Carbohydrate Chemistry

Proven Synthetic Methods

Volume 3

Carbohydrate Chemistry: Proven Synthetic Methods

Series Editor: Pavol Kováč

National Institutes of Health, Bethesda, Maryland, USA

Carbohydrate Chemistry: Proven Synthetic Methods, Volume 1
by Pavol Kováč

Carbohydrate Chemistry: Proven Synthetic Methods, Volume 2
by Gijsbert van der Marel andJeroen Codee

Carbohydrate Chemistry: Proven Synthetic Methods, Volume 3
by René Roy and Sébastien Vidal

Carbohydrate Chemistry | Proven Synthetic Methods Series

Carbohydrate Chemistry

Proven Synthetic Methods

Volume 3

Edited by

René Roy
Sébastien Vidal

CRC Press
Taylor & Francis Group
Boca Raton London New York

CRC Press is an imprint of the
Taylor & Francis Group, an **informa** business

CRC Press
Taylor & Francis Group
6000 Broken Sound Parkway NW, Suite 300
Boca Raton, FL 33487-2742

First issued in paperback 2017

ISBN-13: 978-1-4665-8357-3 (hbk)
ISBN-13: 978-1-138-89424-2 (pbk)

Library of Congress Cataloging-in-Publication Data

Carbohydrate chemistry : proven synthetic methods / edited by Pavol Kovac.
    p. cm.
    ISBN 978-1-4398-6689-4 (hardcover : alk. paper) 1. Carbohydrates--Synthesis. I. Kovác, Pavol, 1938- II. Title.

QD322.S95C37 2011
547'.27--dc22   2011010495
---

*This series is dedicated to Sir John W. Cornforth, the
1975 Nobel Prize winner in chemistry, who was the first
to publicly criticize the unfortunate trend in chemical
synthesis, which he described as "pouring a large volume
of unpurified sewage into the chemical literature."**

* Cornforth, J. W., *Aust. J. Chem.* 1993, 46, 157–170.

Contents

SECTION I   Synthetic Methods

SECTION II   Synthetic Intermediates

Foreword

THE DARK AGES OF PUBLISHING SYNTHETIC ORGANIC CHEMISTRY/CARBOHYDRATE CHEMISTRY: REFLECTION ON THE LAST FEW DECADES

Dear readers: The series Carbohydrate Chemistry, Proven Synthetic Methods is the antithesis of the current trend, virtually a crisis our field currently experiences in executing and publishing synthetic carbohydrate chemistry. You may be surprised that I call crisis what is happening, but I trust that, as you read on, you will largely agree with this description. Volume 3 of this series is the product of the consolidated effort of many authors and a team of three editors who show by their involvement in the series that they share the same concerns. I wish to use this opportunity to thank the contributors of this volume, the Taylor & Francis Group staff involved in its production, and all those who deserve credit for the success of the series. I hope that the unsustainable trend, examples of which I shall show later, can be slowed down and eventually turned around. When you continue reading, you will find evidence that justifies the title of this writing. The works cited are only samples taken at random from carbohydrate literature, but the situation is no different in synthetic organic chemistry in general.

I trust that the contributors and the editors of this series are just a fraction of a large cohort of synthetic carbohydrate chemists who see the situation in publishing synthetic chemistry the same way I did when, years back, I first approached Taylor & Francis Group with the idea to start the series. Sadly, many who we, the editors of the three volumes published to this date, invited to contribute have not responded. It is possible they did so lest their work might not do well when subjected to the checkers' and editors' scrutiny. Be that as it may, there is a large body of carefully executed synthetic work out there, which I document here by arbitrarily citing just a few examples.*[1–40] I hope that the contributions we receive in the future will be from a much wider spectrum of colleagues from all over the world, individuals who trust reproducibility of their protocols and welcome checkers' critical but fair examination, realizing that their experimental procedures may become even more reliable when problems the checkers might find would be corrected.

A question that likely comes to mind when one goes through current organic/carbohydrate synthesis literature is "Are synthetic chemists ever going to wake up and come to their senses? Don't they see that scientific publishing is swarmed with incompetence and dilettantism and is controlled by business interests, which care more about filling pages of an ever-increasing number of unnecessary journals than about the quality of what's in them?" It's like a nightmare to see that, as the number of publications in synthetic chemistry increases, more substandard work is getting published. At the same time, talented young individuals have difficulties getting their

---

* Works quoted here are only a small sample of a large number of papers of similar, high quality.

research funded because some *established* authors suck up resources to produce and publish more substandard work. Some gifted, highly motivated young individuals cannot get into tenure track programs or cannot get tenure or jobs because their CVs are evaluated by committee members whose main qualification is the ability to read Impact Factors of journals. How long can this go on? I am presenting this to the carbohydrate community in the hope that something can be done so that it does not go on like this forever. Unworthy papers are getting published because, evidently, there are not enough qualified people to serve as editors of the steadily increasing number of journals, and the shortage of competent referees to evaluate the huge volume of manuscripts submitted is even more critical. If you have any doubts about this, type the following phrase into your Internet browser: "broken peer review system."

To become an editor of a scientific journal or a member of an editorial board of one, one used to have to be a seasoned, well-established scientist, an individual well known and respected in the field because of one's accomplishments. An editor used to be not only an administrator but also a person who was experienced and knowledgeable and able and willing to be a counselor and mentor to less-experienced authors, for example, young scientists at the early stage of their career who could and would learn from an editor's advice. To illustrate, I shall briefly describe my first encounter with, at that time, a very young, actually the only, journal dealing exclusively with carbohydrate chemistry. The acceptance of a manuscript for publication depended on the quality of data, which had to be collected with the most up-to-date technology, to provide solid evidence in support of the conclusions. When I started my career in synthetic carbohydrate chemistry, merely to have a paper published in *that* journal was *the* objective measure of the quality of work and soundness of science described. I was very disappointed when the referee of the first synthetic paper I submitted to *that* journal did not recommend the manuscript for publication because—what would be unheard of today as a reason for rejecting a paper—the combustion analysis figures for one of the few new compounds I reported, a chromatographically homogeneous, syrupy product, were more than 0.4% off the calculated values. However, Dr. A. B. Foster, the editor who handled the paper, suggested that I distil the substance—again something that would be virtually unheard of to suggest or do today with carbohydrates. I did so, found it relatively easy to do, and the compound in question readily produced correct analytical figures after distillation.[41] From there on, when the compound was amenable to purification in this way, I had been distilling almost routinely syrups that tenaciously held traces of solvents and was surprised that, when pure, they could be distilled at reduced pressure at temperatures way above 200°C without decomposition and, unlike before distillation, virtually never failed to produce correct analytical figures.[42–51] I am pleased to say that I have become a better, more skilled chemist also because of the constructive criticism of the referee and the good advice by a knowledgeable and caring journal editor.

Forty years later, my experience with editors of two well-known chemistry journals was vastly different from the one I just described. In the first occurrence, I contacted the editorial office of a journal, which because of its publisher's name—virtually synonymous with "The Truth in Chemistry"—would be expected to hold itself to higher standards. I asked why they published a paper where proofs of purity and specific optical rotations for a large number of new carbohydrate derivatives were not included. They replied promptly and stated clearly that they agreed that "...all new compounds should be fully characterized and for those with chiral centers the optical rotation should be given." For goodness' sake, when you agree, why don't you put your money where your mouth is and require it? The essence of their response was that what they do "...is in line with standard organic chemistry reporting practice; in any case each reader can make their own judgment as to the merits of results based on the published data." Obviously, they felt very strongly about their policy. For me to respond would mean to potentially start a discussion that an individual cannot win with editors who feel they stand on a very tall pedestal and are not ready to give up even a fraction of their superiority. Nevertheless, here I wish to pose a question: If, as they suggest, each reader is supposed to make their own judgment as to the merits of results based on the published data (or lack of it, I might add), then why do we need peer review? Is it not the purpose of peer review to filter out material that is poor science and not worthy of publication?* I did not respond to the editors' sorry attempt to validate their practices.

In the second episode, I contacted the publisher of the second oldest journal dealing exclusively with carbohydrate chemistry and suggested to them that they ask the journal editor to consider minor revisions to the journal's Instructions for Authors. To be concrete, I suggested the following:

- Instead of requesting that authors provide "melting point for solid product," they should be requested to provide "melting point for crystalline product obtained by crystallization from a suitable solvent."
- "Specific optical rotation for natural product and enantiomerically pure compound should be reported," should be replaced with "specific optical rotation for pure chiral compounds should be reported."
- Instead of "proton and carbon NMR chemical shifts and coupling constants...should be listed," authors should report "assigned proton and carbon NMR chemical shifts, and coupling constants."
- "Evidence for compound purity should include one or more of the following: (1) a 1D proton NMR spectrum or proton-decoupled carbon NMR spectrum; (2) combustion elemental analytical values for carbon and hydrogen; (3) quantitative gas chromatography or HPLC analytical data; (4) for a known compound, a narrow melting point range in close agreement with a cited literature value, and so on" should be replaced with "Evidence for purity of stable, synthetic compounds should include combustion elemental analysis figures for carbon and hydrogen within 0.4% of calculated values, unless absence of such data can be justified in an acceptable way."

After the publisher waited more than two months, the journal editor actually responded but stood firmly by the original wording of the Instructions. He completely

---

* Should I be kind to these editors and assume that when they said "...each reader can make their own judgment..." they actually meant "each reader *must* make their own judgment" because the peer review system is broken? See a footnote on page xviii.

ignored my first three suggestions and offered the following indefensible litany in support of the arguably flawed journal's policy regarding proof of purity of organic compounds (no material changes have been made in the journal editor's arguments cited here; *my comments are in italics*):

- "Use of combustion elemental analysis as the only standard to prove sample purity is not the modern trend." *Sadly, it is true.*
- "High-field NMR spectra can easily tell the purity of any small synthetic molecule." *False.*
- "Elemental analysis consumes large quantities of samples, but many final synthetic products can only be obtained in less than this amount." *I am not suggesting that you should send your deprotected, amorphous decasaccharide that took six months or longer to synthesize to be burned in the process that would likely not produce correct analytical figures anyway, but if you cannot spare 3 mg of a simple, fully protected, synthetic product, go back to the lab, make more, and plan your next work better.*
- "Reviewers definitely can identify problems if there is concern about the product purity." *They should but it is too obvious that they don't.*
- Lastly, he declared: "It has become standard in almost every research area to use HR MS combined with high-field NMR to substitute elemental analyses, which policy is also adopted by all ACS journals." *Sadly, it's true.* And he continued: "Thus, I would like to keep our original guidelines concerning the proof of product purity, with elemental analysis as one of the several options (actually, in many cases NMR and HPLC or GC results are much more reliable than EA reports.)" *Really?* And then he added the punch line: "....people can fake EA data easily but not easy for NMR spectra and HPLC/GC diagrams." *Have you heard of Photoshop?*

It should not come as a surprise to anyone that my jaws dropped having read these pathetic justifications. Inspired by the wording by the previous editor, now I ask you, the readers, to make your own judgment as to the merits of this editor's arguments. It really looks incredible, but this editor suggests that we forget about doing good science and stick to the *modern trend*, which, frankly, stinks. He obviously refuses to accept the simple fact that *there are no options for an objective, scientifically acceptable way to prove purity of organic substances.* If it were not so, and if combustion analysis were not important, Fritz Pregl would not have been awarded the 1923 Nobel Prize in chemistry solely for his contributions to quantitative microanalysis of organic compounds. There is only one *option* to determine the absolute purity of organic substances: combustion analysis. The whole world knows it, and it is, therefore, still regarded worldwide as the touchstone for purity of organic substances,[52] because all other methods have limitations. Some look only at the portion of the sample that is volatile, some only at the portion that moves along a particular column or happens to be detected by a probe. It is only the combustion analysis that enables one to determine the empirical formula (the formula for a compound that contains the smallest set of integer ratios for the elements in the compound) for the *whole* sample analyzed and *not*, for example, only for the solute in solution or for the volatile part of the sample. Having said that, I want to emphasize for those who are not familiar with my writings on this subject in the past[53-55] that there are situations or classes of compounds when correct analytical figures do not have to be included as proof of purity. To borrow wording from the esteemed editor of the second oldest journal dealing exclusively with carbohydrate chemistry, qualified, competent referees and journal editors should be able to make the right judgment about when the foregoing exception should be applied. Not to require elemental analysis data as the proof of purity of new synthetic, stable, small molecules or, as some journals do, namely, require it formally but not to enforce the requirement, is wrong, but it *is* every journal editor's prerogative. However, *to maintain that "in many cases NMR and HPLC or GC results are much more reliable than EA reports" means to defy fundamental principles of chemistry*, and absurdity of that claim equals believing that Earth is the center of the universe. Don't you agree that the situation in publishing organic chemistry is pretty bad when someone with such beliefs manages a scientific, chemistry journal and is entrusted with the task of mentoring the next generation of chemists?

But let me go back to the topic of publishing scientific journals. Now, anybody, regardless of qualification, experience, and competence, who receives the offer and accepts the job can become a member of an editorial board, and invitations to join keep coming because there are too many new journals. Because the semiqualified members of editorial boards are very often peer reviewers, there is no wonder that the peer review system is broken. Similar is the situation with soliciting manuscripts. To be asked by an editor of a renowned journal, who by the sheer nature of the job title used to be a person who meant something in the field, to contribute a paper was an honor and recognition earned over a long time of good work. Now, editors of new journals send letters to names on the lists of attendees of every meeting begging for contributions in the hope to fill and sell all those newly available pages. Many of these new journals are the last resort of those whose science and/or writing ability is even worse than some of those whose papers appear in mainstream journals. Serious authors should exercise careful judgment before responding to requests from these new vehicles for publication to submit a manuscript, because the effect of increasing the number of journals on the value of publications in them is the same as that of inflation on currency, namely, it decreases it.

It seems also that to publish reviews belongs to the ambitions of every editor, regardless of the scope of the journal. The reason is too obvious: reviews generate citations for the journal and potentially increase the impact factor (IF), the golden calf of contemporary scientific publishing. Reviews used to be written by authorities in the field,[56-72] and they wrote them not when it was convenient but when the science progressed to the stage that it was deemed useful. Now, people write reviews whenever an invitation happens to come, even every year to a different journal on the same topic, or pass the job on to their students. Reviews resulting thus serve the students, to fulfill one of the requirements to defend their thesis, and the mentor, to add a number to his or her citation index. However, to do a search confined to the topic of one's thesis, add the material to the previous mentor's review, and then publish the review jointly are not the same as writing an erudite, authoritative review and critically but objectively evaluating the material treated.

The result of the shortage of qualified referees/reviewers is even more obvious.* Lack of qualification of many individuals in these positions is evident because a competent referee would readily notice the flaws and reject or request the revision of manuscripts where[†]

- No proof of identity or purity for numerous compounds synthesized is provided, yields are inflated, NMR data are not assigned, spectra provided as supporting information are of poor quality, and they clearly show the presence of impurities. These are arguably the most frequently occurring problems within publishing synthetic chemistry. Not providing proofs of purity and/or identity reminds one of Benjamin Franklin's famous quotation: "A claim made without proof is merely the mindless blathering of the ignoramus."
- Reproducibility of results is compromised/impossible because physical constants important for the characterization of intermediates and/or final compounds were not included.[74–109] In other words, there is not much to reproduce. Such communication could be considered as belonging to the category "paper chemistry," which, as we know, is easy chemistry to do when identity and/or proof of purity of compounds does not have to be

---

* Apparently this applies not only to publishing synthetic chemistry but generally to publishing science, as is clearly documented by the paper in *Science*[73] where a courageous good Samaritan is reporting on his experience with peer review. He wrote a spoof paper containing grave errors, which a competent peer reviewer should easily identify and not recommend publication. He submitted various versions of it to over 200 journals, which presumably had them peer-reviewed, and 157 journals accepted it for publication. Bohannon feels that the problem lies in open-access journals that collect publication fees when the paper is accepted, and he addressed his merciless but justified criticism in that direction. However, works cited here are not from publication-for-fee journals. Publication of these manuscripts was obviously recommended by poorly qualified reviewers. These are only the tip of the iceberg, and provide convincing evidence that the problem is much more widespread than throughout journals that charge fees for publication.

† Some of the publications cited in the following are so shockingly poor that one wonders how bad the quality of work has to be for the manuscript to be rejected. Nevertheless, it is *not* my intention to offend or embarrass any author. Authors, including me, often make mistakes; authors do not know everything that should be avoided in scientific publishing, but those who accept the task of judging the work of others are expected to know that. I blame referees of papers cited here, which is only a small sample of hundreds of flawed papers in the contemporary synthetic literature, for all violations of good laboratory practices that appear in print and for lowering standards of what can be published in a scientific journal. However, editors of these journals are the main culprits because they pick referees and it is their responsibility to not let poor-quality papers slip through, when Instructions for Authors in their journals clearly require that data allowing reproducibility of results, proof of identity, and proof of purity be included. Therefore, names of authors of the flawed papers are removed from the References, although I realize that whoever is interested in identifying them can easily do so. If you think it took systematic research or a lot of time for a bookworm to do nothing else for months but sift through thousands of papers in the sea of chemical literature to find these references, you could not be farther from the truth. I did not spend even one minute of extra time searching. I encountered these references inadvertently, just doing my everyday reading within my normal daily duties as a researcher and mentor, or when checking references in papers I had been asked to referee/review, before and after I was removed from some journal's list of potential reviewers because I rejected too many papers, thereby potentially harming their impact factor.

provided. Such papers are of little value. An experienced chemist could write many such papers without even going to the laboratory. The purpose they serve is only to boost authors' egos by increasing the size of their bibliography and fill pages of journals, rather than the purpose a good scientific paper should serve, namely, to promote the advancement of science.

- Established trivial names of carbohydrates are mutilated.[110–126]
- Physical constants found for known compounds are compared with literature values, the two sets of numbers are vastly different, and no comment or explanation for the differences is offered.[124,125]
- Amorphous compounds are reported as hydrates to make combustion analysis figures fit theoretical values.[*,8,127,128]
- Specific optical rotation data for chiral compounds are not reported.[83,99, 105,107,114,129–181] How has the art of organic chemistry gotten from State **A**, when chemists were proud of, and wanted to show reproducibility of, their results by measuring $[\alpha]_D$ values in duplicate, and journals actually published description and findings of these two parallel experiments,[182–184] to State **B**, when chemists do not publish $[\alpha]_D$ values at all, is anybody's guess. Is it, perhaps, that when a set of authors first got away with not including $[\alpha]_D$ data for chiral substances in a paper, they created a nasty precedent? When more such papers appeared, editors of almost every chemistry journal jumped on the wagon, not to miss the *new trend*, and continued accepting manuscripts characterized by the same flaw. It allowed them to accept and publish more papers. That helped in the chase for higher IF, which depends, among other things, on the number of papers—good or bad—published. I am not sure which of the *following the trend* came first and which came second, but the *no* $[\alpha]_D$ *trend* resembles another situation that resulted from *following the trend*, and where we also lowered the stick by which we measure the acceptable/unacceptable level of decency. Just for the sake of analogy—and we do not have to go back as far as to Rudolpho Valentino—have you ever heard of or seen in magazines/popular culture journals from the past the revered, handsome, and sexy Cary Grant, Fred Astaire, Gregory Peck, Spencer Tracy, or William Holden showing up in public without a clean shave, or Clark Gable without his mustache perfectly trimmed? One used to have to appear in public either with a clean shave or with trimmed beard or mustache. Anything less than that used to be considered poor taste, disrespectful, and socially unacceptable. But now? After

---

* It is not appropriate to report analytical data for *amorphous* compounds as hydrates, hemihydrates, etc., just because the analytical figures happen to fit that particular formula. In a crystal, *discrete* numbers of solvent molecules may partake in the lattice formation. In an *amorphous* substance, however, this is not so. Therefore, it is likely that it is possible to find analysis fit any compound using solvation. An analysis of an *amorphous* compound is irrelevant if it needs solvation adjustment. In other words, such analysis is no proof of purity, and optical rotations found for such materials should not be considered reliable. We often deal with hygroscopic compounds or substances that tenaciously hold solvents. With other proofs of identity and purity, *reasonably* presented in a paper, it is not such a big deal if *occasionally*, when it can be justified, we cannot obtain, and then do not include in the paper, correct analytical figures for all compounds.

a number of consecutive nights of heavy partying, some male Hollywood celebrities, who have been perceived by the media and advertising agencies as sexy, had no time to shave in the morning. When they appeared in public looking like Clint Eastwood in one of his spaghetti westerns and got away with it, almost every moron in the cohort of their peers felt that it was the *new trend* that must be followed, lest they might lose the sex appeal they have fantasized about having. The *new trend* seems to have caught on, and now when you see a male actor promoting *any* merchandise on the TV or in a magazine, from cars to hotel rooms to erectile dysfunction remedies, he has not shaved for at least two weeks in order to, presumably, look sexy, and thus the product he promotes to appear more appealing and easier to sell. This is where we are heading: the more gaudy and outrageous, the more popular and more marketable. By the same token, authors subscribe enthusiastically to "let's publish as few data as we can get away with, lest someone might find us wrong, so that we can be proud of our long bibliography and feel great about ourselves," and journals are more than happy to cooperate. What is very serious here, however, is that not reporting $[\alpha]_D$ values for chiral substances appears to be a clever way (*new trend?*) to *hide impurity of compounds*, when other proof of purity of newly synthesized substances is not provided. I cannot think of any other reason than that for not reporting $[\alpha]_D$ values for optically pure compounds, when it is an important characteristic, and the data are easy to obtain without destroying the sample. By not reporting specific optical rotation, the authors of such papers not only ignore the proper practice for the characterization of chiral organic substances but are also virtually thumbing their noses at the rest of the synthetic community because now, with the absence of other evidence of purity, nobody can prove that compounds they report are impure. As I mentioned earlier, editors of a leading organic chemistry journal maintain that their publishing policy is "...in line with standard organic chemistry reporting practice." Sadly, I have to agree with that: poor characterization of compounds is currently becoming the norm in publishing synthetic organic/carbohydrate chemistry. Nevertheless, the current standards of publishing synthetic organic chemistry, which allow these violations of good laboratory practices, are flawed, are unfair, and should be resented within the community of organic chemists. They are flawed because they encourage sloppy work and make it possible for sloppy workers to publish extensively. They are unfair because they make conscientious chemists less competitive: it takes them much longer to produce a quality manuscript when they want to adhere to good laboratory practices and characterize products properly.

- Meaningless/unreliable optical rotation was reported for compounds, which judging by the mode of isolation described must have been impure.[185] Similarly, a journal which can pride itself with one of the highest Impact Factors among chemistry journals publishes a synthetic paper [185a] where a one-step preparation of the key reagent, a new compound, is described in a way that it should be obvious to any competent chemist that the product cannot be a pure substance. No clue about the degree of purity is provided, and the latter is used for further conversions without further purification. The referees are satisfied, and the editor accepts publication of combustion analysis figures for two of three products that are off the calculated values by more than 0.4 %, one of these quite off 0.4%. Questions every competent chemist should like to have answered are (1) Are these authors trying to provide credible evidence for their compounds being impure? (2) What is the basis for the high Impact factor of this journal, when quality of work published in it is definitely not it?

- Official carbohydrate nomenclature was ignored.[98,107,110,125,126,129,169,178,186–195]
- $[\alpha]_D$ was reported to four decimal places and concentration in mg/mL.[196]
- Compounds crystallized after the evaporation of solutions or during workup, but crystallization from solvents was not attempted and melting points were not measured and compared with those previously reported.[136,197]
- Melting points were reported for amorphous materials.[127,198]
- A carbohydrate is called *aglycon*.[74,199–201]
- Melting points are reported for crystalline compounds described as *white solids* or variously colored solids, but solvents for crystallization are not mentioned. Thus, it is not clear if m.p.s were measured for residues after the evaporation of solvents or for materials crystallized from suitable solvents and recrystallized to constancy.[31,116,125,202–210] Melting points measured for crystalline residues formed upon the concentration of solutions without crystallization from suitable solvents are virtually worthless. Reporting melting points is important, but it makes sense only if the value is a constant physical characteristic and can be used for identification purposes.
- Methyl α-D-glucopyranoside was subjected in aqueous media and in the presence, among other things, of water-soluble reagents to what was claimed to be the regioselective oxidation. Isolation of the product consisted of partition of the reaction mixture between water and toluene. The water phase was extracted once with Et_2O, filtered, and concentrated (no chromatography or other meaningful purification) "to give the pure methyl-α-D-ribo-hexapyranosid-3-ulose in 96% yield as a dark brown solid."[107] No credible proof of purity or $[\alpha]_D$ was reported to allow comparison with the value determined for the previously synthesized substance, *which was reported as a crystalline substance*.[211,212] Nevertheless, the referees and the journal editor were evidently satisfied because the *dark brown solid*, that is, unpurified sewage (to use the term coined by Sir John Cornforth,[213] the winner of the 1975 Nobel Prize in chemistry), was published in a journal[107] that has earned one of the highest IFs among chemistry journals. This and numerous examples of substandard synthetic works published in high IF journals show that there is no direct relationship between the Golden Calf Factor and the quality of work published in these journals. That, objectively speaking, turns the IF into a farce. I rest my case.

- Compounds were obtained by crystallization from solvents as white or yellow solids or white needles, but melting points were not reported.[140,147,150,185,193,214]
- Instead of supporting the claimed purity of their material by showing correct combustion analysis figures, authors offered the following absurd statement to characterize the synthesized compound: "The resulting colorless oil was analytically pure and was used for the next step without further purification."[215]
- Substantial differences were found when physical constants measured for newly synthesized, known compounds were compared with the data in the literature, but no comments were offered. Convincing evidence of purity of these compounds, which would show that these newly measured physical constants could be used as the reliable means for identification purposes and proof of purity, was not provided either.[202]
- Meaningless m.p. and/or $[\alpha]_D$ values are reported for inseparable anomeric mixtures.[187,192]
- Preparation of known compounds is described, but neither physical nor spectral characteristics are compared with those published. Thus, the identity of products with compounds reported previously is not provided.[139,143,149,193,197,204,216]
- Lactose was treated in refluxing acetic anhydride in the presence of NaOAc, and "the expected α anomer was obtained." No reference is given that would show that α product should be the expected product with lactose when the method applied is known to produce generally 1,2-trans acetates. There are many examples in the literature showing that octaacetyl β lactose was the main product when the parent disaccharide was treated at the conditions described.[217–219] The structure of the product is shown in the scheme as the β anomer. The material obtained was purified by *recrystallization*, but the observed melting point is not reported or compared with the published values.[220]
- The course of a reaction and formation of its product was studied. The phenomenon observed was deemed important enough for the authors to explain it mechanistically in a scheme and verbally in Results and Discussion, but the key experiment, and isolation and proper identification of the key product, described as *the only isolable product*, is not described in the Experimental section at all.[193] It is possible that the authors of this paper did not mean for it to appear in print this way and that the loss of some text resulted from an unfortunate accident. Qualified referees are required and expected to read every manuscript carefully, and also the journal editor for the galley proofs, and should not have missed something so obvious.

If all the examples I have presented are not convincing enough about the notion that we have a crisis to deal with, the best and irrefutable evidence that there are not enough qualified journal editors and people to serve as referees of the huge number of papers submitted to too many journals, a large percentage of which are unnecessary, is this title of a paper published in the leading organic chemistry journal: "One-Pot Conversion of Glycals to *cis*-1,2-Isopropylidene-α-glycosides."[191] One would expect at least the most senior author of a paper to know the difference between an acetal and a glycoside, but if that should not be the case, we have, presumably, referees and journal editors to correct titles as the one cited. Thus, as in other unfortunate cases described here, the final responsibility rests with journal editors because editors choose referees. When a title as the one just cited appears in print, it looks like the editor decided about the referees by pulling the names from a hat instead of entrusting this very responsible task to someone based on his or her qualifications. Passing such a title the referees' and the editor's scrutiny calls the editor's qualifications to edit an organic chemistry journal, as well as the credibility of the journal, into serious question.

Before the publication of *unpurified sewage*[213] became virtually the new norm, material published even in short communications was possible to reproduce because authors included relevant data.[*,211–229] Also, when combustion analysis figures for new compounds did not appear in print, purity of new substances was supported by a statement to that effect.[*,230–235] Finding such a statement in more recent literature is a rarity.[236] Are we now supposed to understand and accept that purity is archaic and, therefore, unimportant when we see in print the description of a large number of new compounds without presenting the evidence of purity and without reporting physical constants?

Among other flaws from which publishing in organic/carbohydrate chemistry currently suffers are not giving due credit to previous workers who published on related topics; erroneous citation of literature; not referencing preparation of starting materials or intermediates; diluting Abstracts with general statements, instead of describing concisely but concretely what was done, the latter being the sole purpose of Abstracts; referring in Abstract to compound numbers without giving the reader any clue about the structure; and, of course, probably the most common, inflated yields. That anybody can claim >95 yields is beyond the comprehension of a qualified chemist, when not only practicing chemists' experience at the bench but also a recent study[237] shows clearly that such yields are unrealistic. Notwithstanding my dislike of the ever-increasing number of journals, the time has, perhaps, come for the organic chemistry community to start *The Journal of Reproducible Results*.[53] Concerning the credit due, writers do not realize that referencing other people's work does not diminish their own contribution. Granted, to find these references means extra work, but it also shows that one thoroughly reviewed what has been done before in the area studied, which is as much a prerequisite to discovering the unknown as it is to avoiding the rediscovery of the already known.

In all fairness, one has to admit that commerce and incompetence could not have usurped scientific publishing without the scientific community participating in the process, because of the pressure by the *publish or perish* environment. Solution

---

* Publications quoted here are only a small sample of a large number of papers of similar, high quality.

to this unsustainable situation is only in our field policing itself.* Journal editors who care about quality over quantity should toughen the requirements for accepting papers for publication and/or enforcing the existing ones. That can only be accomplished when our field decides that having a fewer journals is better than the existing alternative. When these fewer journals are staffed with knowledgeable editors and qualified referees, it will be possible to arrive at the situation when objective criteria, namely, high quality of work, high degree of difficulty, and clear writing, decide the acceptance for publication. High concentration of such papers would then justly make the journal a high-impact journal without having to assign it a numerical value. To publish high-quality science in *such* journals as a means of survival in the *publish or perish* environment will then make sense. It will let the natural selection of scientific publishing take its course, and the journals that only care about selling the pages regardless of their contents will die of natural causes or become the last resort to below mediocrity.

Now what do I mean by quality synthetic work? It is not only a matter of knowledge and laboratory skills but also a question of work ethics. No result of one experiment should be published regardless of how stressed or intoxicated the authors are by their illusion that their work is immensely important. Then nobody would even think about arguing that enough material to run combustion analysis was not available. One should refrain from using the highly popular statement that "…the yields reported were not optimized." Such a statement means nothing less than that the work was executed hastily and published prematurely and that reliability of all other results in the paper is probably also dubious. As I have already indicated elsewhere in this series, a synthetic experiment is finished only when *at least* simple, stable, key intermediates in a long synthesis are purified to pass the test of purity by combustion analysis and are properly characterized. This includes attempting the crystallization of all small synthetic molecules, recrystallization of crystalline compounds to constancy, and reporting reliable physical constants. Crystallization should always be attempted, especially with, but not limited to, compounds that are solid. When reporting melting points, solvents for crystallization should also be reported to make it clear that melting points were measured with materials obtained by crystallization from proper solvents and not of residues obtained after the evaporation of solvents from solutions. Since it can be reasonably assumed that many compounds described these days as *white solids* could be crystallized had such attempts been made, the description of compounds as *white solids* or *powders* or *semisolid mass* without further comment/explanation should not be acceptable either. If attempts at crystallization failed, including a statement

---

* The situation regarding high-impact journals has become so oppressive that some Nobel laureates have decided to boycott some *luxury* journals, the latter term being coined by Randy Scheckman, who recently won the Nobel Prize in physiology and medicine (http://kingsreview.co.uk/magazine/blog/2014/02/24/how-academia-and-publishing-are-destroying-scientific-innovation-a-conversation-with-sydney-brenner). You can learn from the same source that the pressure to publish in these journals leads to *publish rubbish* and has serious consequences on the quality of science published. The community of synthetic/carbohydrate chemists should embrace these views found in the source just cited, expressed by the highest-level authorities in the biomedical sciences, and join this refreshing wave of, hopefully, the beginning of a brighter future of our field.

to that effect is useful; there is no shame in admitting "compound could not be crystallized from common organic solvents." To report or not report $[\alpha]_D$ values for chiral substances claimed to be pure isomers should not be open for discussion. Some beautiful works have been published on the complete structure determination of natural products when only a fraction of a milligram of material was available, for example, Reference 238. On the other hand, a multistep synthesis carried out on only a few-milligram scale without including acceptable evidence of identity and purity of most intermediates, as we often see published, is not equally beautiful. Such work should not be accepted for publication when authors would like to claim priority of the synthesis because, then, the characterization of key intermediates or of the final compound (or its simple derivative) cannot be done due to the insufficient amount of material available. And don't get me started on publishing NMR data without assignment. NMR spectroscopy allows chemists to look at molecules on the atomic level. Spectra where the structurally diagnostic resonances are assigned give the chemist a rare opportunity to provide objective evidence for the structure claimed. That many journals decided to publish NMR data without interpretation/assignment, thereby degrading one of the chemist's most powerful techniques to a tool for a fingerprint comparison, is ludicrous, mind boggling, and inexcusable abomination. As far as the purity of synthetic products is concerned, what we often see in contemporary chemical literature and is documented here by traceable citations is a disgrace to the art of organic synthesis. When we ask in this series that authors comply with certain requirements as far as purity is concerned, we are not asking for anything else than what used to be the norm during the days when I and many of my peers obtained our PhD in organic synthesis. Those standards were tough but in line with proper laboratory practices. We have now chromatography, NMR, and other spectral methods that people 100 years ago (Emil Fischer and his peers) did not have. That lack of sophisticated instrumentation did not prevent those pioneers of modern carbohydrate chemistry from the preparation and reporting of pure substances. The availability of all modern tools we now have does not grant us a mandate for changing the criteria of purity at will; the same still holds: a compound is either pure or impure. And reporting preparation of unpurified sewage[213] should not be allowed to appear in any respectable chemical publication.

In this series, with the help of conscientious authors, we strive to avoid the problems described here, which appear too frequently throughout the carbohydrate literature. We believe that the reproducibility of synthetic protocols is the most important aspect of publications dealing with synthetic work and that the present *modern trend* is unsustainable. It is up to the conscientious chemist to ignore vehicles for publication that publish substandard, irreproducible work and support journals that scrutinize submitted work through the unbiased, constructive criticism of qualified strict-but-fair referees without regard to the number of pages per issue. The recent National Institutes of Health (NIH) leadership's commendable initiative[239] shows what the NIH suggests should be done concerning problems in biomedical sciences, which are in many ways similar to those we have in organic synthesis. Importantly, the director and the deputy director of the NIH realize that their mission can only succeed "with the full engagement of the entire biomedical-research enterprise." This agrees with "solution to this unsustainable situation is only in our field policing itself" mentioned earlier. I dare the synthetic chemistry community to join the directors of the NIH in their effort to clean the Augean stable.*

This series is only a small drop in the vast pool of organic chemistry literature, and it is addressed primarily to carbohydrate chemists. Some of you, the readers, may ask why I have not chosen for the publication of my concerns and findings shown earlier a vehicle that would reach a much wider readership within the organic chemistry community if my intention to publicize these thoughts and findings was to make them a wake-up call. Well, I tried. Before I had the first article addressing this topic accepted for publication by *Chemistry and Biodiversity*,[53] I submitted an article describing similar concerns to a journal where it would be available to the largest possible cohort of organic chemists and where I was hoping to start a discussion on these issues with my peers. The editors did "not believe that (*name of the journal*) was the proper forum." I was surprised to hear that because it was (and is) the leading journal in the organic chemistry field. One can only wonder what they might have thought would be a proper vehicle for such discussion. *The National Enquirer*?

Pavol (Paul) Kováč, Series Editor
Carbohydrate Chemistry: Proven Synthetic Methods

P.S. I realize that by what I have written I may be stepping on some big toes. I plead *guilty* and expect and accept crucifixion. If it should materialize, it should start with bringing up credible arguments in support of why all what I said above is wrong and why and how it would serve the advancement of science if we let the trend described continue. The wrath should not be extended to my young associates who are only guilty of trying to deliver good, reproducible work and do not deserve their career to be harmed because of their mentor's sins.

REFERENCES

1. Bebault, G. M.; Dutton, G. G. S. *Carbohydr. Res.* 1974, *37*, 309–319.
2. Patroni, J. J.; Stick, R. V.; Skelton, B. W.; White, A. H. *Aust. J. Chem.* 1988, *41*, 91–102.
3. Kihlberg, J.; Frejd, T.; Jansson, K.; Magnusson, G. *Carbohydr. Res.* 1986, *152*, 113–130.
4. Uhrig, M. L.; Manzano, V. E.; Varela, O. *Eur. J. Org. Chem.* 2006, 162–168.
5. Khan, S. H.; Compston, C. A.; Palcic, M. M.; Hindsgaul, O. *Carbohydr. Res.* 1994, *262*, 283–295.
6. Nashed, M. A.; Anderson, L. *J. Am. Chem. Soc.* 1982, *104*, 7282–7286.

---

* While this volume was in production, the most credible authors [240] shared with the scientific community their concern regarding the situation in the biomedical field, which is similar to that in publishing synthetic chemistry. In this context, nothing could be more satisfying for the author of these lines than to see concerns similar to his own expressed by the utmost authorities in the life sciences. They criticize hasty publication of large number of papers whose results cannot be reproduced, point out the inflated importance of high impact journals, and the negative impact of unqualified referees and journal editors upon scientific publishing and advancement of science.

7. McGeary, R. P.; Jablonkai, I.; Toth, I. *Tetrahedron* 2001, *57*, 8733–8742.
8. Lowary, T. L.; Eichler, E.; Bundle, D. R. *Can. J. Chem.* 2002, *80*, 1112–1130.
9. Dondoni, A.; Fantin, G.; Fogagnolo, M.; Merino, P. *J. Carbohydr. Chem.* 1990, *9*, 735–744.
10. Balavoine, G.; Berteina, S.; Gref, A.; Fischer, J.-C.; Lubineau, A. *J. Carbohydr. Chem.* 1995, *14*, 1217–1236.
11. Cao, S.; Hernandez-Mateo, F.; Roy, R. *J. Carbohydr. Chem.* 1998, *17*, 609–631.
12. Peri, F.; Nicotra, F.; Leslie, C. P.; Micheli, F.; Seneci, P.; Marchioro, C. *J. Carbohydr. Chem.* 2003, *22*, 57–71.
13. Ziegler, T. *Liebigs Ann. Chem.* 1990, 1125–1131.
14. Kihlberg, J. O.; Leigh, D. A.; Bundle, D. R. *J. Org. Chem.* 1990, *55*, 2860–2863.
15. Xia, J.; Srikrishnan, T.; Alderfer, J. L.; Jain, R. K.; Piskorz, C. F.; Matta, K. L. *Carbohydr. Res.* 2000, *329*, 561–577.
16. Cao, S.; Meunier, S. J.; Anderson, F.; Letellier, M.; Roy, R. *Tetrahedron Asymm.* 1994, *5*, 2303–2231.
17. Nemati, N.; Karapetyan, G.; Nolting, B.; Endress, H.-U.; Vogel, C. *Carbohydr. Res.* 2008, *343*, 1730–1742.
18. Kefurt, K.; Moravcová, J.; Bambasová, S.; Buchalová, K.; Vymetaliková, B.; Kefurtová, Z.; Staněk, J.; Paleta, O. *Collect. Czech. Chem. Commun.* 2001, *66*, 1665–1681.
19. Jana, M.; Misra, A. K. *Beilstein J. Org. Chem* 2013, *9*, 1757–1762.
20. Pozsgay, V.; Dubois, E.; Pannell, L. *J. Org. Chem.* 1997, *62*, 2832–2846.
21. Hirsch, J.; Kóóš, M. *Chem. Pap.* 2005, *59*, 21–24.
22. Gu, G.; Yang, F.; Du, Y.; Kong, F. *Carbohydr. Res.* 2001, *336*, 99–106.
23. Barili, L. P.; Berti, G.; Catelani, G.; D'Andrea, F.; Di Bussolo, V. *Carbohydr. Res.* 1996, *290*, 17–31.
24. Tatai, J.; Fügedi, P. *Org. Lett.* 2007, *9*, 4647–4650.
25. Müller, B.; Blaukopf, M.; Hofinger, A.; Zamyatina, A.; Brade, H.; Kosma, P. *Synthesis* 2010, *2010*, 3143–3151.
26. Evtushenko, E. V. *J. Carbohydr. Chem.* 2010, *29*, 639–678.
27. Hirsch, J.; Kóóš, M.; Tvaroška, I. *Chem. Pap.* 2009, *63*, 329–335.
28. Barath, M.; Petrušová, M.; Hirsch, J.; Petruš, L. *Collect. Czech. Chem. Commun.* 2006, *71*, 1532–1548.
29. Sanchez, S.; Bamhaoud, T.; Prandi, J. *Eur. J. Org. Chem.* 2002, *2002*, 3864–3873.
30. Mandal, P. K.; Dhara, D.; Misra, A. K. *Beilst. J. Org. Chem.* 2014, *10*, 293–299.
31. Dhara, D.; Misra, A. K. *Tetrahedron Asymm.* 2013, *24*, 1488–1494.
32. Giordano, M.; Iadonisi, A.; Pastore, A. *Eur. J. Org. Chem.* 2013, *2013*, 3137–3147.
33. Wang, L. X.; Sakairi, N.; Kuzuhara, H. *J. Carbohydr. Chem.* 1990, *9*, 441–450.
34. Narouza, M. R.; Soliman, S. E.; Bassily, R. W.; -Sokkary, R. I. E.; Nasra, A. Z.; Nashed, M. A. *Synlett* 2013, *24*, 2271–2273.
35. Yan, S.; Liang, X.; Diao, P.; Yang, Y.; Zhang, J.; Wang, D.; Kong, F. *Carbohydr. Res.* 2008, *343*, 3107–3111.
36. Dhara, D.; Kar, R. K.; Bhunia, A.; Misra, A. K. *Eur. J. Org. Chem.* 2014, 4577–4584.
37. Gunter, K. U.; Ziegler, T. *Synthesis* 2014, *46*, 2362–2370.
38. Lipták, A.; Fügedi, P.; Nanasi, P. *Carbohydr. Res.* 1979, *68*, 151–154.
39. Lipták, A.; Fügedi, P.; Kerekgyarto, J.; Nanasi, P. *Carbohydr. Res.* 1983, *113*, 225–231.
40. Santra, A.; Ghosh, T.; Misra, A. K. *Beilstein J. Org. Chem.* 2013, *9*, 74–78, No 79.
41. Kováč, P.; Petríková, M. *Carbohydr. Res.* 1971, *16*, 492–494.
42. Kováč, P. *Carbohydr. Res.* 1973, *31*, 323–330.
43. Kováč, P.; Petríková, M. *Chem. Zvesti.* 1972, *26*, 72–76.
44. Hirsch, J.; Kováč, P.; Kováčik, V. *J. Carbohydr. Nucl. Nucl.* 1974, *1*, 431–448.
45. Kováč, P. *Chem. Zvesti* 1971, *25*, 460–466.
46. Kováč, P. *Carbohydr. Res.* 1972, *22*, 464–466.

47. Kováč, P.; Longauerová, Z. *Chem. Zvesti* 1972, *26*, 179–182.
48. Kováč, P.; Hirsch, J. *Chem. Zvesti* 1973, *27*, 668–675.
49. Kováč, P.; Longauerová, Z. *Chem. Zvesti* 1973, *27*, 415–420.
50. Anderle, D.; Kováč, P. *J. Chromatogr.* 1974, *91*, p. 463–467.
51. Kováč, P.; Alföldi, J.; Košík, M. *Chem. Zvesti* 1974, *28*, 820–832.
52. Harwood, L. M.; Moody, C. J.; Percy, J. M.; 2 ed.; Experimental organic chemistry: Standard and microscale; Harwood, L. M. M., C. J.; Percy, J. M. , Eds.; Blackwell, Malden, MA, 2007, p. 206.
53. Kováč, P. *Chemistry and Biodiversity* 2004, *1*, 606–608.
54. Kováč, P. *An Open Letter to the Community of Organic Chemists*; Medicinal Chemistry Research Progress; Colombo, G. P. and Ricci, S., Eds.; Nova Science Publishers, Hauppauge, NY, 2009, pp. 1–5.
55. Kováč, P. In lieu of introduction; *Carbohydrate Chemistry: Proven Synthetic Methods*; Kováč, P., van der Marel, G. A. and Codée, J. D. C., Eds.; CRC/Taylor & Francis: Boca Raton, FL, 2014; Vol. 2.
56. Barton, D. H. R. *Pure and Appl. Chem.* 1981, *53*, 1081–1099.
57. Paulsen, H. *Angew. Chem. Int. Ed. Engl.* 1982, *21*, 155–173.
58. Glaudemans, C. P. J. *Chem. Rev.* 1991, *91*, 25–35.
59. Roy, R. *Top. Curr. Chem.* 1997, *187*, 241–274.
60. Fügedi, P.; Garegg, P. J.; Lönn, H.; Norberg, T. *Glycoconjugate J.* 1987, *4*, 97–108.
61. Schmidt, R. R. *Angew. Chem. Int. Ed. Engl.* 1986, *25*, 212–235.
62. Shashkov, A. S.; Chizhov, O. S. *Bioorg. Khim.* 1976, *2*, 437–497.
63. Vliegenthart, J. F. G. *FEBS Lett.* 2006, *580*, 2945–2950.
64. Garegg, P. J. *Pure Appl. Chem.* 1984, *56*, 845–858.
65. Kunz, H. *Angew. Chem. Int. Ed. Engl.* 1987, *26*, 294–308.
66. Sharon, N.; Lis, H. *Science* 1972, *177*, 949–959.
67. Gelas, J.; Horton, D. *Heterocycles* 1981, *16*, 1587–1601.
68. Flowers, H. *Methods Enzymol.* 1987, *138*, 359–404.
69. Garegg, P. *Adv. Carbohydr. Chem. Biochem.* 1997, *52*, 179–205.
70. Gyorgydeák, Z.; Szilágyi, L.; Paulsen, H. *J. Carbohydr. Chem.* 1993, *12*, 139–163.
71. David, S.; Hanessian, S. *Tetrahedron* 1985, *41*, 643–663.
72. Danishefsky, S. J.; Bilodeau, M. T. *Angew. Chem. Int. Ed. Engl.* 1996, *35*, 1381–1419.
73. Bohannon, J. *Science* 2013, *342*, 60–65.
74. *Chem. Commun.* 2009, 2505–2507.
75. *Tetrahedron Asymm.* 2006, *17*, 2449–2463.
76. *Tetrahedron Lett.* 2003, *44*, 2785–2787.
77. *Tetrahedron Lett.* 2003, *44*, 1791–1793.
78. *Tetrahedron Lett.* 2003, *44*, 1787–1789.
79. *Tetrahedron Lett.* 1998, *39*, 8681–8684.
80. *Tetrahedron Lett.* 2003, *44*, 2853–2856.
81. *Tetrahedron Lett.* 1998, *39*, 9801–9804.
82. *Tetrahedron Lett.* 2007, *48*, 3061–3064.
83. *Synlett* 2009, 3267–3270.
84. *Org. Biomol. Chem.* 2014, *12*, 376–382.
85. *Chem. Asian J.* 2009, *4*, 386–390.
86. *ChemBioChem* 2008, *9*, 1716–1720.
87. *Chem. Commun.* 2007, 380–382.
88. *Bioorg. Med. Chem. Lett.* 2006, *16*, 6310–6315.
89. *Tetrahedron Lett.* 2006, *47*, 2475–2478.
90. *Org. Biomol. Chem.* 2003, *1*, 3642–3644.
91. *Chem. Commun.* 2004, 1706–1707.
92. *Bioorg. Med. Chem. Lett.* 2003, *13*, 2185–2189.

93. *Tetrahedron Lett.* 2003, *44*, 5247–5249.
94. *Chem. Commun.* 2002, 714–715.
95. *Chem Commun* 2002, 2104–2105.
96. *Tetrahedron Lett.* 2001, *42*, 5283–5286.
97. *Chem. Lett.* 2001, 224–225.
98. *Angew. Chem. Int. Ed.* 2000, *39*, 2727–2729.
99. *Angew. Chem. Int. Ed.* 1998, *37*, 786–789.
100. *Tetrahedron Lett.* 1998, *39*, 1937–1940.
101. *Chem. Commun.* 1997, 2087–2088.
102. *J. Org. Chem.* 1997, *62*, 5660–5661.
103. *Tetrahedron Lett.* 1997, *38*, 4285–4286.
104. *Tetrahedron Lett.* 1997, *38*, 5181–5184.
105. *Tetrahedron Lett.* 1997, *38*, 3885–3888.
106. *Bioorg. Med. Chem. Let.* 1992, *2*, 255–260.
107. *Angew. Chem. Int. Ed.* 2013, *52*, 7809–7812.
108. Goswami, M.; Ellern, A.; Pohl, N. L. B. *Angew. Chem. Int. Ed.* 2013, *52*, 8441–8445.
109. *Angew. Chem. In.t Ed. Engl.* 2013, *52*, 6068–6071.
110. *Angew. Chem. Int. Ed. Engl.* 2007, *46*, 7023–7025.
111. *J. Org. Chem.* 2004, *69*, 7758–7760.
112. *J. Carbohydr. Chem.* 2011, *30*, 165–177.
113. *J. Org. Chem.* 2003, *68*, 5261–5264.
114. *J. Am. Chem. Soc.* 1996, *118*, 9239–9248.
115. *Tetrahedron Lett.* 2007, *48*, 3783–3787.
116. *J. Org. Chem.* 2008, *73*, 3848–3853.
117. *Carbohydr. Res.* 2005, *340*, 2670–2674.
118. *Tetrahedron Lett.* 2003, *44*, 7467–7470,
119. *Chem. Pharm. Bull.* 2004, *52*, 965–971.
120. *J. Org. Chem.* 2009, *74*, 1549–1556.
121. *J. Carbohydr. Chem.* 1998, *17*, 915–922.
122. *Carbohydr. Res.* 1993, *243*, 385–391.
123. *J. Org. Chem.* 2009, *74*, 4982–4991.
124. *Beilstein J. Org. Chem.* 2010, *6*, 699–703.
125. *Angew. Chem. Int. Ed. Engl.* 2009, *48*, 7798–7802.
126. *Russ. Chem. Bull.* 2000, *49*, 1305–1309.
127. *Carbohydr. Res.* 1996, *284*, 207–222.
128. *J. Org. Chem.* 2000, *65*, 801–805.
129. *Beilstein J. Org. Chem* 2012, *8*, 448–455.
130. *J. Am. Chem. Soc.* 2002, *124*, 3198–3199.
131. *J. Org. Chem.* 2010, *6*, 801–809.
132. *J. Am. Chem. Soc.* 1994, *116*, 1766–1775.
133. *Chem. Eur. J.* 2013, *19*, 5259–5262.
134. *Monatsh. Chem.* 2009, *140*, 1251–1256.
135. *Aust. J. Chem.* 2009, *62*, 546–552.
136. *Carbohydr. Res.* 2005, *340*, 1983–1996.
137. *J. Org. Chem.* 2002, 1614–1618.
138. *Org. Chem.* 2009, *74*, 4705–4711.
139. *Tetrahedron Lett.* 2011, *52*, 3912–3915.
140. *Monatsh. Chem.* 2009, *140*, 1257–1260.
141. *Organ. Lett.* 2000, *2*, 3797–3800.
142. *Angew. Chem. Int. Ed. Engl.* 2008, *47*, 6395–6398.
143. *Org. Lett.* 2012, *15*, 2952–2955.
144. *J. Am. Chem. Soc.* 2012, *134*, 15229–15232.

145. *Carbohydr. Res.* 1998, *312*, 61–72.
146. *J. Org. Chem.* 2009, *74*, 8452–8455.
147. *Chem. Commun.* 2009, 5536–5537.
148. *J. Org. Chem.* 2008, *73*, 7574–7579.
149. *J. Org. Chem.* 2008, *73*, 794–800.
150. *Biochemistry* 2002, *41*, 5075–5085.
151. *J. Carbohydr. Chem.* 2007, *26*, 469–495.
152. *J. Org. Chem.* 2006, 5158–5166.
153. *Org. Lett.* 2005, *7*, 3251–3254.
154. *Tetrahedron Lett.* 2004, *45*, 6433–6437.
155. *Tetrahedron Lett.* 2003, *44*, 9051–9054.
156. *Tetrahedron Lett.* 1996, *37*, 1389–1392.
157. *Org. Lett.* 2002, *4*, 281–283.
158. *J. Am. Chem. Soc.* 1999, *121*, 12063–12072.
159. *Org. Lett.* 2008, *10*, 905–908.
160. *Eur. J. Org. Chem.* 2008, 5526–5542.
161. *Nat. Protocol.* 2013, *8*, 1870–1889.
162. *Pharm. Bull.* 1991, *39*, 2883–2887.
163. *Chem. Eur. J.* 2013, *19*, 17425–17431.
164. *Adv. Synth. Catal.* 2011, *353*, 879–884.
165. *Synlett* 2009, 603–606.
166. *J. Am. Chem. Soc.* 2002, *124*, 10036–10053.
167. *Carbohydr. Res.* 2009, *344*, 145–148.
168. *Synlett* 2009, 3111–3114.
169. *J. Am. Chem. Soc.* 2006, *128*, 11906–11915.
170. *Synthesis* 2007, 1412–1420.
171. *Carbohydr. Res.* 2002, *337*, 1247–1259.
172. *Angew. Chem. Int. Ed. Engl.* 2002, *41*, 2360–2362.
173. *Carbohydr. Res.* 2000, *329*, 873–878.
174. *Tetrahedron Lett.* 2011, *52*, 6196–6198.
175. *Steroids* 2004, *69*, 599–604.
176. *J. Med. Chem.* 2006, *49*, 1792–1799.
177. *Organ. Lett.* 2007, *9*, 1465–1468.
178. *J. Am. Chem. Soc.* 2008, *130*, 2928–2929.
179. *Eur. J. Med. Chem.* 2011, *46*, 5959–5969.
180. *Synlett* 1995, 523–524.
181. *Chem. Biodiv.* 2007, *4*, 508–513.
182. Fischer, E.; Oetker, R. *Chem. Ber.* 1914, *46*, 4029–4040.
183. Fischer, E. *Ber.* 1914, 196–210.
184. Zemplén, G.; Gerecs, A. *Ber.* 1930, *63*, 2720–2729.
185. (a) *Chem. Commun.* 2013, *49*, 7815–7817; (b) *Angew. Chem. Int. Ed. Engl.* 2005, *44*, 7605–7607.
186. *J. Org. Chem.* 2008, *73*, 7952–7962.
187. *J. Org. Chem.* 2010, *75*, 1107–1118.
188. *J. Org. Chem.* 2007, *72*, 4663–4672.
189. *J. Org. Chem.* 1997, *62*, 6961–6967.
190. *J. Am. Chem. Soc.* 2008, *130*, 16791–16799.
191. *J. Org. Chem.* 2003, *68*, 7541–7543.
192. *Chem. Eur. J.* 2009, *15*, 10972–10982.
193. *J. Org. Chem.* 2010, *75*, 6747–6755.
194. *Chem. Eur. J.* 2003, *9*, 1909–1921.
195. *J. Org. Chem.* 2014, *10*, 1741–1748.

196. *Bioorg. Med. Chem.* 2010, *18*, 3726–3734.
197. *Synth. Commun.* 2010, *40*, 3378–3383.
198. *Organic Lett.* 2004, *6*, 3797–3800.
199. *Carbohydr. Res.* 1980, *86*, 133–136.
200. *Carbohydr. Res.* 2001, *336*, 107–115.
201. *Tetrahedron Lett.* 1999, *40*, 2217–2220.
202. *Carbohydr. Res.* 2013, *380*, 1–8.
203. *Tetrahedron Asymm.* 2008, *19*, 1919–1933.
204. *Angew. Chem. Int. Ed. Engl.* 2009, *48*, 2723–2726.
205. *J .Org. Chem.* 2006, *71*, 1390–1398.
206. *Can. J. Chem.* 2002, *80*, 555–558.
207. *Carbohydr. Res.* 2001, *337*, 87–91.
208. *Tetrahedron* 2013, *69*, 542–550.
209. *Carbohydr. Res.* 2010, *345*, 559–564.
210. *Carbohydr. Res.* 1996, *290*, 233–237.
211. *Acta Chem. Scand.* 1967, *21*, 910–914.
212. *Carbohydr. Res.* 1976, *49*, 201–207.
213. Cornforth, J. W. *Austr. J. Chem.* 1993, *46*, 157–170.
214. *Carbohydr. Res.* 2006, *341*, 2469–2477.
215. *Chem. Eur. J.* 2007, *13*, 4510–4522.
216. *Tetrahedron Lett.* 2009, *50*, 4536–4540.
217. Hudson, C. S.; Johnson, J. M. *J. Am. Chem. Soc.* 1915, *37*, 1270–1275.
218. Hronowski, L. J. J.; Szarek, W. A.; Hay, G. W.; Krebs, A.; Depew, W. T. *Carbohydr. Res.* 1989, *190*, 203–218.
219. Xu, P.; Yang, J. Y.; Kováč, P. *J. Carbohydr. Chem.* 2012, *31*, 711–720.
220. Xue, J.; Pan, Y.; Guo, Z. *Tetrahedron Lett.* 2002, *43*, 1599–1602.
221. Feather, M. S.; Whistler, R. L. *Tetrahedron Lett.* 1962, 667–668.
222. Hanessian, S. *Tetrahedron Lett.* 1967, *8*, 1549–1552.
223. Paulsen, H.; Koebernick, H.; Stenzel, W.; Köll, P. *Tetrahedron Lett.* 1975, *18*, 1493–1494.
224. Holy, A.; Soucek, M. *Tetrahedron Lett.* 1971, *12*, 185–188.
225. Brewster, K.; Harrison, J. M.; Inch, T. D. *Tetrahedron Lett.* 1979, *20*, 5051–5054.
226. Whistler, R. L.; Nayak, U. G.; Perkins, A. W., Jr. *Chem. Commun.* 1968, 1339–1340.
227. Pacák, J.; Točík, Z.; Černý, J. *Chem. Commun.* 1969, 77.
228. Lubineau, A.; Bienayme, H.; Gallic, J. L. *J. Chem. Soc. Chem. Commun.* 1989, 1918–1919.
229. Roy, R.; Laferriere, C. A. *J. Chem. Soc. Chem. Commun.* 1990, 1709–1711.
230. Hanessian, S. *Chem. Commun.* 1966, 796–798.
231. Nicolaou, K. C.; Dolle, R. E.; Chucholowski, A.; Randall, J. L. *J. Chem. Soc. Chem. Commun.* 1984, 1153–1154.
232. Windholz, T. B.; Johnston, D. B. R. *Tetrahedron Lett.* 1967, 2555–2557.
233. Chaudhary, S. K.; Hernandez, O. *Tetrahedron Lett.* 1979, 95–98.
234. Hanessian, S.; Banoub, J. *Carbohydr. Res.* 1977, *53*, C13–C16.
235. Byramova, N. E.; Ovchinnikov, M. V.; Backinowsky, L. V.; Kochetkov, N. K. *Carbohydr. Res.* 1983, *124*, C8–C11.
236. Tóth, A.; Medgyes, A.; Bajza, I.; Lipták, A.; Batta, G.; Kontrohr, T.; Peterffy, K.; Pozsgay, V. *Bioorg. Med. Chem. Lett.* 2000, *10*, 19–21.
237. Wernerová, M.; Hudlický, T. *Synlet* 2010, 2701–2707.
238. Daubenspeck, J. M.; Zeng, H.; Chen, P.; Dong, S.; Steichen, C. T.; Krishna, N. R.; Pritchard, D. G.; Turnbough, C. L. *J. Biol. Chem.* 2004, *279*, 30945–30953.
239. Collins, F. S.; Tabak, L. A. *Nature* 2014, *505*, 612–613.
240. Alberts, B.; Kirschner, M. W.; Tilghman, S.; Varmus, H. *PNAS* 2014, *111*, 5773–5777.

Introduction

The influence of sciences on the everyday life of people has greatly increased in the past decades, improving life expectancy and survival and curing diseases or dealing with epidemics. Breakthroughs have only been made possible through reliable and, more important, reproducible data reported by scientists. Biomolecules such as carbohydrates have found numerous applications, and proper reporting of the results obtained in carbohydrate chemistry has therefore been of prime importance.

Such careful and accurate reporting of data starts with the correct depiction of the chirality of carbohydrates, since modern software include *flip horizontal* and *flip vertical* tools to help the design of complex graphical artworks. Nevertheless, an organic chemist has to consider that these symmetry operations are producing enantiomers of the original molecule. Therefore, efforts from the scientific community to properly illustrate their reports are required. This comment applies not only to students but also to their mentors, the reviewers, and the editors of prestigious scientific reviews.

To continue down the list of problems we currently face in publishing quality carbohydrate/organic chemistry, the description of experimental protocols to synthesize a carbohydrate derivative often suffers from lack of detail, which sometimes compromises the positive outcome when reproducing the results. The very first information that an organic chemist looks for when analyzing the reaction mixture by thin layer chromatography is the retention factor (R_f value). This information is typically not reported in most recent publications. Also, the detailed description of the reaction (e.g., color change, the number of compounds present, and the consistency of the reaction mixture) is often omitted due to limited space, mostly imposed by editorial guidelines. In this context, the new investigators are strongly encouraged to read the excellent introductory chapter in Volume 2 by series editor Pavol (Paul) Kováč titled "For Young Readers Starting Their Careers in Synthetic Carbohydrate Chemistry: Tips and Tricks from up My Sleeve." In this very useful writing "In lieu of Introduction," Paul, with his more than 40 years of experience in carbohydrate chemistry, provides key advice on everyday manipulations that can ultimately afford better yields and high-purity products.

Our book series is intended to help organic chemists but also biologists and biochemists who aim to obtain synthetic compounds with limited efforts and assurance of success when following every step of the synthetic process described in the chapters. Volume 3 of the series contains 33 chapters written by experts in the discipline. As established within this series, each reaction has been thoroughly checked by independent research laboratories. Where checkers could not reach similar yields for a given reaction, the range of obtained values has been provided, thus giving some clues about the difficulty of a given process. The synthetic methods provided in Volume 3 describe either new compounds, known compounds obtained through simplified/improved procedures, or even known compounds but for which full characterization could not be found by scanning the literature.

After the pioneering Volume 1 edited by Paul Kováč and the succeeding and successful Volume 2 edited by Jeroen C. Codée and Gijs van der Marel, we now present to the carbohydrate community Volume 3 as a continuation, where we have compiled the most recent aspects of carbohydrate chemistry.

René Roy
Montréal, Québec, Canada

Sébastien Vidal
Lyon, France

Editors

René Roy has held since 2004 a Canadian Research Chair in therapeutic chemistry in the Department of Chemistry at the Université du Québec à Montréal (Québec, Canada). He has more than 40 years of experience in carbohydrate chemistry. He earned a PhD in carbohydrate chemistry in 1980 from the Université de Montréal under the expert guidance of Prof. Stephen Hanessian. He joined the National Research Council of Canada in Ottawa (Canada), where he was active from 1980 till 1985, and acquainted himself with carbohydrate-based vaccines. He then served as a professor at the Department of Chemistry, University of Ottawa, from 1985 to 2002. He was the recipient of the 2003 Melville L. Wolfrom Award from the ACS Division of Carbohydrate Chemistry for his contributions in the design of vaccines and glycodendrimers. He has more than 310 publications and has contributed to the development of two commercial carbohydrate-based vaccines against meningitis. His current interests are in multivalent carbohydrate protein interactions, medicinal chemistry, and nanomaterials.

Sébastien Vidal is a Centre National de la Recherche Scientifique (CNRS) researcher at the University of Lyon, Lyon, France. He earned a PhD in organic chemistry (2000) at the University of Montpellier (France) under the guidance of Prof. Jean-Louis Montero, where he synthesized mannose-6-phosphate analogues for drug delivery applications. He then joined the group of Sir J. Fraser Stoddart at the University of California, Los Angeles (UCLA) as a postdoctoral fellow. During the following two and a half years, he studied the synthesis and characterization of glycodendrimers and also the design of pseudorotaxanes. In 2003, he moved to the National Renewable Energy Laboratory (NREL, Golden, Colorado) and studied with Prof. Joseph J. Bozell the combination of organometallic and carbohydrate chemistries for the design of new reactions involving these two aspects of modern organic synthesis. After one year, he joined the group of Prof. Peter G. Goekjian at the University of Lyon in 2004. After successfully applying for a CNRS position the same year, Dr. Vidal started his own research projects dealing with carbohydrate chemistry and applications in biology. His main interests are in the design of glycoclusters for antiadhesive strategy against bacterial infections but also in enzyme inhibitors targeting glycogen phosphorylase with applications in type 2 diabetes (hypoglycemic drugs) or glycosyltransferases such as OGT. For his outstanding research, he was recently endowed the young investigator award, the 2014 *Prix du Groupe Français des glycosciences.*

Series Editor

Pavol Kováč, PhD, Dr. h.c., with more than 40 years of experience in carbohydrate chemistry and more than 300 papers published in refereed scientific journals, is a strong promoter of good laboratory practices and a vocal critic of publication of experimental chemistry lacking data that allow reproducibility. He earned an MSc in chemistry at Slovak Technical University in Bratislava (Slovakia) and a PhD in organic chemistry at the Institute of Chemistry, Slovak Academy of Sciences, Bratislava. After postdoctoral training at the Department of Biochemistry, Purdue University, Lafayette, Indiana (R. L. Whistler, advisor), he returned to the Institute of Chemistry and formed a group of synthetic carbohydrate chemists, which had been active mainly in oligosaccharide chemistry. After relocating to the United States in 1981, he first worked at Bachem, Inc., Torrance, California, where he established a laboratory for the production of oligonucleotides for automated synthesis of DNA. He joined the National Institutes of Health in 1983, where he is currently one of the principal investigators and chief of the section on carbohydrates (NIDDK, Laboratory of Bioorganic Chemistry), which was originally established by the greatest American carbohydrate chemist Claude S. Hudson and which is arguably the world's oldest research group continuously working on the chemistry, biochemistry, and immunology of carbohydrates. Dr. Kováč's primary interest is in the development of conjugate vaccines for infectious diseases from synthetic and bacterial carbohydrate antigens.

Contributors

Irene Agnolin
Max Mousseron Institute for
  Biomolecules
National Graduate School of Chemistry
Montpellier, France

Károly Ágoston
Carbosynth Ltd.
Compton, United Kingdom

Guillaume Anquetin
Sorbonne Paris Cité
University Paris Diderot
Paris, France

Abdelkrim Azzouz
Department of Chemistry
Université du Québec à Montréal
Montréal, Québec, Canada

Robert N. Ben
Department of Chemistry
University of Ottawa
Ottawa, Ontario, Canada

Sayantan Bhaduri
Department of Chemistry
Indiana University
Bloomington, Indiana

Markus Blaukopf
Department of Chemistry
University of Natural Resources and
  Life Sciences, Vienna
Vienna, Austria

Marie Bøjstrup
Carlsberg Laboratory
Copenhagen, Denmark

Andrea V. Bordoni
Gerencia Química
Centro Atómico Constituyentes
Comisión Nacional de Energía Atómica
Buenos Aires, Argentina

Omar Boutureira
Departament de Química Analítica i
  Química Orgànica
Universitat Rovira i Virgili
Tarragona, Spain

Walter J. Boyko
Department of Chemistry
Villanova University
Villanova, Pennsylvania

Jennie G. Briard
Department of Chemistry
University of Ottawa
Ottawa, Ontario, Canada

Yoan Brissonnet
Chimie Et Interdisciplinarité, Synthèse,
  Analyse, Modélisation
UFR des Sciences et des Techniques
Nantes, France

Stephen P. Brown
Department of Chemistry
University of Pennsylvania
Philadelphia, Pennsylvania

Giorgio Catelani
Department of Pharmacy
University of Pisa
Pisa, Italy

Contributors

Vincent Chagnault
Laboratoire LG2A
Centre National de la Recherche
    Scientifique
and
Institut de Chimie de Picardie
Université de Picardie Jules Verne
Amiens, France

Thibaut Chalopin
Laboratoire Chimie Et
    Interdisciplinarité, Synthèse,
    Analyse, Modélisation
Centre National de la Recherche
    Scientifique
Université de Nantes
Nantes, France

Felicia D'Andrea
Department of Pharmacy
University of Pisa
Pisa, Italy

Rosa M. de Lederkremer
Facultad de Ciencias Exactas y
    Naturales
Departamento de Química Orgánica
    Centro
de Investigaciones en Hidratos de
    Carbono
Universidad de Buenos Aires
Buenos Aires, Argentina

Alexei V. Demchenko
Department of Chemistry and
    Biochemistry
University of Missouri, St. Louis
St. Louis, Missouri

Catarina Dias
Faculty of Sciences
Department of Chemistry and
    Biochemistry
Center for Chemistry and Biochemistry
University of Lisbon
Lisbon, Portugal

Nicolas Drillaud
Laboratoire LG2A
Centre National de la Recherche
    Scientifique
and
Institut de Chimie de Picardie
Université de Picardie Jules Verne
Amiens, France

Christophe Dussouy
Institut de Chimie Moléculaire
    de Reims
Centre National de la Recherche
    Scientifique
Reims, France

Olof Engström
Arrhenius Laboratory
Department of Organic Chemistry
Stockholm University
Stockholm, Sweden

Carsten Fleck
Department of Chemistry,
    Pharmaceutical and Medicinal
    Chemistry
University of Hamburg
Hamburg, Germany

Maxime B. Fusaro
Laboratoire LG2A
Centre National de la Recherche
    Scientifique
and
Institut de Chimie de Picardie
Université de Picardie Jules Verne
Amiens, France

Denis Giguère
Département de Chimie
Université Laval
Québec City, Québec, Canada

Robert M. Giuliano
Department of Chemistry
Villanova University
Villanova, Pennsylvania

David Goyard
Department of Chemistry
Université du Québec à Montréal
Montréal, Québec, Canada

Tiziana Gragnani
Department of Pharmacy
University of Pisa
Pisa, Italy

Lorenzo Guazzelli
School of Chemistry and Chemical
  Biology
Centre for Synthesis and Chemical
  Biology
University College Dublin
Dublin, Ireland
and
Department of Pharmacy
University of Pisa
Pisa, Italy

David Gueyrard
Laboratoire LG2A
Centre National de la Recherche
  Scientifique
and
Institut de Chimie de Picardie
Université de Picardie Jules Verne
Amiens, France

Laure Guillotin
Institut de Chimie Organique
  et Analytique
Université d'Orléans
Orléans, France

András Guttman
Horváth Laboratory of Bioseparation
  Sciences
University of Debrecen
Debrecen, Hungary

Scott J. Hasty
Department of Chemistry and
  Biochemistry
University of Missouri, St. Louis
St. Louis, Missouri

Fernando Hernandez-Mateo
Faculty of Sciences
Department of Organic Chemistry
University of Granada
Granada, Spain

Ján Hirsch
Institute of Chemistry
Slovak Academy of Sciences
Bratislava, Slovakia

Ralph Hollaus
Department of Chemistry
University of Natural Resources and
  Life Sciences, Vienna
Vienna, Austria

Sławomir Jarosz
Institute of Organic Chemistry
Polish Academy of Sciences
Warsaw, Poland

Anushka B. Jayasuiya
Alberta Glycomics Centre
and
Department of Chemistry
Gunning-Lemieux Chemistry Centre
University of Alberta
Edmonton, Alberta, Canada

Xiao G. Jia
Department of Chemistry and
  Biochemistry
University of Missouri, St. Louis
St. Louis, Missouri

Solen Josse
Laboratoire LG2A
Centre National de la Recherche
    Scientifique
and
Institut de Chimie de Picardie
Université de Picardie Jules Verne
Amiens, France

László Kalmár
Horváth Laboratory of Bioseparation
    Sciences
University of Debrecen
Debrecen, Hungary

W. Scott Kassel
Department of Chemistry
Villanova University
Villanova, Pennsylvania

Abhijeet K. Kayastha
Department of Chemistry and
    Biochemistry
University of Missouri, St. Louis
St. Louis, Missouri

János Kerékgyártó
Faculty of Sciences and Technology
Bioorganic Laboratory
Department of Ecology
University of Debrecen
Debrecen, Hungary

Leonid O. Kononov
N. K. Kochetkov Laboratory of
    Carbohydrate Chemistry
N. D. Zelinsky Institute of Organic
    Chemistry
Russian Academy of Sciences
Moscow, Russian Federation

Paul Kosma
Department of Chemistry
University of Natural Resources and
    Life Sciences, Vienna
Vienna, Austria

Pavol Kováč
Laboratory of Bioorganic Chemistry
National Institute of Diabetes and
    Digestive and Kidney Diseases
National Institutes of Health
Bethesda, Maryland

José Kovensky
Laboratoire LG2A
Centre National de la Recherche
    Scientifique
and
Institut de Chimie de Picardie
Université de Picardie Jules Verne
Amiens, France

Michał Kowalski
Institute of Organic Chemistry
Polish Academy of Sciences
Warsaw, Poland

Martina Lahmann
School of Chemistry
University of Bangor
Bangor, United Kingdom

Katarzyna Łęczycka
Institute of Organic Chemistry
Polish Academy of Sciences
Warsaw, Poland

Reko Leino
Laboratory of Organic Chemistry
Åbo Akademi University
Åbo, Finland

Tianlei Li
Laboratory of Bio-Organic Chemistry
Department of Chemistry
University of Namur
Namur, Belgium

Cédric Epoune Lingome
Laboratoire LG2A
Centre National de la Recherche
    Scientifique
and
Institut de Chimie de Picardie
Université de Picardie Jules Verne
Amiens, France

Arkadiusz Listkowski
Institute of Physical Chemistry
Polish Academy of Sciences
Warsaw, Poland

Marcos J. Lo Fiego
Departamento de Química
Instituto de Química del Sur
Universidad Nacional del Sur
Buenos Aires, Argentina

Todd L. Lowary
Alberta Glycomics Centre
and
Department of Chemistry
Gunning-Lemieux Chemistry Centre
University of Alberta
Edmonton, Alberta, Canada

Xiaowei Lu
Laboratory of Bioorganic Chemistry
National Institute of Diabetes and
    Digestive and Kidney Diseases
National Institutes of Health
Bethesda, Maryland

Caroline Maierhofer
Fachbereich Chemie and Konstanz
    Research School of Chemical
    Biology
Universität Konstanz
Konstanz, Germany

Michał Malik
Institute of Organic Chemistry
Polish Academy of Sciences
Warsaw, Poland

Carla Marino
Facultad de Ciencias Exactas y
    Naturales
Departamento de Química Orgánica
Centro de Investigaciones en Hidratos
    de Carbono
Universidad de Buenos Aires
Buenos Aires, Argentina

Nicolò Marnoni
Department of Pharmaceutical Sciences
University of Piemonte Orientale "A.
    Avogadro"
Novara, Italy

Alice Martins
Faculty of Sciences
Department of Chemistry and
    Biochemistry
Center for Chemistry and Biochemistry
University of Lisbon
Lisbon, Portugal

Daniel Matzner
Life and Medical Sciences Institute
University of Bonn
Bonn, Germany

Anup Kumar Misra
Division of Molecular medicine
Bose Institute
Kolkata, India

Hani Mobarak
Arrhenius Laboratory
Department of Organic Chemistry
Stockholm University
Stockholm, Sweden

Chinmoy Mukherjee
Department of Chemistry
Indiana University
Bloomington, Indiana

Bernhard Müller
Department of Chemistry
University of Natural Resources and
    Life Sciences, Vienna
Vienna, Austria

Kottari Naresh
Department of Chemistry
Université du Québec à Montréal
Montréal, Québec, Canada

Isabelle Opalinski
Institut des Milieux et des Matériaux
    de Poitiers
Université de Poitiers
Poitiers, France

Weidong Pan
The Key Laboratory of Chemistry for
    Natural Products
Chinese Academy of Sciences
Guiyang, Guizhou, People's Republic
    of China

Luigi Panza
Dipartimento di Scienze del Farmaco
Università degli Studi del Piemonte
    Orientale "Amedeo Avogadro"
Novara, Italy

Sébastien Papot
Institut des Milieux et des Matériaux de
    Poitiers
Université de Poitiers
Poitiers, France

Wenjie Peng
Department of Cell and Molecular
    Biology
The Scripps Research Institute
La Jolla, California

Nicholas A. Piro
Department of Chemistry
Villanova University
Villanova, Pennsylvania

Nikita M. Podvalnyy
N. K. Kochetkov Laboratory of
    Carbohydrate Chemistry
N. D. Zelinsky Institute of Organic
    Chemistry
Russian Academy of Sciences
Moscow, Russian Federation

Nicola L.B. Pohl
Department of Chemistry
Indiana University
Bloomington, Indiana

Denis Postel
Laboratoire LG2A
Centre National de la Recherche
    Scientifique
and
Institut de Chimie de Picardie
Université de Picardie Jules Verne
Amiens, France

Jani Rahkila
Laboratory of Organic Chemistry
Åbo Akademi University
Åbo, Finland

Boddu Venkateswara Rao
Department of Chemistry
Indian Institute of Science Education
    and Research
Pune, India

Jun Rao
College of Chemistry and Life Sciences
Zhejiang Normal University
Jinhua, Zhejiang, People's Republic of
    China

Stéphanie Rat
Laboratoire LG2A
Centre National de la Recherche
    Scientifique
and
Institut de Chimie de Picardie
Université de Picardie Jules Verne
Amiens, France

Amélia P. Rauter
Faculty of Sciences
Department of Chemistry and
    Biochemistry
Center for Chemistry and Biochemistry
University of Lisbon
Lisbon, Portugal

Brigitte Renoux
Institut des Milieux et des Matériaux de
    Poitiers
Université de Poitiers
Poitiers, France

Sylvain Rocheleau
Department of Chemistry
McGill University
Montréal, Québec, Canada

René Roy
Department of Chemistry
Université du Québec à Montréal
Montréal, Québec, Canada

Sarah Roy
Département de Chimie
Université Laval
Québec City, Québec, Canada

Deepak Sail
Laboratory of Bioorganic Chemistry
National Institute of Diabetes and
    Digestive and Kidney Diseases
National Institutes of Health
Bethesda, Maryland

Francisco Santoyo-Gonzalez
Faculty of Sciences
Department of Organic Chemistry
University of Granada
Granada, Spain

Torben Seitz
Fachbereich Chemie and Konstanz
    Research School of Chemical
    Biology
Universität Konstanz
Konstanz, Germany

Radia Sennour
Department of Chemistry
Université du Québec à Montréal
Montréal, Québec, Canada

Tze Chieh Shiao
Department of Chemistry
Université du Québec à Montréal
Montréal, Québec, Canada

M. Soledade Santos
Faculty of Sciences
Department of Chemistry and
    Biochemistry
Center for Chemistry and Biochemistry
University of Lisbon
Lisbon, Portugal

Sameh E. Soliman
Laboratory of Bioorganic Chemistry
National Institute of Diabetes and
    Digestive and Kidney Diseases
National Institutes of Health
Bethesda, Maryland

Zoltán Szurmai
Faculty of Sciences and Technology
Bioorganic Laboratory
Department of Ecology
University of Debrecen
Debrecen, Hungary

Mikael Thomas
Institut des Milieux et des Matériaux de
    Poitiers
Université de Poitiers
Poitiers, France

Abdellatif Tikad
Laboratory of Bio-Organic Chemistry
Department of Chemistry
University of Namur
Namur, Belgium

Melissa Barrera Tomas
Department of Chemistry
Université du Québec à Montréal
Montréal, Québec, Canada

Mohamed Touaibia
Department of Chemistry and
    Biochemistry
Université de Moncton
Moncton, New Brunswick, Canada

Purav P. Vagadia
Department of Chemistry
Villanova University
Villanova, Pennsylvania

Monica Varese
Department of Pharmaceutical Sciences
University of Piemonte Orientale "A.
    Avogadro"
Novara, Italy

Juan A. Ventura
Instituto de Quimica Orgánica
General Consejo Superior de
    Investigaciones Científicas
Madrid, Spain

Stéphane P. Vincent
Laboratory of Bio-Organic Chemistry
Department of Chemistry
University of Namur
Namur, Belgium

Anne Wadouachi
Laboratoire LG2A
Centre National de la Recherche
    Scientifique
and
Institut de Chimie de Picardie
Université de Picardie Jules Verne
Amiens, France

Göran Widmalm
Department of Organic Chemistry
Arrhenius Laboratory
Stockholm University
Stockholm, Sweden

Grzegorz Witkowski
Institute of Organic Chemistry
Polish Academy of Sciences
Warsaw, Poland

Valentin Wittmann
Fachbereich Chemie and Konstanz
    Research School of Chemical
    Biology
Universität Konstanz
Konstanz, Germany

Peng Xu
Laboratory of Bioorganic Chemistry
National Institute of Diabetes and
    Digestive and Kidney Diseases
National Institutes of Health
Bethesda, Maryland

Anna Zawisza
Department of Organic and Applied
    Chemistry
University of Lodz
Lodz, Poland

Xiaojun Zeng
College of Chemistry and Life Sciences
Zhejiang Normal University
Jinhua, Zhejiang, People's Republic of
	China

Gaolan Zhang
State Key Laboratory of Natural and
	Biomimetic Drugs
and
School of Pharmaceutical Sciences
Peking University
Haidian, Beijing, People's Republic of
	China

Xiangming Zhu
College of Chemistry and Life Sciences
Zhejiang Normal University
Jinhua, Zhejiang, People's Republic of
	China

Alexander I. Zinin
N. K. Kochetkov Laboratory of
	Carbohydrate Chemistry
N. D. Zelinsky Institute of Organic
	Chemistry
Russian Academy of Sciences
Moscow, Russian Federation

Deanna L. Zubris
Department of Chemistry
Villanova University
Villanova, Pennsylvania

Section I

Synthetic Methods

1 Synthesis of Higher-Carbon Sugars Using the Phosphonate Methodology

Part I—Synthesis of Methyl (methyl 2,3,4-Tri-O-benzyl-α-D-glucopyranosid)uronate

Grzegorz Witkowski, Anna Zawisza,[†]
Michał Malik, and Sławomir Jarosz*

CONTENTS

The chiron approach represents a useful methodology in the synthesis of optically pure complex derivatives.[1] Synthesis of such targets with numerous hydroxyl groups can be conveniently carried out from sugar chirons. These starting materials are also used for the preparation of compounds possessing wide therapeutic applications such as iminosugars (e.g., nojirimycin).[2] Carbohydrate chirons may be also applied

* Corresponding author; e-mail: slawomir.jarosz@icho.edu.pl.
† Checker; e-mail: aniazawisza@poczta.onet.pl.

in the synthesis of rare *normal-size* carbohydrates (C5–C9) as well as the so-called higher-carbon sugars having more than 10 carbon atoms in the chain.

The synthetic efforts in our laboratory have been focused on the preparation of such aforementioned demanding targets.[3] The most convenient way for their preparation consists of a coupling of two properly activated carbohydrate subunits. From several methods proposed by us,[3] the *phosphonate methodology* seems to be the most applicable.[4] It requires conversion (Scheme 1.1) of one carbohydrate into the phosphonate (**I**) and the other into an ulose (**II**) followed by coupling of these two synthons under mild conditions, which provides precursor (enone **III**).

In this chapter, we describe the synthesis of the precursor **6** of the sugar phosphonate of type **I** (Scheme 1.2).

Preparation of the methyl uronate **6** started from the readily available methyl α-D-glucopyranoside (**1**) according to the standard carbohydrate methodology[8] and

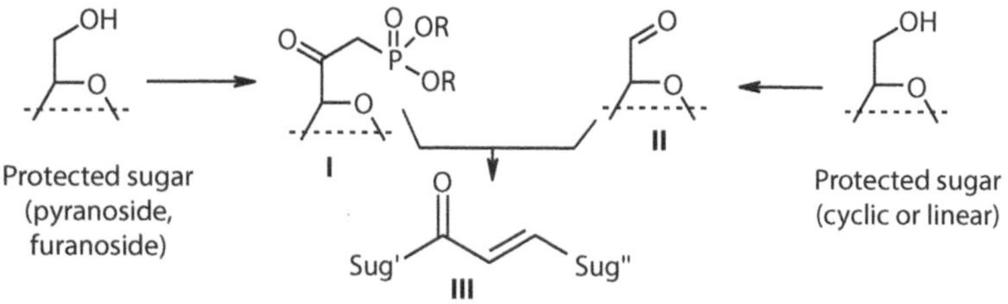

SCHEME 1.1 Coupling of two monosaccharide subunits using a Horner–Wadsworth–Emmons approach.

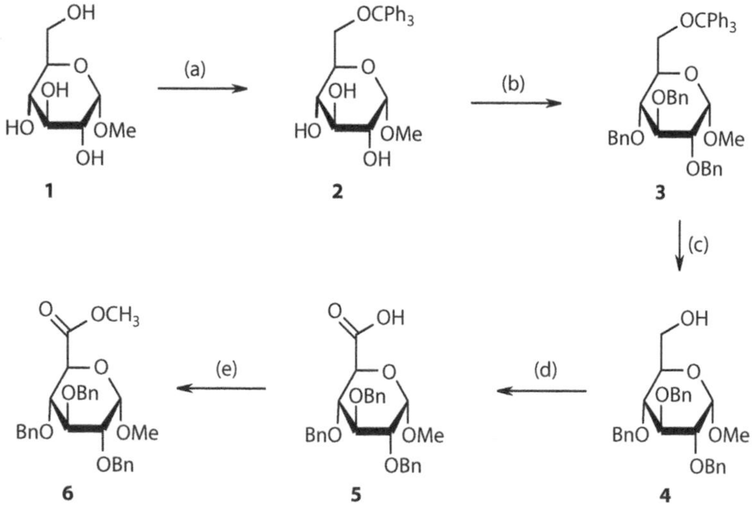

SCHEME 1.2 Preparation of methyl(methyl-2,3,4-tri-*O*-benzyl-α-D-glucopyranosid)uronate (**6**): (a) Ph₃C–Cl, py, Δt, 18 h, 86%; (b) BnBr, NaH, DMF, 6 h; (c) CH₂Cl₂/MeOH, pTsOH, rt., 5 h, 87% (from **2**); (d) acetone, Jones' reagent, 16 h; and (e) MeOH/THF, H₂SO₄ (cat.), rt., 8 h, 67% (from **4**).

proceeded with minor modifications. Protection of the primary hydroxyl group as a trityl ether left the secondary positions free. The resulting trityl ether **2** was then benzylated with benzyl bromide/sodium hydride in DMF, which furnished the fully protected glucoside **3**.

Removal of the trityl protection from the primary position provided the mono-alcohol **4** in high yield. This compound (CAS No. 53008–65–4) is commercially available, however, only in rather small quantities. According to the preparation described in this chapter, multigram quantities of glucoside **4** can be easily obtained.

The primary alcohol of derivative **4** was oxidized with Jones' reagent[5,8] to provide the uronic acid **5**. The preparation described here is superior to other methods in terms of the yield/cost ratio. The two-step approach combining the use of Dess–Martin periodinane and oxidation of the resulting aldehyde with sodium chlorite is experimentally less convenient.[6] Reaction of alcohol **4** with PDC in DMF often gave a product isolation of which from precipitate is difficult on a large scale.[7] Finally, the acid was esterified under standard conditions with acidic methanol avoiding the use of dangerous, especially on a large scale, diazomethane.[8] The overall yield for the synthesis of ester **6** from methyl α-D-glucopyranoside **1** described here is nearly 50% on a multigram scale. This preparation requires only two chromatographic purifications, intermediate **4** and the final ester **6**.

EXPERIMENTAL METHODS

GENERAL METHODS

NMR spectra were recorded in $CDCl_3$ (internal Me_4Si) with a Varian AM-600 (600 MHz ^1H, 150 MHz ^{13}C) at room temperature. Chemical shifts (δ) are reported in ppm relative to Me_4Si (δ 0.00) for ^1H and residual chloroform (δ 77.00) for ^{13}C. Structurally significant resonances were assigned by COSY (^1H–^1H), HSQC (^1H–^{13}C), and HMBC (^1H–^{13}C) correlations. Optical rotations were measured with a Jasco P-1020 apparatus using sodium light. Elemental analyses were performed with an Elementar vario EL III. Reagents were purchased from Sigma-Aldrich and Alfa Aesar and used without further purification. Commercial DMF and pyridine were dried over freshly activated 3 Å molecular sieves for at least 3 days. Hexanes (65°C–80°C fraction from petroleum) and EtOAc were purified by distillation. Other solvents were used without further purification. Thin-layer chromatography was carried out on silica gel 60 F_{254} (Merck). Flash chromatography was performed on Büchi glass columns packed with silica gel 60 (400–600 mesh, Merck), using Knauer Smartline system with a Büchi fraction collector. The organic solutions were dried over $MgSO_4$.

Methyl 6-O-(Triphenylmethyl)-α-D-glucopyranoside (2)[9]

Methyl α-D-glucopyranoside (**1**, 9.71 g, 50.0 mmol) was dissolved in dry pyridine (100 mL) to which DMAP (61 mg, 0.5 mmol) and trityl chloride were added (16.73 g, 60.0 mmol), and the mixture was stirred overnight (ca. 18 h) in oil bath at 60°C. After this time, TLC (5:1 CH_2Cl_2–MeOH) indicated formation of the product (R_f 0.9). Water (10 mL) was added and (after 15 min) the solution was concentrated to dryness in vacuum. The resulting solid material was dissolved in EtOH (25 mL) to which xylene (300 mL) was added and the solvents were evaporated under reduced pressure (most of pyridine was removed at this stage). The residue was dissolved in EtOAc (250 mL), washed with water (100 mL) and brine (100 mL) in order to remove traces of compound 1, and dried and concentrated to about 50 mL. Diethyl ether (250 mL) was added, which induced precipitation of the solid. It was filtered off, washed with Et$_2$O (5×30 mL), and dried (first on air then under high vacuum), providing methyl 6-O-(triphenylmethyl)-α-D-glucopyranoside (2; 18.77 g white amorphous solid), which was sufficiently pure for the next.

Methyl 2,3,4-Tri-O-benzyl-6-O-(triphenylmethyl)-α-D-glucopyranoside (3)[8,9]

Methyl 6-O-(triphenylmethyl)-α-D-glucopyranoside (2, 18.77 g, 43.0 mmol) was dissolved in dry DMF (215 mL) and cooled in water-ice bath. Sodium hydride (60% dispersion in mineral oil, 6.70 g, 167.7 mmol) was added to a vigorously stirred solution in 10 portions over 1 h. After 30 min, benzyl bromide (16.9 mL, 141.9 mmol) was added dropwise over 1 h via a syringe pump. After an additional 1 h, the cooling bath was removed and the mixture was stirred at rt. for 6 h. At this point, TLC (3:1 hexanes–EtOAc) indicated disappearance of the starting material (R_f 0.1) and formation of a new product (R_f 0.8). Methanol (50 mL) was carefully added to decompose the excess of hydride, the mixture was stirred overnight and concentrated, and the residue was dissolved in Et$_2$O (430 mL). The organic solution was washed with 1 M H$_2$SO$_4$ (200 mL), water (3×200 mL), and brine (100 mL), dried, and concentrated. Crude 2,3,4-tri-O-benzyl-6-O-(triphenylmethyl)-α-D-glucopyranoside (3) thus obtained was used in the next step without further purification.

Methyl 2,3,4-Tri-O-benzyl-α-D-glucopyranoside (4)[9,10]

Compound 3 obtained in the previous step was dissolved in CH$_2$Cl$_2$/MeOH (250 mL; 1:1 v/v) to which p-TsOH (1.64 g) was added and the mixture was stirred at room temperature for 3 h. After this time, TLC (2:1 hexane–EtOAc) indicated disappearance of the starting material (R_f 0.7) and formation of a new and more polar product (R_f 0.3). Triethylamine (10 mL) was added, the mixture was concentrated, and the product was purified by flash chromatography (4:1 → 1:1 hexanes–EtOAc) to afford methyl 2,3,4-tri-O-benzyl-α-D-glucopyranoside (4) (17.36 g, 87% yield over two steps) as colorless oil, [α]$_D$ +23.3 (c 1.0, CHCl$_3$); lit.[10]: 66.5°C–67°C, [α]$_D$ + 23.5 (c 1.0, chloroform). ^{1}H NMR δ: 7.38–7.27 (m, 15H, arom.), 4.99 (d, J = 10.9 Hz, 1H, OCH$_2$Ph), 4.88 (d, J = 11.0 Hz, 1H, OCH$_2$Ph), 4.84 (d, J = 10.9 Hz, 1H, OCH$_2$Ph), 4.80 (d, J = 12.1 Hz, 1H, OCH$_2$Ph), 4.65 (m, 2H, 2×OCH$_2$Ph), 4.57 (d, J = 3.5 Hz, 1H, H-1), 4.01 (t, J = 9.3 Hz, 1H, H-3), 3.76 (ddd, J = 11.7, 5.3, 2.7 Hz, 1H, H-6), 3.72–3.62 (m, 2H, H-4 and H-6), 3.54–3.48 (m, 2H, H-2 and H-5), 3.36 (s, 3H, OMe), 1.64 (dd, J = 7.5, 5.4 Hz, 1H, OH). ^{13}C NMR δ: 138.7 and 2×138.1 (quat. benzyl), 128.5–127.6 (arom.), 98.2 (C-1), 81.9 (C-3), 80.0 (C-2), 77.4 (C-5), 75.7 (OCH$_2$Ph),

75.0 (OCH_2Ph), 73.4 (OCH_2Ph), 70.6 (C-4), 61.9 (C-6), 55.2 (OMe). Anal. calcd for $C_{28}H_{32}O_7$: C, 72.39; H, 6.94. Found: C, 72.55; H 7.07.

Methyl 2,3,4-Tri-O-benzyl-α-D-glucopyranosiduronic Acid (5)[11]

To a solution of compound **4** (17.36 g, 37.4 mmol) in acetone (190 mL) was added Jones' reagent (prepared from CrO_3 [7.48 g, 74.8 mmol] and H_2SO_4 [12 mL] in water to 40 mL) in five portions over 4 h. After stirring for an additional 12 h, TLC (1:1 hexanes–EtOAc) indicated disappearance of the starting material (R_f 0.5) and formation of a new and more polar product (R_f 0.2). 2-Propanol (50 mL) was added and the mixture was stirred for 1 h, to decompose excess of the oxidant (the color turned dark green). Toluene (150 mL) was added and most of acetone was removed under vacuum (175 mbar, 40°C). Ether (400 mL) and water (100 mL) were added, the layers were separated, and the aqueous layer was extracted with CH_2Cl_2 (3 × 75 mL). The combined organic solutions were washed with water (2 × 200 mL) and brine (100 mL), dried, and concentrated providing methyl 2,3,4-tri-O-benzyl-α-D-glucopyranosiduronic acid (**5**), which was used in next step without further purification.

Methyl (methyl-2,3,4-tri-O-benzyl-α-D-glucopyranosid)uronate (6)[12,8]

To a solution of crude **5** in MeOH (150 mL) and THF (30 mL) was added conc. sulfuric acid (0.4 mL) and the mixture was stirred at room temperature for 8 h. After this time, TLC (1:1 hexanes–EtOAc) indicated disappearance of the starting material (R_f 0.2) and formation of a new, less polar product (R_f 0.8). Triethylamine (10 mL) was added, the mixture was concentrated, and the product was isolated by flash chromatography (7:1 → 4:1 hexanes–EtOAc) to afford methyl (methyl 2,3,4-tri-O-benzyl-α-D-glucopyranosid)uronate (**6**, 12.37 g, 67% over two steps) as a colorless oil, $[\alpha]_D$ +12.0 (c 1.0, $CHCl_3$); lit.,[8] $[\alpha]_D$ +13.0 (c 1, $CHCl_3$); lit.,[13] $[\alpha]_{578}$ +17.6 (c 1, $CHCl_3$). ^{1}H NMR δ: 7.37–7.20 (m, 15H, arom.), 4.95 (d, J = 10.9 Hz, 1H, OCH_2Ph), 4.84–4.78 (m, 3H, 3 × OCH_2Ph), 4.64 (d, J = 12.1 Hz, 1H, OCH_2Ph), 4.60 (d, J = 3.5 Hz, 1H, H-1), 4.57 (d, J = 10.9 Hz, 1H, OCH_2Ph), 4.19 (d, J = 10.0 Hz, 1H, H-5), 3.99 (t, J = 9.3 Hz, 1H, H-3), 3.72 (dd, J = 10.0, 9.3 Hz, 2H, H-4), 3.70 (s, 3H, $COOCH_3$), 3.57 (dd, J = 9.3, 3.5 Hz, 1H, H-2), 3.39 (s, 3H, OMe). ^{13}C NMR δ: 170.0 ($COOCH_3$), 138.5 and 2 × 137.9 (quat. benzyl), 128.5–127.7 (arom.), 98.7 (C-1), 81.4 (C-3), 79.54 (C-4), 79.3 (C-2), 75.9 (OCH_2Ph), 75.1 (OCH_2Ph), 73.6 (OCH_2Ph), 70.1 (C-5), 55.6 (OMe), 52.4 ($COOCH_3$).

Anal. calcd for $C_{29}H_{32}O_7$: C, 70.71; H 6.55. Found: C, 70.83, H 6.57.

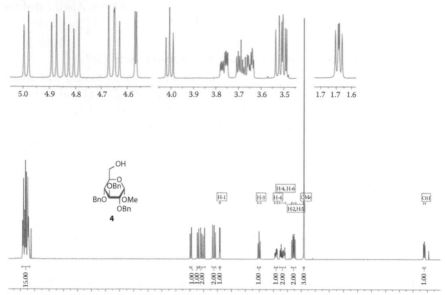

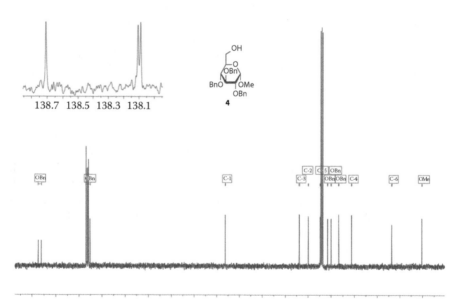

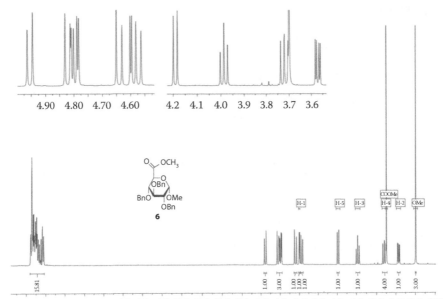

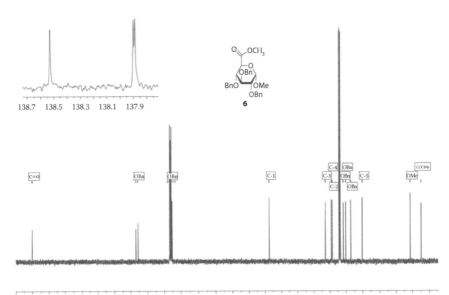

REFERENCES

1. (a) Hanessian, S. *Total Synthesis of Natural Products: The Chiron Approach*, Pergamon Press, New York, **1983**; (b) Fraser-Reid, B. *Acc. Chem. Res.* **1996**, *29*, 57–66 and references therein.
2. Tyler, P.C.; Winchester, B.G. Chapter 7, Synthesis and biological activity of castanospermine and close analogs. in *Iminosugars as Glycosidase Inhibitors: Nojirimycin and Beyond*, 1st edn.; Stütz, A.E. Ed., Wiley, Weinheim, Germany, **1999**, pp. 125–156.
3. Review: Jarosz, S. *Curr. Org. Chem.* **2008**, *12*, 985–994; see also: Cieplak, M.; Jarosz, S. *Tetrahedron: Asymmetry* **2011**, *22*, 1757–1762.
4. Jarosz, S.; Mach, M. *J. Chem. Soc. Perkin Trans. 1* **1998**, 3943–3948.
5. Bowden, K.; Heilbron, I.M.; Jones, E.R.H.; Weedon, B.C.L. *J. Chem. Soc.* **1946**, 39–45.
6. Chambers, D.J.; Evans, G.R.; Fairbanks, A.J. *Tetrahedron* **2005**, *61*, 7184–7192.
7. de Raadt, A.; Stütz, A.E. *Carbohyd. Res.* **1991**, *220*, 101–116.
8. Kováč, P.; Alfodi, J.; Košík, M. *Chem. Zvesti* **1974**, *28*, 820–832.
9. Dorgeret, B.; Khemtemourian, L.; Correia, I.; Soulier, J.-L.; Lequin, O.; Ongeri, S. *Eur. J. Med. Chem.* **2011**, *46*, 5959–5969.
10. Eby, R.; Schuerch, C. *Carbohydr. Res.* **1974**, *34*, 79–90.
11. Cheng, K.; Liu, J.; Liu, X.; Li, H.; Sun, H.; Xie, J. *Carbohydr. Res.* **2009**, *344*, 841–850.
12. De Raadt, A.; Stutz, A.E. *Carbohydr. Res.* **1991**, *220*, 101–115.
13. Schmidt, R.R.; Ruecker, E. *Tetrahedron Lett.* **1980**, *21*, 1421–1424.

2 Synthesis of Higher-Carbon Sugars Using the Phosphonate Methodology

Part II—Synthesis of Dimethyl (methyl 2,3,4-Tri-O-benzyl-α-D-gluco-heptopyranos-6-ulos-7-yl)phosphonate and Application for Carbon Chain Elongation

Michał Malik, Anna Zawisza,[†]
Grzegorz Witkowski, and Sławomir Jarosz*

CONTENTS

* Corresponding author; e-mail: slawomir.jarosz@icho.edu.pl.
[†] Checker; e-mail: a.m.zawisza@wp.pl.

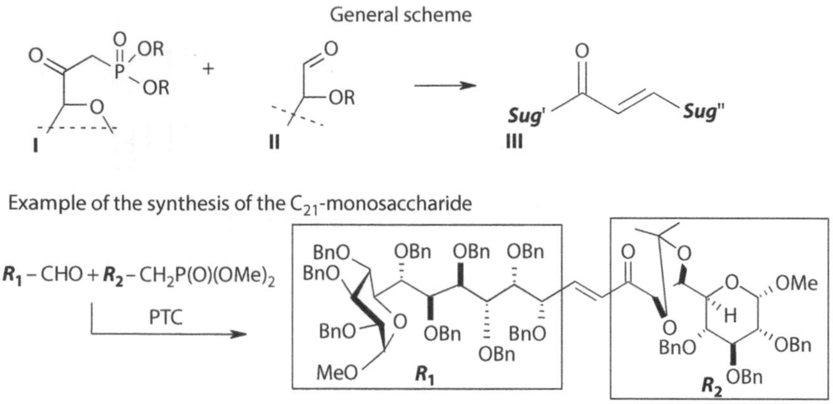

General scheme

Example of the synthesis of the C_{21}-monosaccharide

SCHEME 2.1   Coupling of two monosaccharide subunits by a phosphonate methodology.

Higher-carbon sugars can be conveniently synthesized according to the methodology shown in Scheme 2.1. It is based on a coupling of a carbohydrate phosphonate **I** with a carbohydrate aldehyde **II**, which provides the higher carbohydrate enone **III** in good yield and with (almost exclusively) *E*-configuration across the double bond.[1] The method is applicable to making higher carbohydrates of almost unrestricted chain length. We have already reported the synthesis of the C_{21}-monosaccharide by coupling of a C_{12}-aldehyde and the C_9-phosphonate **2**.[2]

In this chapter, we present a model synthesis of the higher carbohydrate enone composed of two units derived from D-glucose and D-glyceraldehyde. In the procedure, we will also present an efficient method of the synthesis of phosphonate **2**[3] (derivative of D-glucose) and its application in the preparation of higher-carbon sugars.

Methyl (methyl 2,3,4-tri-*O*-benzyl-α-D-glucopyranosid)uronate (see Chapter 1) **1** was treated with the lithiated anion of dimethyl methylphosphonate to afford the desired Horner–Wadsworth–Emmons reagent **2** in excellent yield (Scheme 2.2).

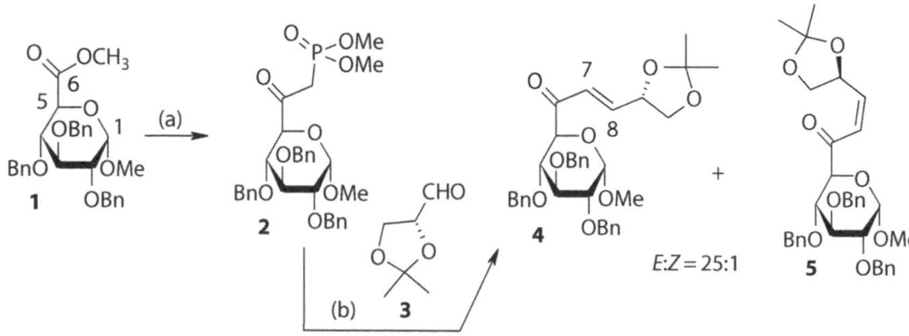

SCHEME 2.2   Synthesis of methyl 2,3,4-tri-*O*-benzyl-7,8-dideoxy-7(*E*)-eno-9,10-*O*-isopropylidene-D-*glycero*-α-D-*gluco*-dec-1,5-pyranoside (**4**): (a) LiCH$_2$P(O)(OMe)$_2$, THF, −78°C to rt., 2 h (93%); (b) **3**, 18-*c*-6, K$_2$CO$_3$, toluene, 24 h (90%).

This synthon was subjected to treatment with 2,3-di-O-isopropylidene-D-glyceraldehyde under mild solid phase transfer catalysis conditions (as proposed by Makosza[4]) to afford the title compound **4** with the E-geometry (as established by ¹H NMR: $J_{7,8} = 15.7$ Hz) across the double bond. We were able to isolate also small amounts (ca. 3%) of the isomeric Z-olefin **5**.

GENERAL METHODS

NMR spectra were recorded at room temperature for solutions in CDCl₃ (internal Me₄Si) with a Varian AM-600 (600 MHz ¹H, 150 MHz ¹³C). Chemical shifts (δ) are reported in ppm relative to Me₄Si (δ 0.00) for ¹H and residual chloroform (δ 77.00) for ¹³C. Carbohydrate ring nuclei-NMR signals were assigned by COSY (¹H–¹H), HSQC (¹H–¹³C), and HMBC (¹H–¹³C) correlations. Reagents were purchased from Sigma-Aldrich and Alfa Aesar and used without further purification. THF (predried over KOH) was distilled over benzophenone/sodium and stored over freshly activated molecular sieves 3 Å. Toluene (predried over CaH₂) was distilled and stored over freshly activated molecular sieves 3 Å. Hexanes (65°C–80°C fraction from petroleum) and EtOAc were purified by distillation. Thin-layer chromatography was carried out on silica gel 60 F₂₅₄ (Merck). Flash chromatography was performed on Büchi glass columns packed with silica gel 60 (400–600 mesh, Merck), using a Knauer Smartline system with a Büchi fraction collector. Solutions in organic solvents were dried over MgSO₄. Optical rotations were measured with a Jasco P-1020 polarimeter at room temperature ($c = 1$, CH₂Cl₂). MS spectra were recorded with a SYNAPT G2-S HDMS spectrometer. Melting points were measured with a SRS EZ-Melt apparatus and are uncorrected.

DIMETHYL (METHYL 2,3,4-TRI-O-BENZYL-α-D-GLUCO-HEPTOPYRANOS-6-ULOS-7-YL)PHOSPHONATE (2)

The reaction was conducted under argon. Dimethyl methylphosphonate (4.7 mL, 44.0 mmol, 4.2 equiv.) was dissolved in anhydrous THF (100 mL) and the mixture was cooled to −78°C. n-Butyllithium (16.9 mL of a 2.5 M solution in hexanes, 4 equiv.) was added under vigorous stirring over a period of 30 min with a syringe pump. Then, a solution of ester **1** (5.15 g, 10.5 mmol) in anhydrous THF (20 mL) was added at −78°C *via* a syringe pump during 30 min. The cooling bath was removed and the reaction mixture was allowed to reach room temp. (about 1 h). After this time, TLC indicated total consumption of the starting material (hexanes–EtOAc, 2:1) and formation of a new product (5% (v/v) methanol in 1:1 hexanes–EtOAc). The mixture was transferred to a separatory funnel, then water (100 mL), brine (50 mL), and ethyl acetate (50 mL) were added. The content was shaken and the layers were separated. The aqueous one was extracted with 5% (v/v) methanol in ethyl acetate (4×30 mL). Combined organic solutions were washed with brine (20 mL), dried, and concentrated *in vacuo*. Flash chromatography (linear gradient, 1:1 hexanes–EtOAc → EtOAc) afforded the title compound **2** (5.7 g, 93%). The compound is unstable at ambient temperature and should not be stored for prolonged periods of time. HRMS: found, $m/z = 607.2074$. Calcd for $C_{31}H_{37}O_9PNa$ (M + Na⁺): 607.2073; $R_f = 0.36$ (5% (v/v) methanol in 1:1 hexanes–EtOAc).

¹H NMR δ: 7.36–7.23 (m, arom.), 4.98 (d, $J = 10.9$ Hz, 1H, OCH_2Ph), 4.81 (m, 3H, OCH_2Ph), 4.63 (m, 3H, 2 × OCH_2Ph, H-1), 4.36 (d, $J = 9.9$ Hz, 1H, H-5), 4.03 (~t, $J = 9.3$ Hz, 1H, H-3), 3.72 (~t, $J = 11$ Hz, 6H, 2 × POCH_3), 3.64 (~t, $J = 9.0$ Hz, 1H, H-4), 3.52 (dd, $J = 9.7$, 3.5 Hz, 1H, H-2), 3.43 (s, 3H, OCH_3), 3.27 (dd, $J = 14.0$, 22.6 Hz, 1H, H-7a), 3.17 (dd, $J = 14.0$, 22.4 Hz, 1H, H-7b). ¹³C NMR δ: 198.4 (d, $J = 6.8$ Hz, C-6), 138.4, 137.9, 137.8 (3 × quat. benzyl), 128.5–127.6 (arom.), 98.6 (C-1), 81.8 (C-3), 79.3 (C-2), 78.8 (C-4), 75.8 (OCH_2Ph), 75.0 (OCH_2Ph), 73.8 (C-5), 73.5 (OCH_2Ph), 55.7 (OCH_3), 53.1 (d, $J = 6.4$ Hz, POCH_3), 52.9 (d, $J = 6.4$ Hz, POCH_3), 39.4 (d, $J = 129$ Hz, C-7).

METHYL 2,3,4-TRI-*O*-BENZYL-7,8-DIDEOXY-7(E)-ENO-9,10-*O*-ISOPROPYLIDENE-D-*GLYCERO*-α-D-GLUCO-DEC-1,5-PYRANOSID-6-ULOSE (4)

The reaction was conducted under argon.* To a solution of phosphonate **2** (5.7 g, 9.76 mmol) in anhydrous toluene (140 mL), 2,3-*O*-isopropylidene-D-glyceraldehyde (**3**; ~2 equiv., freshly prepared† from 1,2:5,6-di-*O*-isopropylidene-D-mannitol) was added followed by finely pulverized potassium carbonate (4.4 g, 31.9 mmol, 3.3 equiv.) and 18-crown-6 ether (60 mg, 0.23 mmol, 2% mol) and the mixture was stirred at room temperature for 24 h. TLC indicated total consumption of the starting material (2:1 hexanes–EtOAc) and formation of two products slightly differing in polarities (3:1 hexanes–EtOAc).‡ The mixture was partitioned between water (150 mL) and diethyl ether (75 mL), the layers were separated, and the aqueous one was extracted with diethyl ether (3 × 40 mL). Combined ethereal solutions were washed with brine (20 mL), dried, and concentrated *in vacuo*. Flash chromatography (linear gradient, hexanes → 7:3 hexanes–EtOAc) provided both geometric isomers of the desired product as pure substances:

4 (isomer *E*) as a pale yellow thick oil (5.01 g, 87%, more polar), HRMS: found, $m/z = 611.2623$. Calcd for $C_{35}H_{40}O_8Na$ (M + Na⁺): 611.2621. Anal. calcd for $C_{35}H_{40}O_8$: C, 71.4; H, 6.85. Found: C, 71.18; H, 6.82%; [α]$_D$ = +42.7; $R_f = 0.37$ (hexanes–EtOAc, 3:1).

¹H NMR δ: 7.36–7.2 (m, arom.), 6.88 (dd, $J = 15.7$, 5.5 Hz, 1H, H-8), 6.60 (dd, $J = 15.7$, 1.4 Hz, 1H, H-7), 4.97 (d, $J = 10.9$ Hz, 1H, OCH_2Ph), 4.80 (m, 3H, OCH_2Ph),

---

* This reaction may be carried out without such precautions and in commercial toluene, but the yield is decreased by ~15%.

† Preparation of **3** (Schmid, R.C. et al., *Org. Synth. Coll.* 9, 450, 1998): 1,2:5,6-di-*O*-isopropylidene-D-mannitol⁵ (2.85 g, 10.9 mmol) was dissolved in DCM (28.5 mL) to which saturated sodium bicarbonate (1.1 mL) was added under vigorous stirring. Then NaIO₄ (4.7 g, 22 mmol, 2 equiv.) was added in portions (temperature was kept below 35°C). After 3 h, TLC (hexanes-EtOAc, 2:1) indicated disappearance of the starting material and formation of a new, slightly less polar product. MgSO₄ was added (1.4 g) and the mixture was stirred for additional 15 min. The solid material was filtered off and rinsed with DCM (3 × 15 mL). The solvent was evaporated under reduced pressure and the aldehyde was used without further purification.

‡ If at this point TLC indicates incomplete conversion of the starting material, another portion of finely pulverized potassium carbonate (3 g) should be added and stirring continued for another few hours.

4.61 (m, H-1, H-9, OC*H*₂Ph), 4.53 (d, *J* = 10.6 Hz, 1H, OC*H*₂Ph), 4.37 (d, *J* = 9.9 Hz, 1H, H-5), 4.06 (m, 2H, H-3, H-10*a*), 3.65 (dd, *J* = 9.8, 9.0 Hz, 1H, H-4), 3.56 (m, 2H, H-2, H-10*b*), 3.40 (s, 3H, OC*H*₃), 1.41 (s, 3H, C(C*H*₃)₂), 1.39 (s, 3H, C(C*H*₃)₂).

¹³C NMR δ: 195.0 (C-6), 144.3 (C-8), 138.5, 137.9, 137.7 (3×quat. benzyl), 128.5–127.9 (arom.), 127.3 (C-7), 110.2 (*C*(CH₃)₂), 98.7 (C-1), 81.7 (C-3), 79.32 (C-2), 79.29 (C-4), 75.9 (O*C*H₂Ph), 75.1 (C-9), 75.0 (O*C*H₂Ph), 73.6 (O*C*H₂Ph), 73.0 (C-5), 68.6 (C-10), 55.8 (O*C*H₃), 26.4 (C(*C*H₃)₂), 25.7 (C(*C*H₃)₂).

5 (isomer *Z*), (198 mg, 3%, less polar), which crystallized on standing, m.p. 87–89°C; *R*ᶠ = 0.42 (hexanes:ethyl acetate, 3:1). [α]ᴅ = +118.6; HRMS: found, *m/z* = 611.2625. Calcd for C₃₅H₄₀O₈Na (M + Na⁺): 611.2621. ¹H NMR δ: 7.36–7.19 (m, arom.), 6.39 (m, 2H, H-7, H-8), 5.34 (m, 1H, H-9), 4.97 (d, *J* = 10.9 Hz, 1H, OC*H*₂Ph), 4.79 (m, 3H, OC*H*₂Ph), 4.64 (d, *J* = 12.1 Hz, 1H, OC*H*₂Ph), 4.62 (d, *J* = 3.5 Hz, 1H, H-1), 4.54 (d, *J* = 10.5 Hz, 1H, OC*H*₂Ph), 4.46 (dd, *J* = 8.3, 7.3 Hz, 1H, H-10*a*), 4.22 (d, *J* = 10.0 Hz, 1H, H-5), 4.04 (app. t, *J* = 9.3 Hz, 1H, H-3), 3.60 (dd, *J* = 9.8, 9.1 Hz, 1H, H-4), 3.54 (m, 2H, H-2, H-10*b*), 3.39 (s, 3H, OC*H*₃), 1.44 (s, 3H, C(C*H*₃)₂), 1.39 (s, 3H, C(C*H*₃)₂).¹³C NMR δ: 195.7 (C-6), 150.8 (C-8), 138.5, 137.9, 137.6 (3× quat. benzyl), 128.5–127.9 (arom.), 124.5 (C-7), 109.8 (*C*(CH₃)₂), 98.6 (C-1), 81.7 (C-3), 79.4 (C-2), 78.9 (C-4), 75.9 (O*C*H₂Ph), 75.0 (O*C*H₂Ph), 74.6 (C-9), 74.2 (C-5), 73.6 (O*C*H₂Ph), 69.4 (C-10), 55.8 (O*C*H₃), 26.5 (C(*C*H₃)₂), 25.3 (C(*C*H₃)₂). Anal. calcd for C₃₅H₄₀O₈: C, 71.41; H, 6.85. Found: C, 71.58; H, 6.72.

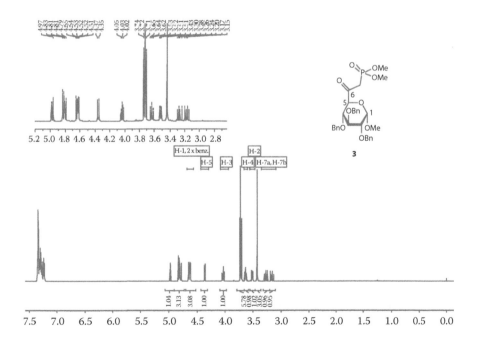

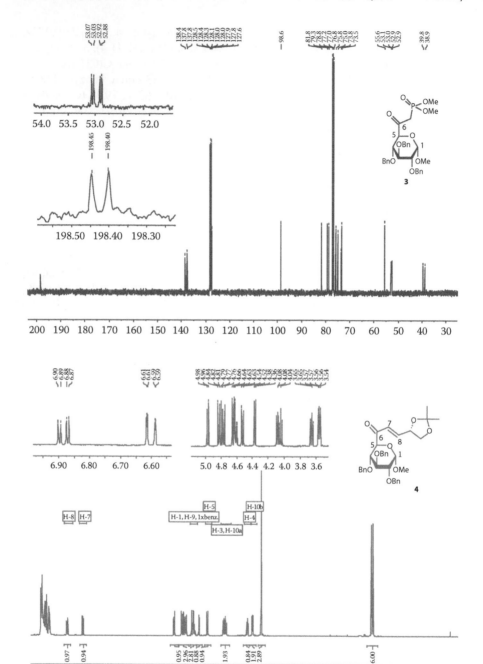

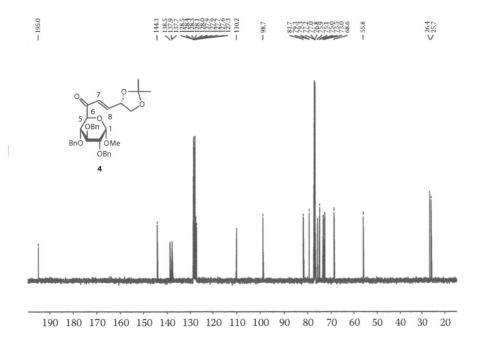

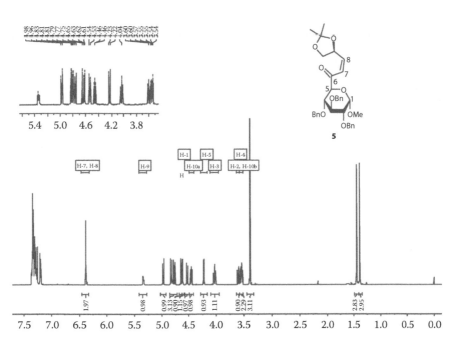

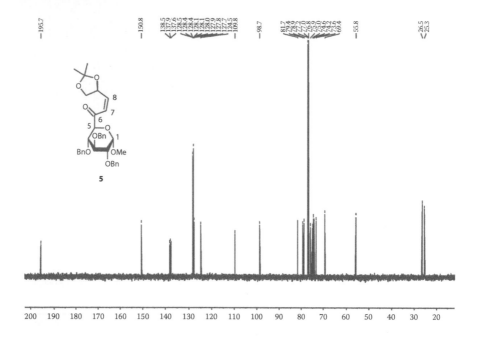

REFERENCES

1. *Review*: Jarosz, S. *Curr. Org. Chem.*, **2008**, *12*, 985–994; see also: Cieplak, M.; Jarosz, S. *Tetrahedron: Asymmetry*, **2011**, *22*, 1757–1762; Jarosz, S.; Mach, M. *J. Chem. Soc. Perkin Trans. 1*, **1998**, 3943–3948.
2. Jarosz, S. *Tetrahedron Lett.*, **1994**, *35*, 7655–7658; Jarosz, S.; Sałański, P.; Mach, M. *Tetrahedron*, **1998**, *54*, 2583–2594.
3. Jarosz, S.; Mach, M. *J. Chem. Soc. Perkin Trans. 1*, **1998**, 3943.
4. Fedorynski, M.; Wojciechowski, K.; Matacz, Z.; Makosza, M. *J. Org. Chem.*, **1978**, *43*, 4682–4864.
5. Kierstead R.W.; Faraone, A.; Mennona, F.; Mullin, J.; Guthrie, R.W.; Crowley, H.; Simko, B.; Blaber, L.C. *J. Med. Chem.*, **1983**, *26*, 1561–1569; this compound is also commercially available.

3 Preparation of Methyl, Butyl, Hexyl, and Octyl 2,3,4-Tri-O-acetyl-D-glucopyranuronates Using Microwave Irradiation

Cédric Epoune Lingome,
David Gueyrard,[†] Stéphanie Rat,
José Kovensky, and Anne Wadouachi [*]

CONTENTS

* Corresponding author; e-mail: anne.wadouachi@u-picardie.fr.
† Checker; e-mail: david.gueyrard@univ-lyon1.fr.

R = methyl 89%
    butyl 81%
    hexyl 90%
    octyl 90%

Uronic acids are carbohydrates present in relevant biologically active compounds as glycoconjugates and oligosaccharides. The uronic acid units can be linked through their ester linkage involving carboxylic acid function[1] or their anomeric carbon atom in glycosidic bond. Alkyl 2,3,4-tri-O-acetyl-D-glucuronates are key intermediate compounds for the synthesis of glycosyl donors as glucuronyl trichloroacetimidates for the synthesis of the target molecules.[2,3]

Glucurono-6,3-lactone has been used as precursor for the synthesis of the title compound 2a.[4] A three-step reaction involving a ring-opening reaction with methanol was followed by acetylation with acetic anhydride in pyridine and finally cleavage of the anomeric acetyl group with an equimolar amount of tributyltin methoxide in dichloroethane[5,6] or THF[7] or with ammonium acetate[8,9] in DMF. Compound 2a can be also obtained from peracetylated methyl glucopyranuronyl bromide in the presence of cadmium carbonate and water.[10–12] All the previously reported procedures require a selective deprotection of the anomeric position and these steps are performed using organic solvents.

We have developed a microwave-assisted glycosidation and/or esterification of glucuronic acid in the presence of different promoters.[13,14] Here, we report the use of FeCl$_3$, as selective Lewis acid catalyst for the microwave-assisted transesterification of the 2,3,4-tri-O-acetyl-D-glucopyranurono-6,1-lactone 1[13] with methanol, butanol, hexanol, or octanol. The corresponding compounds 2a–d are obtained in a short reaction time (10 min) and the chemoselectivity directs the reaction to the formation of esterified products exclusively. The synthesis of alkyl 2,3,4-tri-O-acetyl-D-glucuronates was performed in the presence of 20 mol% of FeCl$_3$ and 6 equivalents of the alcohol. In the case of methanol, an excess of the alcohol (20 equiv.) was necessary due to its high volatility.

EXPERIMENTAL METHODS

GENERAL METHODS

All chemicals were purchased from Aldrich or Acros (France). Thin-layer chromatography (TLC) was performed on silica gel 60 F$_{254}$ (Merck) plates with visualization by UV light (254 nm) and/or charring with a vanillin–H$_2$SO$_4$ or a 5% alcoholic molybdophosphoric acid reagent. Preparative column chromatography was performed using 230–400 mesh Merck silica gel (purchased from Aldrich). Optical rotations were determined with a Jasco Dip-370 electronic micropolarimeter (10 cm cell). ^{1}H and ^{13}C NMR spectra were recorded on a Bruker 300 WB spectrometer at

300 and 75 MHz, respectively. Chemical shifts are given as δ values with reference to tetramethylsilane (TMS). High-resolution electrospray experiments (ESI-HRMS) were performed on a Waters-Micromass Q-TOF *Ultima Global* hybrid quadrupole time-of-flight instrument, equipped with an electrospray (Z-Spray) ion source (Waters-Micromass, Manchester, United Kingdom). All solvents were distilled before use. Microwave irradiation was performed in a CEM-Discover® system or in a Biotage initiator 2.0 system.

GENERAL PROTOCOL

To a mixture of 2,3,4-tri-O-acetyl-D-glucopyranurono-6,1-lactone (**1**)[12] (100 mg, 0.33 mmol) and alcohol (20 equiv. for methanol or 6 equiv. for butanol, hexanol, and octanol), anhydrous $FeCl_3$ (10.7 mg, 0.2 equiv.) was added. The mixture was placed in the microwave reactor equipped with a condenser under argon atmosphere and irradiated at 60°C (methanol) or 65°C (butanol, hexanol, and octanol) for 10 min, using the dynamic method. The crude product was purified by column chromatography (7:3 to 5:5 cyclohexane–ethyl acetate) to give the desired product **2**.

Methyl 2,3,4-Tri-O-acetyl-D-glucopyranuronate (2a)

Eluent: petroleum ether–ethyl acetate 1:1. Mixture of α,β-isomers: colorless syrup. Yield: 97 mg, 89%. α/β: 4/1.* IR: 3440, 2958, 1743, 1367, 1220, 1035 cm^{-1}. α-isomer, ^{1}H NMR (CDCl$_3$) δ: 2.02 (s, 6H, 2×CH_3CO), 2.07 (s, 3H, CH_3CO), 3.60 (s, 1H, OH-1), 3.74 (s, 3H, CH_3OCO), 4.58 (d, 1H, $J_{5,4}$ 10.1 Hz, H-5), 4.91 (dd, 1H, $J_{2,3}$ 9.6 Hz, $J_{2,1}$ 3.6 Hz, H-2), 5.17 (t, 1H, $J_{4,3}$ 9.6 Hz, H-4), 5.53 (d, 1H, $J_{1,2}$ 3.6 Hz, H-1), 5.57 (t, 1H, $J_{3,4}$ 9.6 Hz, H-3). ^{13}C NMR (CDCl$_3$) δ: 20.8 (3 x CH$_3$CO), 53.2 (CH$_3$OCO), 68.3 (C-5), 69.2 (C-3), 69.7 (C-4), 70.9 (C-2), 90.5 (C-1), 168.5 (C-6), 169.8 (CO), 170.3 (2×CO). HRMS [M-H] calcd for C$_{13}$H$_{17}$O$_{10}$: 333.0822. Found [M-H]: 333.0829.

Butyl 2,3,4-Tri-O-acetyl-D-glucopyranuronate (2b)

Eluent: petroleum ether–ethyl acetate 3:2. Mixture of α,β-isomers: colorless syrup. Yield: 100 mg, 81%. α/β: 85/15.† IR: 3469, 2963, 1751, 1370, 1219, 1038 cm^{-1}. α-isomer, ^{1}H NMR (CDCl$_3$) δ: 0.91 (t, 3H, J 7.3 Hz, CH_3CH$_2$), 1.29–1.41(m, 2H, CH_2), 1.54–1.64 (m, 2H, CH$_2$), 2.02 (s, 6H, 2×CH_3CO), 2.06 (s, 3H, CH_3CO), 4.03–4.21 (m, 2H, CO$_2$CH_2), 4.58 (d, 1H, $J_{5,4}$ 10.1 Hz, H-5), 4.90 (dd, 1H, $J_{2,3}$ 9.5 Hz, $J_{2,1}$ 3.5 Hz, H-2), 5.18 (t, 1H, $J_{4,3}$ 9.5 Hz, H-4), 5.51 (d, 1H, $J_{1,2}$ 3.5 Hz, H-1), 5.55 (t, 1H, $J_{3,4}$ 9.5 Hz, H-3). ^{13}C NMR (CDCl$_3$) δ: 13.8 (CH$_3$), 19.1 (CH$_2$), 20.8 (3×CH$_3$CO), 30.5 (CH$_2$), 66.2 (CO$_2$CH$_2$), 68.3 (C-5), 69.3 (C-3), 69.5 (C-4), 70.9 (C-2), 90.4 (C-1), 168.3 (C-6), 169.6 (CO), 170.7 (2×CO). HRMS [M+Na] calcd for C$_{16}$H$_{24}$O$_{10}$Na: 399.1257. Found [M+Na]: 399.1267.

Hexyl 2,3,4-Tri-O-acetyl-D-glucopyranuronate (2c)

Eluent: cyclohexane–ethyl acetate 7:3 then 5:5. Mixture of α,β-isomers: colorless syrup. Yield: 117 mg, 90%. α/β: 80/20. IR: 3442, 2933, 1743, 1367, 1228, 1035 cm^{-1}. α-isomer, ^{1}H NMR (CDCl$_3$) δ: 0.87 (t, 3H, J 7.0 Hz, CH_3CH$_2$), 1.23–1.33 (m, 8H, CH$_2$),

---

* The ratio of isomers may vary.
† The ratio of isomers may vary.

1.55–1.66 (m, 2H, CH_2), 2.01 (s, 6H, 2×CH_3CO), 2.06 (s, 3H, CH_3CO), 4.02–4.19 (m, 2H, CO$_2$CH_2), 4.57 (d, 1H, $J_{5,4}$ 10.1 Hz, H-5), 4.91 (dd, 1H, $J_{2,3}$ 9.6, $J_{2,1}$ 3.5 Hz, H-2), 5.18 (t, 1H, $J_{4,3}$ 9.8 Hz, H-4), 5.52 (d, 1H, $J_{1,2}$ 3.5 Hz, H-1), 5.55 (t, 1H, $J_{3,4}$ 9.8 Hz, H-3). ^{13}C NMR (CDCl$_3$) δ: 14.1 (CH_3CH_2), 20.8 (3×CH_3CO), 22.6 (CH_2), 25.5 (CH_2), 28.5 (CH_2), 31.5 (CH_2), 66.5 (CO$_2$CH_2), 68.4 (C-5), 69.5 (C-3), 69.7 (C-4), 70.9 (C-2), 90.4 (C-1), 168.3 (C-6), 169.6 (CO), 170.2 (2×CO). HRMS [M+Na] calcd for C$_{18}$H$_{28}$O$_{10}$Na: 427.1580. Found [M + Na]: 427.1571.

Octyl 2,3,4-Tri-O-acetyl-D-glucopyranuronate (2d)

Eluent: cyclohexane–ethyl acetate 7:3. Mixture of α,β-isomers: colorless syrup. Yield: 127 mg, 90%. α/β: 76/24. IR: 3473, 2924, 1743, 1229, 1368, 1057 cm^{-1}. α-isomer, ^{1}H NMR (CDCl$_3$) δ: 0.88 (t, 3H, J 7.0 Hz, CH_3CH_2), 1.22–1.36 (m, 12H, CH_2), 1.57–1.64 (m, 2H, CH_2), 2.03 (s, 6H, 2×CH_3CO), 2.08 (s, 3H, CH_3CO), 3.38 (s, 1H, OH-1), 4.03–4.21 (m, 2H, CO$_2$CH_2), 4.59 (d, 1H, $J_{5,4}$ 10.0 Hz, H-5), 4.93 (dd, 1H, $J_{2,3}$ 9.6, $J_{2,1}$ 3.5 Hz, H-2), 5.20 (t, 1H, $J_{4,3}$ 9.6 Hz, H-4), 5.53 (d, 1H, $J_{1,2}$ 3.5 Hz, H-1), 5.57 (t, 1H, $J_{3,4}$ 9.6 Hz, H-3). ^{13}C NMR (CDCl$_3$) δ: 13.8 (CH_3CH_2), 20.8 (3×CH_3CO), 22.8 (CH_2), 25.9 (CH_2), 28.5 (CH_2), 29.3 (CH_2), 31.9 (CH_2), 66.5 (CO$_2$CH_2), 68.4 (C-5), 69.1 (C-3), 69.4 (C-4), 70.8 (C-2), 90.5 (C-1), 168.1 (C-6), 169.6 (CO), 170.2 (2×CO). HRMS [M+Na] calcd for C$_{20}$H$_{32}$O$_{10}$Na: 455.1893. Found [M + Na]: 455.1883.

ACKNOWLEDGMENTS

This chapter was financially supported by the Regional Council of Picardie.

Methyl 2,3,4-tri-O-acetyl-D-glucopyranuronate (**2a**) (mixture of anomers, 80:20 (α:β) ratio) ^{1}H (300 MHz) and ^{13}C (75 MHz) in CDCl$_3$.

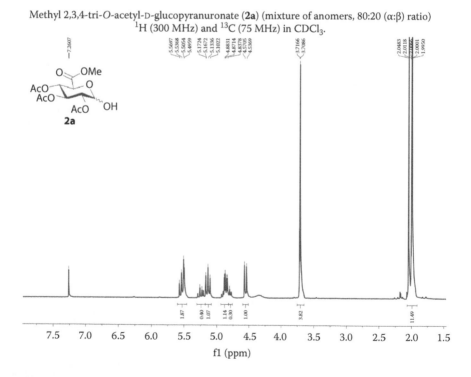

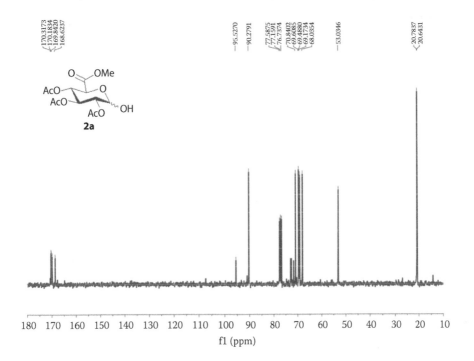

Butyl 2,3,4-tri-*O*-acetyl-D-glucopyranuronate (**2b**) (mixture of anomers, 85:15 (α:β) ratio) ¹H (300 MHz) and ¹³C (75 MHz) in CDCl₃.

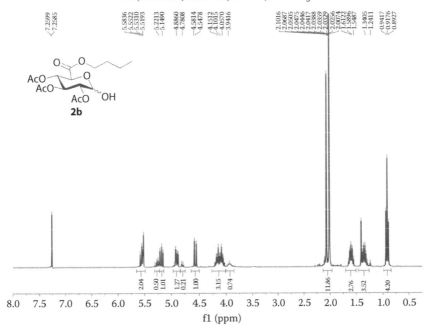

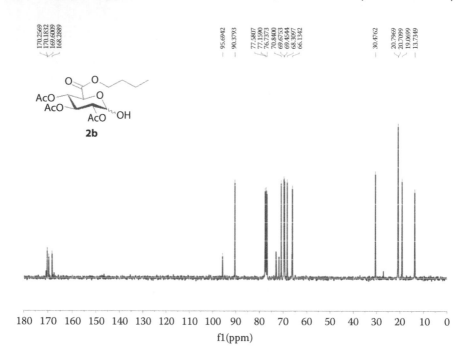

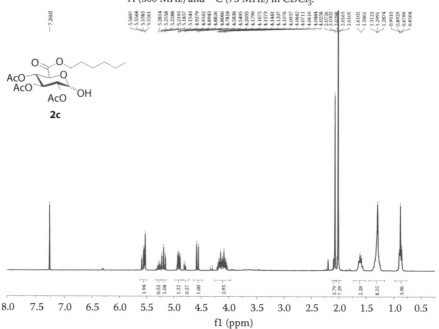

Hexyl 2,3,4-tri-*O*-acetyl-D-glucopyranuronate (**2c**) (mixture of anomers, 80:20 (α:β) ratio)
^{1}H (300 MHz) and ^{13}C (75 MHz) in CDCl$_3$.

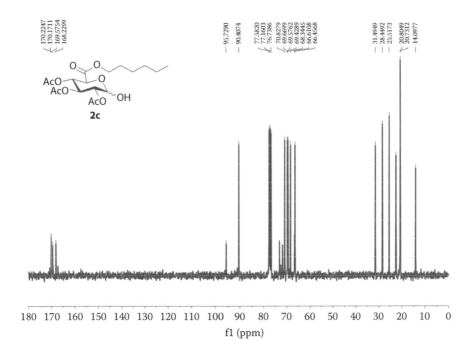

Octyl 2,3,4-tri-*O*-acetyl-D-glucopyranuronate (**2d**) (mixture of anomers, 76:24 (α:β) ratio) ^{1}H (300 MHz) and ^{13}C (75 MHz) in CDCl$_3$.

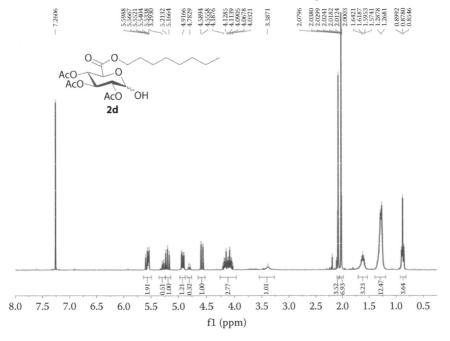

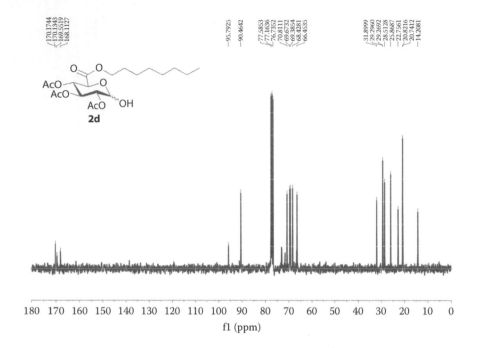

REFERENCES

1. Takahashi, N; Koshijima, T. *Wood Sci. Technol.* **1988**, *22*, 213–241.
2. Wadouachi, A.; Kovensky, J. *Molecules* **2011**, *16*, 3933–3968.
3. Codee, J.D.C.; Christina, A.E.; Walvoort, M.T.C.; Overkleeft, H.S.; van der Marel, G.A. *Top. Curr. Chem.* **2011**, *301*, 253–289.
4. Bollenback, G.N.; Long, J.W.; Benjamin, D.G.; Lindquist, J.A. *J. Am. Chem. Soc.* **1955**, *77*, 3310–3315.
5. Nudelman, A.; Herzig, J.; Gottlieb, H.E.; Keinan, E.; Sterling, J. *Carbohydr. Res.* **1987**, *162*, 145–152.
6. Nakajima, R.; Ono, M.; Aiso, S.; Akita, H. *Chem. Pharm. Bull.* **2005**, *53*, 684–687.
7. Cavalluzzi, M.M.; Lentini, G.; Lovece, A.; Bruno, C.; Catalano, A.; Carocci, A.; Franchini, C. *Tetrahedron Lett.* **2010**, *51*, 5265–5268.
8. Chittaboina, S.; Hodges, B.; Wang, Q. *Lett. Org. Chem.* **2006**, *3*, 35–38.
9. Lu, Y.J.; Liu, Y.; Prashad, M.; Shieh, W.C. *Adv. Chem. Eng. Sci.* **2012**, *2*, 379–383.
10. Iyer, S.S; Rele, S.M.; Baskaran, S.; Chaikof, E.L. *Tetrahedron* **2003**, *59*, 631–638.
11. Pfefferkorn, J.A.; Larsen, S.D.; Van Huis, C.; Sorenson, R.; Barton, T.; Winters, T.; Auerbach, B. et al. *Bioorg. Med. Chem. Lett.* **2008**, *18*, 546–553.
12. Casati, S.; Ottria, R.; Ciuffreda, P. *Steroids* **2009**, *74*, 250–255.
13. Rat, S.; Mathiron, D.; Michaud, P.; Kovensky, J.; Wadouachi, A. *Tetrahedron* **2007**, *63*, 12424–12428.
14. Bosco, M.; Rat, S.; Dupre, N.; Hasenknopf, B.; Lacote, E.; Malacria, M.; Remy, P.; Kovensky, J.; Thorimbert, S.; Wadouachi, A. *ChemSusChem* **2010**, *3*, 1249–1252.

4 Metal-Free, Diamine-Mediated, Oxidative Monoamidation of Benzylated Carbohydrates

Maxime B. Fusaro, Vincent Chagnault,*
Christophe Dussouy,† and Denis Postel

CONTENTS

The amide group is an important functionality widely found in natural products, pharmaceuticals, and polymers. This link between amino acids in peptides and proteins plays an important role in the structure of many biological systems.

Some carbohydrates containing amide functional groups at the anomeric position (glyconamides) can be prepared[1] as key intermediates for the synthesis of high value-added molecules, such as iminosugars and N-alkyliminosugars, which are lead compounds for the treatment of a variety of diseases.[1–4]

---

* Corresponding author; e-mail: vincent.chagnault@u-picardie.fr.
† Checker, under supervision of Jean Bernard Berh; e-mail: jb.behr@univ-reims.fr.

Martin et al. described the oxidative amidation reaction using a variety of primary amines in order to form N-alkylated gluconamides in one step.[1] The reaction, performed in t-BuOH at 90°C in the presence of molecular iodine, leads to modest to good yields depending on the nature of the starting materials (45%–74%). Being aware of the interest in such compounds and the difficulties involved in the synthesis of glyconamides from benzylated carbohydrates, we optimized conditions for this oxidation.[5] It consists of the direct conversion of aldose hemiacetals to amides, which can be directly performed using iodine as the oxidant. The main advantage of this metal-free reaction is that both aldehyde oxidation and C–N bond formation are performed in a single step.[6,7] We found that at room temperature, using methanol (analytical grade) instead of t-BuOH at 90°C, the yields of glyconamides were improved (80%–98%). In the case of diamines, we found that a monoamidation is carried out, which leads to functionalized glyconamides.

Based on our previous work, here, we describe synthesis of N-(3-aminopropyl)-2,3,4-tri-O-benzyl-D-xylonamide **1** and N-(3-aminopropyl)-2,3,4,6-tetra-O-benzyl-D-gluconamide **2**.

EXPERIMENTAL METHODS

GENERAL METHODS

Reagent-grade chemicals were obtained from commercial suppliers and were used as received. Optical rotations were recorded for $CHCl_3$ solutions. FTIR spectra were obtained using ATR and are reported in cm^{-1}. ^1H NMR (300 MHz) and ^{13}C NMR (75 MHz) spectra were recorded in $CDCl_3$. The proton and carbon signal assignments were determined by decoupling, COSY, and HSQC experiments. TLC were performed on silica F_{254} and detection by UV light at 254 nm or by charring with cerium molybdate reagent.* Column chromatography was effected on silica gel 60 (230 mesh). High-resolution electrospray mass spectra in the positive ion mode were obtained with a Q-TOF Ultima Global hybrid quadrupole/time-of-flight instrument, equipped with a pneumatically assisted electrospray (Z-Spray) ion source and an additional sprayer (LockSpray) for the reference compound.

N-(3-AMINOPROPYL)-2,3,4-TRI-O-BENZYL-D-XYLONAMIDE (1)

2,3,4-Tri-O-benzyl-D-xylose[8] (200 mg, 0.476 mmol) was dissolved in methanol (4.8 mL, analytical grade) and 1,3-diaminopropane (79 μL, 0.952 mmol) was added. The mixture was stirred at room temperature for 1 min, and K_2CO_3 (197 mg, 1.43 mmol) and iodine (242 mg, 0.952 mmol) were added successively. The solution was stirred at room temperature for 20 h and saturated aqueous solution of $Na_2S_2O_3$ was added until dark-brown color disappeared (~1 mL). The mixture was diluted with water (5 mL), the desired product was extracted into CH_2Cl_2 (3×20 mL), and the

---

* Preparation: To 235 mL of distilled water was added 12 g of ammonium molybdate, 0.5 g of ceric ammonium molybdate, and 15 mL of concentrated sulfuric acid. Storage is possible.

organic phases were combined, dried over $MgSO_4$, and concentrated. The residue was chromatographed (95:5:1 CH_2Cl_2–MeOH–NH_4OH; R_f=0.23) to give the desired gly-conamide **1** (89%) as colorless oil; the compound can be kept at 4°C for weeks without decomposition. $[\alpha]_D$ = +0.9 (c=1, $CHCl_3$); IR (ATR) 1651 cm^{-1}. ^1H NMR (300 MHz, $CDCl_3$) δ: 7.41–7.20 (m, 15H, Ph), 7.15 (t, J=5.8 Hz, 1H, CONH), 4.75–4.44 (m, 6H, **CH₂Ph**), 4.12 (d, J=2.7 Hz, 1H, H-2), 4.07 (dd, J=6.8, 2.7 Hz, 1H, H-3), 3.76 (ddd, J=6.7, 5.2, 4.1 Hz α, 1H, H-4), 3.68 (dd, J=11.9, 4.0 Hz, 1H, H-5), 3.50 (dd, J=11.9, 5.3 Hz, 1H, H-5), 3.43–3.29 (m, 1H, H-6), 3.29–3.14 (m, 1H, H-6), 2.64 (td, J=6.5, 2.1 Hz, 2H, H-8), 1.60–1.46 (m, 2H, H-7); ^{13}C NMR (75 MHz, $CDCl_3$) δ: 171.2 (**CONH**), 138.3, 138.1, 136.7 (Cq-Ar), 128.8, 128.6, 128.5, 128.2, 128.1, 128.0 (Ar), 80.0, 79.8 (C-2, C-3, C-4), 75.3, 73.9, 73.3 (**CH₂Ph**), 61.7 (C-5), 39.6 (C-8), 37.3 (C-6), 32.0 (C-7). HRMS calcd for $C_{29}H_{37}N_2O_5$: 493.2702. Found m/z: 493.2697 $[M+H]^+$.

N-(3-Aminopropyl)-2,3,4,6-tetra-O-benzyl-D-gluconamide (2)

2,3,4,6-Tetra-O-benzyl-D-glucose* (257 mg, 0.476 mmol) was dissolved in methanol (4.8 mL, analytical grade) and 1,3-diaminopropane (79 µL, 0.952 mmol) was added. The reaction mixture was stirred at room temperature for 1 min, K_2CO_3 (197 mg, 1.43 mmol) and iodine (242 mg, 0.952 mmol) were added successively, and the solution was stirred at room temperature for 20 h. Saturated aqueous solution of $Na_2S_2O_3$ was added until complete disappearance of the dark-brown color (~1 mL), and the mixture was diluted with water (5 mL). The mixture was extracted with CH_2Cl_2 (3×20 mL) and the organic phases were combined, dried over $MgSO_4$, and concentrated under reduced pressure. The residue was chromatographed (95:5:1 CH_2Cl_2–MeOH–NH_4OH) to give the desired glyconamide **2** (89%) as colorless oil; the somewhat unstable compound can be kept at 4°C for weeks without noticeable decomposition; $[\alpha]_D$ +15.9 (c=1, $CHCl_3$); IR (ATR) 1648 cm^{-1}. ^1H NMR (300 MHz, $CDCl_3$) δ: 7.43–7.16 (m, 20H, Ph), 7.10 (t, J=5.9 Hz, 1H, CONH), 4.76–4.45 (m, 8H, **CH₂Ph**), 4.27 (d, J=3.3 Hz, 1H, H-2), 4.08 (dd, J=5.5, 3.4 Hz, 1H, H-3), 3.98–3.90 (m, 1H, H-5), 3.87 (dd, J=7.5, 5.6 Hz, 1H, H-4), 3.66 (dd, J=9.9, 3.0 Hz, 1H, H-6), 3.60 (dd, J=9.8, 5.3 Hz, 1H, H-6), 3.41–3.27 (m, 1H, H-7), 3.27–3.14 (m, 1H, H-7), 2.62 (td, J=6.5, 2.3 Hz, 2H, H-9), 1.58–1.46 (m, 2H, H-8). ^{13}C NMR (75 MHz, $CDCl_3$) δ: 171.3 (**CONH**), 138.2, 138.1, 138.0, 136.9 (Cq-Ar), 128.7, 128.4, 128.3, 128.1, 128.0, 127.9, 127.8, 127.7 (Ar), 80.7, 80.2 (C-2, C-3), 77.6 (C-4), 75.1, 74.1, 74.0, 73.4 (**CH₂Ph**), 71.4, 71.2 (C-5, C-6), 39.5 (C-9), 37.1 (C-7), 32.1 (C-8). HRMS calcd for $C_{37}H_{45}N_2O_6$: 613.3278. Found m/z: 613.3256 $[M+H]^+$.

ACKNOWLEDGMENTS

We thank CNRS (France) for financial support and the Region Picardie and Europe (FEDER) for a grant (to M. B. Fusaro).

---

* Commercially available but can be prepared using the procedure described for 2,3,4-tri-O-benzyl-D-xylose[8].

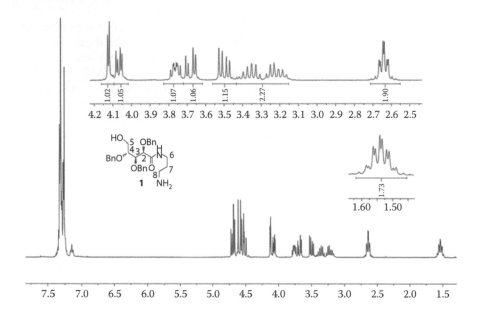

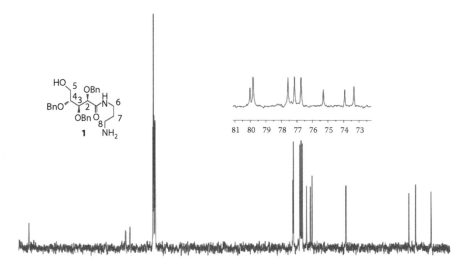

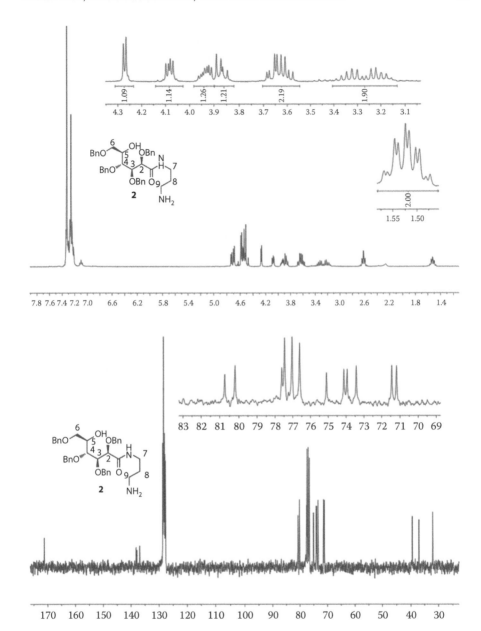

REFERENCES

1. Colombeau, L.; Traore, T.; Compain, P.; Martin, O. R. *J. Org. Chem.* **2008**, *73*, 8647–8650.
2. Decroocq, C.; Laparra, L. M.; Rodríguez-Lucena, D.; Compain, P. *J. Carbohydr. Chem.* **2011**, *30*, 559–574.
3. Compain, P.; Decroocq, C.; Iehl, J.; Holler, M.; Hazelard, D.; Mena Barragán, T.; Ortiz Mellet, C.; Nierengarten, J.-F. *Angew. Chem. Int. Ed.* **2010**, *49*, 5753–5756.
4. Compain, P.; Martin, O. R. In *Iminosugars: From Synthesis to Therapeutic Applications*; Compain, P.; Martin, O. R., Eds.; Wiley-VCH: Weinheim, Germany, **2008**; pp. 1–6.
5. Fusaro, M. B.; Chagnault, V.; Postel, D. *Tetrahedron* **2013**, *69*, 542–550.
6. Cho, C.-C.; Liu, J.-N.; Chien, C.-H.; Shie, J.-J.; Chen, Y.-C.; Fang, J.-M. *J. Org. Chem.* **2009**, *74*, 1549–1556.
7. Chen, M.-Y.; Hsu, J.-L.; Shie, J.-J.; Fang, J.-M. *J. Chin. Chem. Soc.* **2003**, *50*, 129–133.
8. Nadein, O. N.; Kornienko, A. *Org. Lett.* **2004**, *6*, 831–834.

5 Metal-Free Oxidative Lactonization of Carbohydrates Using Molecular Iodine

Maxime B. Fusaro, Vincent Chagnault,*
Solen Josse, Nicolas Drillaud,
Guillaume Anquetin,† and Denis Postel

CONTENTS

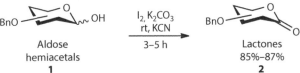

Diversely protected carbohydrate lactones are useful intermediates for synthesis of natural products, in particular for an easy access to *C*-glycosyl compounds.[1–5] Concerning the oxidation of a hemiacetal function to produce lactones, only few high-yielding methods are known. Oxidative lactonization of free carbohydrates can be effected using bromine water and a $BaCO_3$ buffer to form in situ the hypobromite/bromate oxidative agents.[6,7] Under these conditions, oxidation is selective at the anomeric position but suitable mainly to water-soluble compounds. Moreover, this methodology, when applied to unprotected sugars, generally provides a mixture of δ- and γ-lactones and glyconic acids. A similar procedure with molecular iodine has been developed and has the same disadvantages.[8] The procedure used by Cordova and

* Corresponding author; e-mail: vincent.chagnault@u-picardie.fr.
† Checker, under the supervision of Benoit Piro; e-mail: piro@univ-paris-diderot.fr.

coworkers employs large amounts of MnO$_2$ (15 eq.).[9] Under these conditions, the hemiacetal was selectively oxidized in presence of secondary hydroxyl groups.

We previously described an efficient procedure for the direct oxidative lactonization of protected carbohydrates.[10] The method allows quick access, in one step, to lactones **2** using 3 eq. of molecular iodine and 3 eq. of potassium carbonate in dichloromethane or *t*-BuOH at room temperature. The main advantages of this method are the operational simplicity and use of simple reagents. The reaction has also been tested on partially protected compounds and yields of the lactones obtained indicate clearly a chemoselective oxidation. Lactones were obtained in good yields without need of any purification. Reaction times were 2–5 h for most of the carbohydrates used in our tests, except for benzylated derivatives that required 20 h under the same conditions.

We thus describe herein a new procedure for the oxidative lactonization of benzylated carbohydrates (2,3,4-tri-*O*-benzyl-D-xylopyranose **1a**[11] and 2,3,4,6-tetra-*O*-benzyl-D-glucopyranose **1b***), which leads to the lactones **2a** and **2b** in good yields in only 3 and 5 h, respectively.

EXPERIMENTAL METHODS

GENERAL METHOD

All reagent-grade chemicals were obtained from commercial suppliers and were used as received. Melting points are uncorrected. Optical rotations were recorded in CDCl$_3$ or CH$_2$Cl$_2$ solutions. FTIR spectra were obtained using ATR and are reported in cm^{-1}. ^{1}H NMR (300 MHz) and ^{13}C NMR (75 MHz) spectra were recorded in CDCl$_3$. The proton and carbon signal assignments were determined from decoupling experiments, COSY spectra, and HSQC spectra. TLC were performed on Silica F$_{254}$ and detection by UV light at 254 nm or by charring with cerium molybdate reagent.[†] Column chromatography was effected on silica gel 60 (230 mesh). High-resolution electrospray mass spectra in the positive ion mode were obtained on a Q-TOF Ultima Global hybrid quadrupole/time-of-flight instrument, equipped with a pneumatically assisted electrospray (Z-Spray) ion source and an additional sprayer (LockSpray) for the reference compound.

2,3,4-TRI-*O*-BENZYL-D-XYLONO-1,5-LACTONE 2A

In a 10 mL round-bottom flask,[‡] 2,3,4-tri-*O*-benzyl-D-xylose[11] **1a** (200 mg, 0.476 mmol) was dissolved in dichloromethane (analytical grade; 4.8 mL), and K$_2$CO$_3$ (198 mg, 3 eq.), iodine (362 mg, 3 eq.), and KCN (93 mg, 3 eq.) were added successively. The solution was stirred at room temperature, without inert atmosphere, and monitored by TLC. After 3 h, a saturated aqueous solution of Na$_2$S$_2$O$_3$ was

---

* Commercially available but can be prepared using the procedure described for compound **1a**.
† Preparation: To 235 mL of distilled water was added 12 g of ammonium molybdate, 0.5 g of ceric ammonium molybdate, and 15 mL of concentrated sulfuric acid. Storage is possible.
‡ The reaction mixture is dark and creates some poorly soluble material at the surface. A small volume for the flask is better to achieve higher conversion since then less material sticks to the wall of the flask above the level of the solvent.

added under vigorous stirring until complete disappearance of the dark-brown color (~1 mL). The mixture was diluted with water (10 mL) and the desired product was extracted with dichloromethane (3×20 mL). The organic phases were combined, dried over $MgSO_4$, and concentrated under reduced pressure.* The residue was chromatographed (7:3 cyclohexane–EtOAc), to give the desired lactone **2a** (174 mg, 87%), mp 118°C–119°C (EtOAc); $[\alpha]_D^{20}$ +4 (c=0.13, CH_2Cl_2); R_f=0.55 (7:3 cyclohexane–EtOAc). ^{1}H NMR (300 MHz, $CDCl_3$) δ: 7.46–7.27 (m, 15H, Ar), 5.04 (d, J=11.6 Hz, 1H, **CH$_2$**Ph), 4.68 (d, J=11.5 Hz, 2H, **CH$_2$**Ph), 4.59 (d, J=11.6 Hz, 1H, **CH$_2$**Ph), 4.58 (d, J=11.5 Hz, 1H, **CH$_2$**Ph), 4.52 (d, J=11.5 Hz, 1H, **CH$_2$**Ph), 4.41 (ddd, $J_{5a,5b}$=12.3 Hz, $J_{5a,4}$=3.3 Hz, $J_{5a,3}$=1.5 Hz, 1H, H-5a), 4.30 (dd, $J_{5b,5a}$=12.3 Hz, $J_{5b,4}$=2.0 Hz, 1H, H-5b), 4.17 (d, $J_{2,3}$=6.6 Hz, 1H, H-2), 3.92 (ddd, $J_{3,2}$=6.6 Hz, $J_{3,4}$=2.1 Hz, $J_{3,5a}$=1.5 Hz, 1H, H-3), 3.79 (dt, $J_{4,5a}$=3.3 Hz, $J_{4,3}$=2.1 Hz, $J_{4,5b}$=2.1 Hz, 1H, H-4). ^{13}C NMR ($CDCl_3$, 75 MHz) δ: 169.8 (C-1), 137.4 (Cq-Ar), 137.2 (Cq-Ar), 137.1 (Cq-Ar), 128.6–127.9 (Ar), 81.4 (C-3), 78.2 (C-2), 75.3 (C-4), 73.4 (CH$_2$Ph), 72.8 (CH$_2$Ph), 70.6 (CH$_2$Ph), 65.8 (C-5). IR (neat): 1744 cm^{-1}. Anal. calcd for $C_{26}H_{26}O_5$: C, 74.62; H, 6.26. Found: C, 74.44; H, 6.14.

2,3,4,6-Tetra-*O*-benzyl-d-glucono-1,5-lactone 2b

In a 10 mL round-bottom flask,† 2,3,4,6-tetra-*O*-benzyl-D-glucose‡ **1b** (257 mg, 0.476 mmol) was dissolved in dichloromethane (analytical grade; 4.8 mL), and K_2CO_3 (197 mg, 3 eq.), iodine (362 mg, 3 eq.), and KCN (93 mg, 3 eq.) were added successively. The solution was stirred at room temperature, without inert atmosphere, and monitored by TLC. After 5 h, a saturated aqueous solution of $Na_2S_2O_3$ was added under vigorous stirring until complete disappearance of the dark-brown color (~1 mL). The mixture was diluted with water (10 mL) and the desired product was then extracted with dichloromethane (3×20 mL). The organic phases were combined, dried over $MgSO_4$, and evaporated under reduced pressure.§ The residue was purified by chromatography on silica gel (8:2 cyclohexane–EtOAc) to give the desired lactone **2b** (215 mg, 85%) as a viscous colorless oil. $[\alpha]_D^{20}$ +81 (c=0.16, CH_2Cl_2) (Lit.^{12} $[\alpha]_D^{25}$ +79 (c=1, $CHCl_3$)); R_f=0.5 (4:1 cyclohexane:EtOAc). ^{1}H NMR (300 MHz, $CDCl_3$) δ: 7.44–7.20 (m, 20H, Ar), 5.03 (d, J=11.4 Hz, 1H, **CH$_2$**Ph), 4.77 (d, J=11.4 Hz, 1H, **CH$_2$**Ph), 4.75 (d, J=11.2 Hz, 1H, **CH$_2$**Ph), 4.68 (d, J=11.4 Hz, 1H, **CH$_2$**Ph), 4.63 (d, J=11.4 Hz, 1H, **CH$_2$**Ph), 4.60 (d, J=12.1 Hz, 1H, **CH$_2$**Ph), 4.56 (d, J=11.5 Hz, 1H, **CH$_2$**Ph), 4.51 (d, J=12.0 Hz, 1H, **CH$_2$**Ph), 4.49 (m, 1H, H-5), 4.17 (d, $J_{2,3}$=6.6 Hz, 1H, H-2), 3.98 (m, 2H, H-3, H-4), 3.77 (dd, $J_{6a,6b}$=11.0 Hz, $J_{6a,5}$=2.5 Hz, 1H, H-6a), 3.75 (dd, $J_{6b,6a}$=11.0 Hz, $J_{6b,5}$=3.3 Hz, 1H, H-6b). ^{13}C NMR ($CDCl_3$, 75 MHz) δ: 169.4 (C-1), 137.7 (Cq-Ar), 137.6 (2×Cq-Ar), 137.1 (Cq-Ar), 128.6–128.0 (Ar), 81.1 (C-2),

---

* The crude product was analyzed by ^{1}H NMR and provided data identical to the purified sample. The quality of the crude material should therefore be suitable for further reactions without chromatography.

† The reaction mixture is dark and creates some poorly soluble material at the surface. A small volume for the flask is better to achieve higher conversion since then less material sticks to the wall of the flask above the level of the solvent.

‡ Commercially available but can be prepared using the procedure described for compound **1a**.

§ The crude product was analyzed by ^{1}H NMR and provided data identical to the purified sample. The quality of the crude material should therefore be suitable for further reactions without chromatography.

78.3 (C-5), 77.5 (C-3), 76.2 (C-4), 74.0 (**CH₂Ph**), 73.8 (2×O-**CH₂Ph**), 73.7 (O-**CH₂Ph**), 68.4 (C-6). IR (neat): 1755 cm⁻¹. Anal. calcd for C₃₄H₃₄O₆: C, 75.82; H, 6.36. Found: C, 75.97; H, 6.26.

ACKNOWLEDGMENTS

We thank CNRS (France) for financial support and the Region Picardie and Europe (FEDER) for a grant (to M. B. Fusaro) and M. Lucas Fontaine for his kind help.

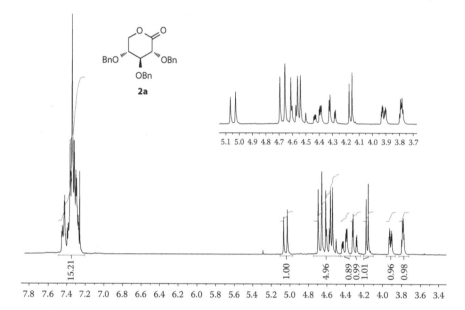

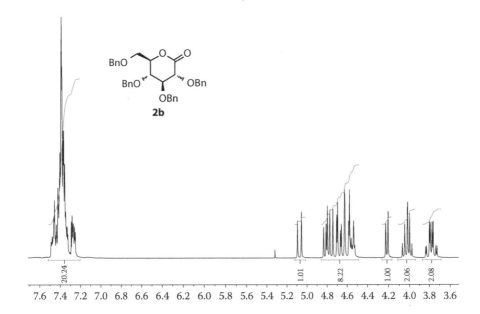

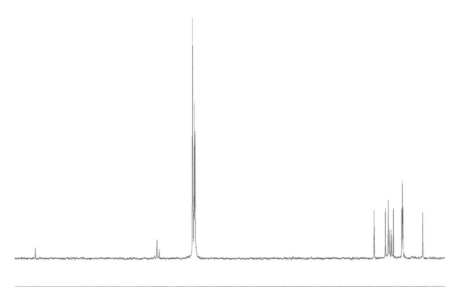

REFERENCES

1. Takahashi, S.; Nakata, T. *J. Org. Chem.* **2002**, *67*, 5739–5752.
2. Sonnet, P.; Izard, D.; Mullie, C. *Int. J. Antimicrob. Agents* **2012**, *39*, 77–80.
3. Kang, S. Y.; Song, K.-S.; Lee, J.; Lee, S.-H.; Lee, J. *Bioorg. Med. Chem.* **2010**, *18*, 6069–6079.
4. Plet, J. R. H.; Porter, M. J. *Chem. Commun.* **2006**, *11*, 1197–1199.
5. Bowen, E. G.; Wardrop, D. J. *Org. Lett.* **2010**, *12*, 5330–5333.
6. Mantell, S. J.; Ford, P. S.; Watkin, D. J.; Fleet, G. W. J.; Brown, D. *Tetrahedron Lett.* **1992**, *33*, 4503–4506.
7. Isbell, H. S. *Methods Carbohydr. Chem.* **1963**, *2*, 13–14.
8. Jeong, Y.-I.; Seo, S.-J.; Park, I.-K.; Lee, H.-C.; Kang, I.-C.; Akaike, T.; Cho, C.-S. *Int. J. Pharm.* **2005**, *296*, 151–161.
9. Córdova, A.; Ibrahem, I.; Casas, J.; Sundén, H.; Engqvist, M.; Reyes, E. *Chem. Eur. J.* **2005**, *11*, 4772–4784.
10. Fusaro, M. B.; Chagnault, V.; Josse, S.; Postel, D. *Tetrahedron* **2013**, *69*, 5880–5883.
11. Nadein, O. N.; Kornienko, A. *Org. Lett.* **2004**, *6*, 831–834.
12. Rajanikanth, B.; Seshadri, R. *Tetrahedron Lett.* **1989**, *30*, 755–758.

6 Synthesis of Glycosyl Vinyl Sulfones for Bioconjugation

*Francisco Santoyo-Gonzalez,**
Fernando Hernandez-Mateo,
and Omar Boutureira[†]

CONTENTS

The covalent coupling of two biomolecules to each other (bioconjugation) or to a solid support (immobilization) is one of the cornerstones of *omic* sciences.[1] Among numerous chemical strategies to attain this goal, the versatile Michael-type addition of amine and thiol groups to vinyl sulfones is an attractive methodology.[2] The latter are excellent Michael acceptors because of the electron-poor nature of their double bond, owed to the sulfone's electron-withdrawing capability that makes them good electrophiles.[3,4] All the conjugate additions with vinyl sulfones share a similar reaction pattern, namely, the addition to the β-position of the sulfone. Accordingly, these reactions are well-established methods for creating β-heterosubstituted

---

* Corresponding author; e-mail: fsantoyo@ugr.es.
† Checker; e-mail: omar.boutureira@urv.cat.

sulfones. Prominent characteristics of this methodology are the water stability of the vinyl sulfone function, the possibility to perform the reactions in physiological conditions (aqueous media, slightly alkaline pH, and room temperature) that preserves the biological function of the biomolecules, the absence of catalysts and by-products, the almost theoretical yields, and the stability of the linkage formed. For these reasons, vinyl sulfones have found application in most of the subdomains of modern proteomics.[5]

Accordingly, and in the context of carbohydrate research, the vinyl sulfone functionalization of the anomeric carbon has proved to be a general strategy for subsequent chemical glycosylation of proteins and for the covalent linkage of a saccharide to amine- and thiol-functionalized supports.[6,7] This strategy has found applications in glycoscience to explore protein–carbohydrate interactions.[6] The authors' group has developed a reliable and simple two-step high-yielding method for the derivatization of saccharides at the anomeric carbon with a vinyl sulfone group spanned by an ethylthio linker.[8] For that purpose, easily accessible or commercially available 1-halo sugars are used as starting materials. The method is based on the preparation of S-glycosyl N-alkyl dithiocarbamates by treatment of glycosyl halides with salts of alkyl dithiocarbamates[9] in, for example, anhydrous acetone at room temperature. In this way, the formed sugar dithiocarbamates act as masked 1-thiol saccharides. The thiolate sugars are easily generated in a second step by treatment with a common organic base, such as triethylamine, and trapped in situ by commercial divinyl sulfone (DVS) present in the reaction media. The glycosyl vinyl sulfones isolated are ready to be used in any conjugation for preparation of glycosylated materials. The procedure is exemplified by reaction of glucosyl bromide **1** with sodium N-benzyldithiocarbamate as a model alkyl dithiocarbamate salt. An improved method for preparation of sodium N-benzyl dithiocarbamate compared to that reported in literature[10] is also described. The synthetic approach described herein is generally applicable for the preparation of any glycosyl vinyl sulfone.

EXPERIMENTAL METHODS

GENERAL METHODS

Commercially available reagents (2,3,4,6-tetra-O-acetyl-α-D-glucopyranosyl bromide (**1**), benzylamine, triethylamine, and divinyl sulfone) and solvents were used without further purification. Thin-layer chromatographies (TLCs) were performed on Merck silica gel 60 F254 aluminum sheets. Detection was effected by charring with sulfuric acid (5% v/v in ethanol), potassium permanganate (1% w/v), and ninhydrin (0.3% w/v) in ethanol and UV light when applicable. Flash column chromatography was performed on Merck silica gel (230–400 mesh, ASTM). Optical rotations were recorded with a PerkinElmer 141 polarimeter at room temperature. IR spectra were recorded with a Satellite Mattson FTIR. ^{1}H and ^{13}C NMR spectra were recorded at room temperature with a Varian DirectDrive (300, 400, and 500 MHz) spectrometer. Chemical shifts are given in ppm and referenced to internal CDCl$_3$. J values are given in Hz. Electrospray ionization (ESI) mass spectra were recorded on an LCT premier spectrometer.

Sodium N-Benzyldithiocarbamate

An aqueous 1M solution of NaOH (20 mL, 20 mmol) was cooled by means of an ice bath. Benzylamine (2.2 mL, 20 mmol) and carbon disulfide (1.2 mL, 20 mmol) were then added and the reaction mixture stirred vigorously for 2 h. After this time, the majority of the solvent was removed under reduced pressure. Isopropanol/diethyl ether (1:5 v/v, 60 mL) was added to the residue giving a white amorphous precipitate that was collected by filtration. After drying in a desiccator under vacuum over anhydrous phosphorous pentoxide at room temperature for 18 h, 3.2 g (78%) of the sodium salt of N-benzyldithiocarbamate was obtained. The compound showed spectroscopic data identical to those reported by Zhang et al.[10]

S-(2,3,4,6-Tetra-O-acetyl-β-D-glucopyranosyl)-N-benzyldithiocarbamate (2)

A mixture of glucopyranosyl bromide **1** (0.411 g, 1 mmol) and sodium N-benzyldithiocarbamate (0.410 g, 2.0 mmol) in anhydrous acetone (30 mL) was magnetically stirred at room temperature until TLC (1:1 EtOAc–hexane) showed complete conversion of the starting material (2–4 h). The mixture was neutralized with a few drops of aqueous 5% HCl and the organic solvent was removed under reduced pressure. Water (40 mL) was added and the mixture was extracted with dichloromethane (2×60 mL). The combined organic extracts were dried (Na_2SO_4) and concentrated and the residue was chromatographed (1:2 EtOAc–hexane) yielding the glucopyranosyl dithiocarbamate derivative **2** as syrup (0.42 g, 82%). R_f (1:1 EtOAc–hexane) 0.49; $[\alpha]_D$ +12.2° (c 1, $CHCl_3$); ν_{max}(film)/cm^{-1} 3291, 2942, 1753, 1512, 1372, 1227, 1046, and 914. ^{1}H NMR ($CDCl_3$, 400 MHz) δ: 7.67 (t, 1H, J 5.1 Hz, NH), 7.39–7.26 (m, 5H), 5.67 (d, 1H, J 10.5 Hz), 5.31 (t, 1H, J 9.3 Hz), 5.18 (t, 1H, J 10.1 Hz), 5.06 (t, 1H, J 9.7 Hz), 4.86 (d, 2H, J 5.0 Hz), 4.20 (dd, 1H, J 12.5, 4.8 Hz), 4.04 (dd, 1H, J 12.7 and 2.3 Hz), 3.82 (ddd, 1H, J 10.1, 4.7 and 2.2 Hz), 2.01, 2.00, 2.00, and 1.99 (4s, 12H). ^{13}C NMR ($CDCl_3$, 100 MHz) δ: 192.9, 170.8, 170.2, 169.7, 169.6, 135.8, 129.1, 128.5, 128.5, 86.2, 76.5, 74.1, 68.8, 68.2, 61.8, 51.6, 20.9, 20.8, 20.7. HRMS (m/z) (ESI) calcd for $C_{22}H_{27}NO_9S_2Na$ [M+Na]$^+$: 536.1025. Found: 536.1019.

[2-(Ethenesulfonyl)ethyl] 2,3,4,6-tetra-O-acetyl-1-thio-β-D-glucopyranoside (3)

To a solution of the glucosyl dithiocarbamate **2** (0.513 g, 1 mmol) in anhydrous acetone (30 mL) was added divinyl sulfone (0.3 mL, 3 mmol) and triethylamine (0.28 mL, 2 mmol). The reaction mixture was magnetically stirred at room temperature until TLC (1:1 EtOAc–hexane) showed complete disappearance of the starting material (1 h). After concentration, chromatography (1:1 EtOAc–hexane → EtOAc) yielded the glycosyl vinyl sulfone **3** as syrup. Total yield: 0.43 g (90%). R_f (1:1 EtOAc–hexane) 0.23; $[\alpha]_D$ −22.5° (c 1, $CHCl_3$). ν_{max}(film)/cm^{-1}: 2943, 1752, 1431, 1374, 1226, 1136, 1039, and 914. ^{1}H NMR ($CDCl_3$, 400 MHz) δ: 6.67 (dd, 1H, J 16.6, 9.8 Hz), 6.45 (d, 1H, J 16.6 Hz), 6.21 (d, 1H, J 9.9 Hz), 5.21 (t, 1H, J 9.4 Hz), 5.03 (t, 1H, J 9.8 Hz), 5.00 (t, 1H, J 9.7 Hz), 4.54 (d, 1H, J 10.0 Hz), 4.16 (d, 2H, J 3.7 Hz), 3.72 (dt, 1H, J 10.1, 3.8 Hz), 3.40–3.27 (m, 2H), 3.07 (m, 1H), 2.91 (m, 1H), 2.09, 2.03, 2.01, and 1.99 (4s, 12H). ^{13}C NMR ($CDCl_3$, 100 MHz) δ: 170.8, 170.2, 169.5, 136.1, 131.4, 84.0, 76.3, 73.6, 69.6, 68.3, 62.1, 55.3, 23.0, 20.8, 20.8, 20.7, 20.7. HRMS (m/z) (ESI) calcd for $C_{18}H_{26}O_{11}S_2Na$ [M+Na]$^+$: 505.0814. Found: 505.0819.

ACKNOWLEDGMENTS

This chapter was funded by Ministerio de Ciencia e Innovación of the Government of Spain (CTQ2011-29299-CO2-01). O.B. thanks the Ministerio de Ciencia e Innovación, Spain (Juan de la Cierva Fellowship) and the European Commission (Marie Curie Career Integration Grant).

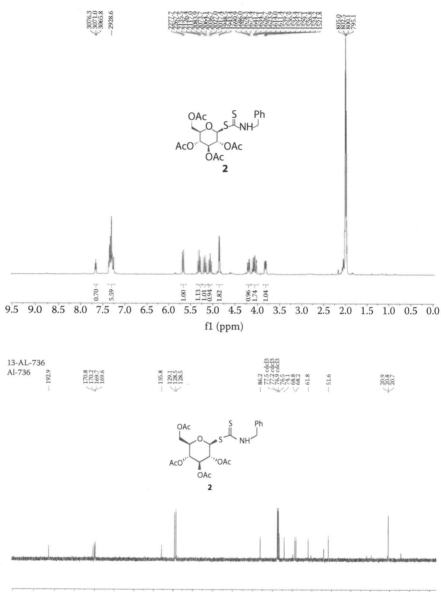

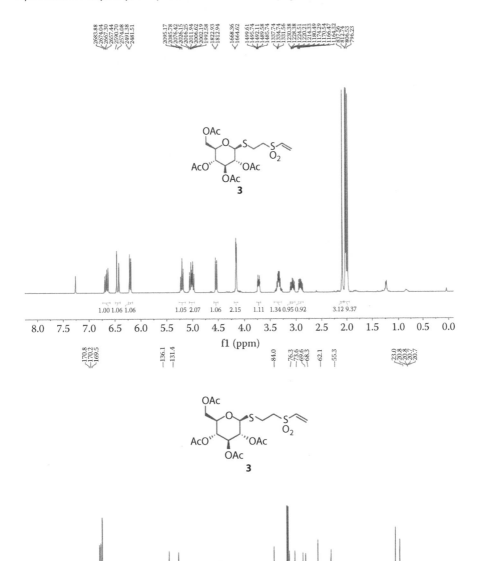

REFERENCES

1. Hermanson, G. T., Ed. *Bioconjugate Techniques*, 2nd edn. Academic Press, San Diego, CA, 2008.
2. Meadows, D. C.; Gervay-Hague, J. *Med. Res. Rev.* **2006**, *26*, 793–814.
3. Forristal, I. *J. Sulfur Chem.* **2005**, *26*, 163–195.
4. Simpkins, N. S. *Tetrahedron* **1990**, *46*, 6951–6984.
5. Lopez-Jaramillo, F. J.; Hernandez-Mateo, F.; Santoyo-Gonzalez, F. Vinyl sulfone: A multi-purpose function for protemics, In *Integrative Proteomics*, Eastwood Leung, H.-C., Ed. InTech, Rijeka, Croatia, 2012, pp. 301–326.
6. Lopez-Jaramillo, F. J.; Ortega-Munoz, M.; Megia-Fernandez, A.; Hernandez-Mateo, F.; Santoyo-Gonzalez, F. *Bioconjug. Chem.* **2012**, *23*, 846–855.
7. Morales-Sanfrutos, J.; Lopez-Jaramillo, J.; Ortega-Muñoz, M.; Megia-Fernandez, A.; Perez-Balderas, F.; Hernandez-Mateo, F.; Santoyo-Gonzalez, F. *Org. Biomol. Chem.* **2010**, *8*, 667–675.
8. Megia-Fernandez, A.; de la Torre-Gonzalez, D.; Parada-Aliste, J.; Lopez-Jaramillo, F. J.; Hernandez-Mateo, F. Santoyo-Gonzalez, F. *Chem. Asian J.* **2014**, 9, 620–631.
9. Fernandez, J. M. G.; Mellet, C. O. *Adv. Carbohydr. Chem. Biochem.* **2000**, *55*, 35–135.
10. Zhang, J.; Lin, X.; Ren, J.; Liu, J.; Wang, X. *Appl. Radiat. Isot.* **2009**, *68*, 101–104.

7 Synthesis of 5-Deoxy-β-D-galactofuranosides (5-Deoxy-α-L-*arabino*-hexofuranosides) Starting from D-Galacturonic Acid Using Photoinduced Electron Transfer Deoxygenation

Carla Marino,* Andrea V. Bordoni,
Marcos J. Lo Fiego,[†] and Rosa M. de Lederkremer

CONTENTS

Developing specific deoxygenation methods for carbohydrates is important as deoxy sugars are useful not only for enzymatic characterization but also for the synthesis of natural products from carbohydrate precursors as chiral templates.[1–3] Since its development in 1975, the Barton–McCombie deoxygenation method has been

---

* Corresponding author; e-mail: cmarino@qo.fcen.uba.ar.
† Checker; e-mail: marcoslf@hotmail.com.

extensively employed.[4] However, this procedure involves the use of organotin compounds, which are toxic, expensive, and difficult to remove. Thus, many modifications have been later developed using other hydride donors.[5]

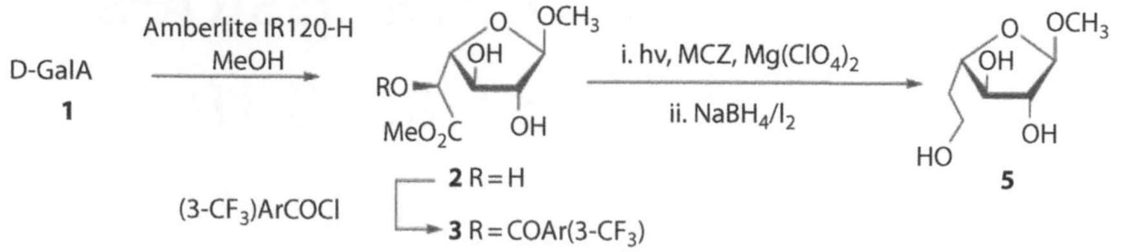

Photochemical reactions are attractive in the modern context of green chemistry. In the frame of our project dedicated to the characterization of β-D-galactofuranosidases,[6] we found very useful the photoinduced electron transfer (PET) reduction[7] of 3-(trifluoromethyl)benzoyl esters, using 9-methylcarbazole (MCZ) as photosensitizer. This reaction was first used by Rizzo et al. for the synthesis of deoxyribonucleosides (**A**, Scheme 7.1). We investigated the PET deoxygenation on D-galactono-1,4-lactone (**B**),[8,9]

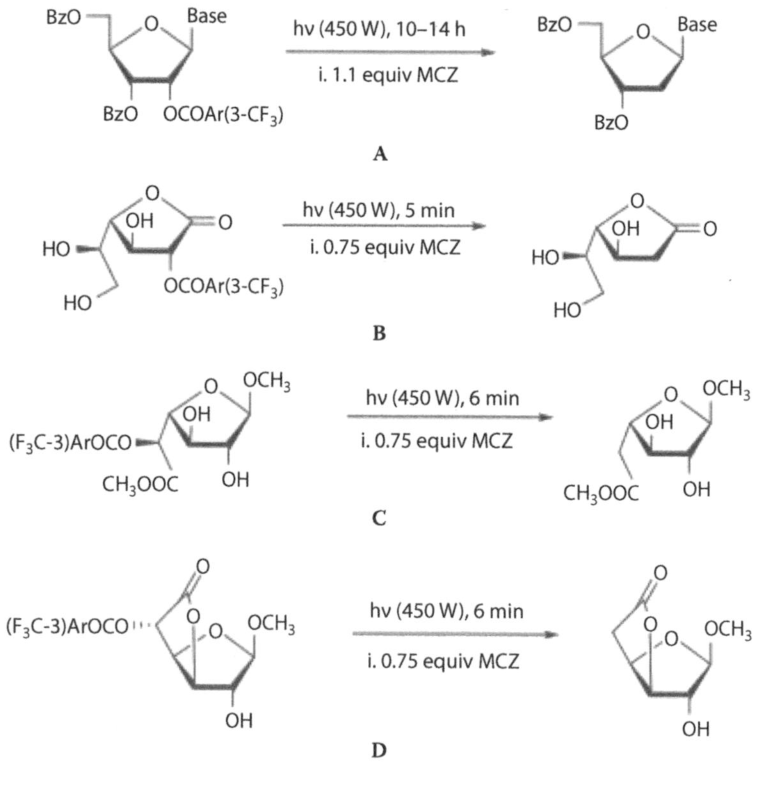

SCHEME 7.1   PET α-deoxygenation. (i) 1.4 equiv $Mg(ClO_4)_2$, 9:1 2-PrOH/H_2O.

D-galacturonic acid (**C**), and D-glucofuranosidurono-6,3-lactone (**D**) derivatives.[10] The reaction is particularly efficient for reduction of the HO vicinal to the carbonyl group of lactones or esters (**B-D**, Scheme 7.1) allowing to significantly reduce the irradiation time and the MCZ amount. The effectiveness of the reaction relies on the stabilization by the carbonyl group of the intermediate radical **I**, formed by homolytic cleavage of the hydroxyl group conveniently derivatized (Scheme 7.2).

We describe here the synthesis of methyl 5-deoxy-β-D-galactofuranoside (**5**) from D-galacturonic acid (**1**) through PET deoxygenation (Scheme 7.3). The furanosyl precursor **2**, obtained by the resin catalyzed reaction of **1** with methanol,[11] is

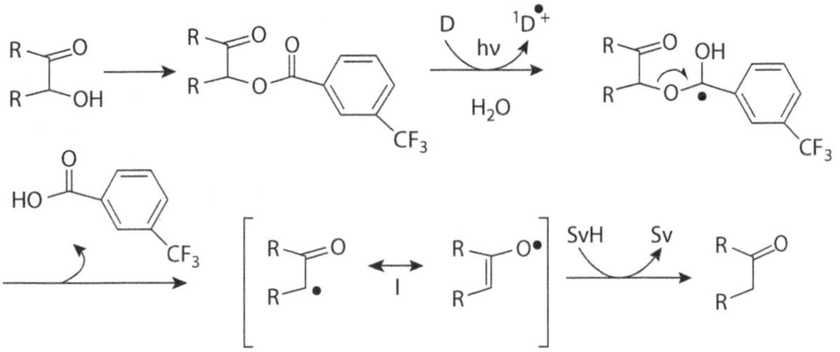

SCHEME 7.2 Proposed mechanism for the PET α-deoxygenation of carbonyl derivatives (D: electron donor, in this case, MCZ).

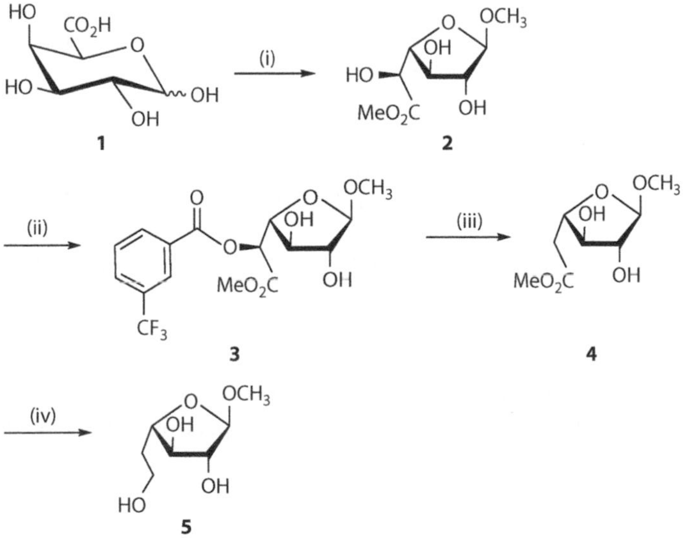

SCHEME 7.3 Synthesis of methyl 5-deoxy-β-L-*arabino*-hexofuranoside (**5**). (i) Amberlite IR120-H, MeOH; (ii) (3-CF$_3$)PhCOCl, 2:1 CH$_2$Cl$_2$/pyridine; (iii) $h\nu$, MCZ, Mg(ClO$_4$)$_2$, 9:1 2-PrOH/H$_2$O; (iv) NaBH$_4$/I$_2$, THF.

converted to the photoreductible ester **3** and subsequently irradiated to afford the deoxygenated compound **4**. Finally, using the $NaBH_4$–I_2 system,[12] 5-deoxy glycofuranoside **5** is obtained. The straightforward access to the furanosyl precursor **2**, combined with the regioselectivity of the acylation and the easiness of the deoxygenation, significantly shortens the pathway to **5**, with respect to previous reports.[13]

EXPERIMENTAL METHODS

GENERAL METHODS

Thin-layer chromatography (TLC) was performed on 0.2 mm silica gel 60 F254 (Merck) aluminum-supported plates. Detection was effected by UV light and by charring with 10% (v/v) H_2SO_4 in EtOH. Column chromatography was performed on silica gel 60 (230–400 mesh, Merck). The ^1H and ^{13}C spectra were recorded with a Bruker AM 500 spectrometer at 500 MHz (^1H) and 125.8 MHz (^{13}C). Assignments were supported by COSY and HSQC experiments. High-resolution mass spectra (HRMS) were recorded on an Agilent LC-TOF with Windows XP–based OS and APCI/ESI ionization high-resolution mass spectrometer analyzer, with Opus V3.1 and DEC 3000 AlphaStation. Optical rotations were measured with a PerkinElmer 343 polarimeter, with a path length of 1 dm. Melting points were determined with a Fisher–Johns apparatus and are uncorrected.

Methyl (methyl β-D-galactofuranosid)uronate (2)

A 250 mL, single-neck round-bottom flask equipped with a rubber septum was charged with D-galacturonic acid (**1**, 1.00 g, 5.15 mmol), MeOH* (20.0 mL), and Amberlite IR-120 H$^+$ resin (1.0 g). The suspension was stirred in an orbital shaker at 35°C and 180 rpm. After 3 days, another portion of resin (~0.3 g) was added, and the stirring was continued for four more days. TLC showed presence of a main product with R_f 0.5 (EtOAc, twice developed) and a slower moving product (R_f 0.4). After filtration of the resin and removal of the solvent under diminished pressure, the residue (1.11 g) was purified by silica gel column chromatography (49:1 EtOAc–toluene). Fractions of R_f 0.5 afforded syrupy **2** (0.64–0.68 g, 56%–59%, $[\alpha]_D$ −125 (c 1, CH_3OH); lit.[14] $[\alpha]_D$ −112 (c 1.4, methanol). ^1H NMR (500 MHz, D_2O) δ: 4.89 (d, J = 1.9 Hz, 1 H, H-1), 4.50 (d, J = 2.7 Hz, 1 H, H-5), 4.29 (dd, J = 2.7, 6.5 Hz, 1 H, H-4), 4.18 (dd, J = 3.9, 6.5 Hz, 1 H, H-3), 4.03 (dd, J = 1.9, 3.9 Hz 1 H, H-2), 3.81 (s, 3 H, OCH_3), 3.38 (s, 3 H, CO_2CH_3). ^{13}C NMR (50.3 MHz, D_2O) δ: 174.2 (C-6).109.0 (C-1), 84.1, 81.0 (C-2,4), 76.3 (C-3), 70.0 (C-5), 55.6, 53.5 (2 OCH_3). Fractions of R_f 0.4 afforded methyl (methyl α-D-galactofuranosid)uronate (0.15–0.19 g, 13%–17%)[†] with physical and spectroscopic data identical to those reported in the literature.[14]

---

* MeOH was distilled.
[†] Longer reaction time increases the amount of the β-anomer.

Methyl [methyl 5-*O*-(3-trifluoromethyl)benzoyl-β-D-galactofuranosid]uronate (3)

A solution of **2** (0.50 g, 2.25 mmol) in CH_2Cl_2* (10.0 mL) containing pyridine[†] (5 mL) was placed in a 50 mL round-bottom flask. The reaction vessel was flushed with argon and the solution was magnetically stirred with external cooling in an ice-water bath. 3-(Trifluoromethyl)benzoyl chloride (0.39 mL, 2.64 mmol) dissolved in CH_2Cl_2* (5.0 mL) was added in three aliquots during 1.5 h. After 1 h of stirring at 0°C, TLC showed a main product of R_f 0.6 (9:1 EtOAc–toluene) and small amounts of a product of R_f 0.9 and unchanged starting material. The NMR spectrum analysis indicated that compound of R_f 0.9 was the tribenzoylated sugar. The solution was diluted with CH_2Cl_2 (10 mL) and successively washed with HCl (5%), water, satd $NaHCO_3$, and water (20 mL each). The organic layer was dried (Na_2SO_4), filtered, and concentrated. The syrup obtained was chromatographed (3:2 toluene–EtOAc) to afford **3** (R_f 0.6), 0.55–0.58 g (62%–65%); $[\alpha]_D$ −66 (*c* 1, CH_3OH). ¹H NMR (500 MHz, $CDCl_3$) δ: 8.37–7.58 (4 H, CH_{Ar}), 5.53 (d, *J*=Hz,1 H, H-5), 4.92 (s, 1 H, H-1), 4.48 (apparent t, *J*=4.5 Hz, 1 H, H-4), 4.11 (m, 2 H, H-2,3), 3.79 (s, 3 H, CO_2CH_3), 3.36 (s, 3 H, OCH_3). ¹³C NMR (125 MHz, D_2O) δ: 168.1 (C-6), 164.8 (COAr), 133.1–122.5 (6 C, C_{Ar}), 108.6 (C-1), 83.1 (C-4), 81.1 (C-2), 77.7 (C-3), 72.3 (C-5), 55.2 (OCH_3), 53.0 (CO_2CH_3). Anal. calcd for $C_{16}H_{17}F_3O_8$: C, 48.74; H, 4.35. Found: C, 49.01; H, 4.48.

Methyl (methyl 5-deoxy-β-L-*arabino*-hexofuranosid)uronate (4)

The PET deoxygenation was performed in a custom-made Pyrex immersion well apparatus equipped with a quartz cold finger and a Hanovia 450 W medium-pressure Hg lamp (λ_{exc}>300 nm) (see Figure 7.1). A solution containing **3** (0.29 g, 0.75 mmol), $Mg(ClO_4)_2$ (33 mg, 0.3 mM), and MCZ (13 mg, 0.075 mmol) in 500 mL of 9:1 2-PrOH–H_2O was placed in the reaction vessel. The solution was degassed by bubbling UHP argon through it during 30 min, while the temperature was maintained at 25°C with the aid of a circulating water bath. Then, the solution was irradiated (~7 min), the lamp was turned off, and an aliquot of 2 or 3 mL was withdrawn, concentrated, and analyzed by TLC. The plate showed conversion of the starting material (R_f 0.63) into a product with R_f 0.47 (9:1 EtOAc–toluene) and faster moving components (R_f 0.95) corresponding to MCZ and a photoproduct of MCZ (TLC analysis showed a minor component, R_f: 0.24). NMR analysis showed that the latter was the free acid (hydrolysis of the ester occurred). When the reaction was complete, generally after 6–7 min of irradiation,[‡] the solvent was removed under reduced pressure and the residue was partitioned between water and CH_2Cl_2. The aqueous layer contained only the deoxygenated product **4** that, after chromatography (97:3 EtOAc–toluene) (0.080–0.086 g, 52%–56%), showed $[\alpha]_D$ −68 (*c* 1, $CHCl_3$). ¹H NMR (500 MHz, D_2O) δ: 4.89 (d, *J*=0.5 Hz, 1 H, H-1), 4.30 (m, 1 H, H-4), 4.04 (m, 1 H, H-2), 3.89 (ddd, *J*=4.4, 5.3, 0.4 Hz,1 H, H-3), 3.73 (CO_2CH_3),

---

* CH_2Cl_2 was dried by refluxing with P_2O_5 followed by distillation.
[†] Pyridine was distilled from NaOH pellets and stored under nitrogen over NaOH.
[‡] If necessary, the irradiation was continued for 2 or 3 min, until total transformation.

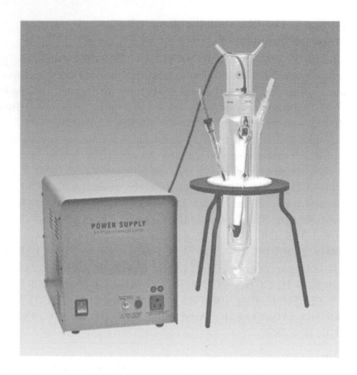

FIGURE 7.1 The quartz cold finger and a Hanovia 450 W medium-pressure Hg lamp (λ_{exc}>300 nm) used in preparation of **4**.

3.36 (OCH_3), 2.86 (dd, J=4.4, 16.0 Hz, 1 H, H-5), 2.72 (dd, J=8.7, 16.0 Hz, 1 H, H-5′). ^{13}C NMR (125 MHz, D$_2$O) δ: 174.3 (C-6), 109.1 (C-1), 81.4 (C-4), 80.5 (C-2), 80.2 (C-3), 55.4 (CO$_2$CH_3), 53.2 (OCH_3), 38.6 (C-5). HRMS (ESI/APCI) calcd for C$_8$H$_{18}$NO$_6$, [M+NH$_4$]$^+$: 224.1929. Found: 224.2234.

Methyl 5-Deoxy-α-ʟ-*arabino*-hexofuranoside (5)

The reaction setup consisted of a 50 mL, two-necked, round-bottom flask with a magnetic stirrer, having a reflux condenser with a calcium chloride drying tube connected to one neck and the other one capped with a rubber septum. A suspension of NaBH$_4$ (0.126 g, 3.34 mmol) in dry THF (6.7 mL) was placed in the flask and cooled to 0°C under argon atmosphere. Under a stream of argon, a solution of I$_2$ (0.34 g, 1.34 mmol) in THF (3.2 mL) was added slowly with a cannula, drop by drop, waiting for the brown color to disappear before adding the next drop (~30 min). Fifteen minutes after the last addition of I$_2$, a solution of compound **4** (0.08 g, 0.38 mmol) in THF (2.0 mL) was transferred with a cannula and under a stream of argon, from a 5 mL, septum-capped, round-bottom flask, into the NaBH$_4$–I$_2$ suspension. The mixture was vigorously stirred under reflux, and when TLC showed complete conversion of the starting material (R_f 0.70, 9:1 EtOAc–MeOH, ~2–5 h) into a product R_f 0.36

(9:1 EtOAc–MeOH), 5% HCl was carefully added, to neutralize the excess of $NaBH_4$ (dropwise, ~2 mL), and the mixture was partitioned between $CHCl_3$ and H_2O. The aqueous layer was thoroughly washed with $CHCl_3$, to remove the iodine, and deionized by elution from a column of mixed bed ion-exchange resin (1.0×6.0 cm) (e.g., IWT® TMD-8 or Amberlite MB-3A) with water. Evaporation of the solvent under reduced pressure and coevaporation with methanol (5 mL×3) afforded **5**. Yield, 0.047–0.051 g, 70%–76%, $[\alpha]_D$ −141 (*c* 1, CH_3OH); lit. 16 $[\alpha]_D$ −137 (*c* 1, methanol). ^1H NMR (500 MHz, D_2O) δ: 4.89 (d, *J* = 1.4 Hz, 1 H, H-1), 4.03 (m, 2 H, H-2,3), 3.83 (ddd, *J* = 0.5, 3.4, 7.1 Hz, 1 H, H-4), 3.75 (ddd, *J* = 5.7, 6.9, 11.1 Hz, 1 H, H-6), 3.71 (ddd, *J* = 6.4, 7.4, 11.1 Hz, 1 H, H-6′), 3.39 (OCH_3), 1.96 (dddd, *J* = 4.8, 7.2, 9.3, 14.2 Hz, 1 H, H-5), 1.87 (dddd, *J* = 5.7, 6.3, 8.3, 14.2 Hz,1 H, H-5′). ^{13}C NMR (125 MHz, D_2O) δ: 108.7 (C-1), 81.6, 81.3 (C-2,3), 80.8 (C-4), 59.0 (C-6), 55.4 (OCH_3), 35.8 (C-5).

ACKNOWLEDGMENTS

We are indebted to CONICET and Universidad de Buenos Aires for financial support. C. Marino, A.V. Bordoni, and R.M. de Lederkremer are research members of CONICET.

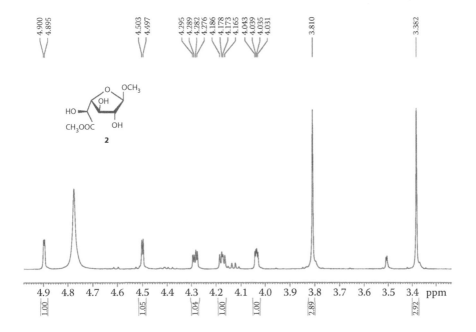

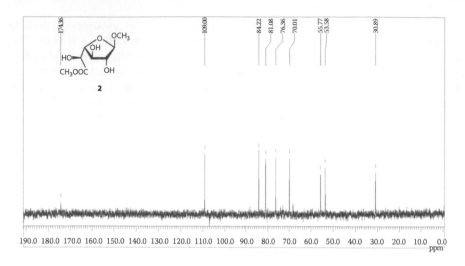

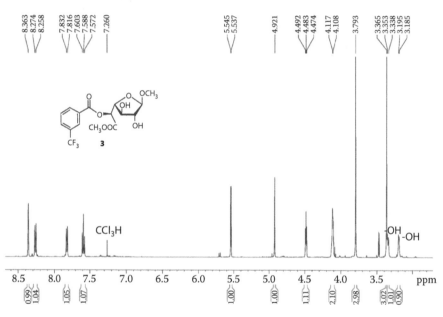

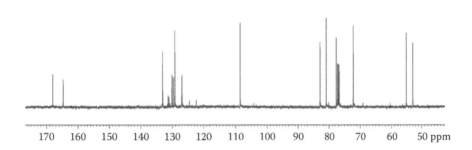

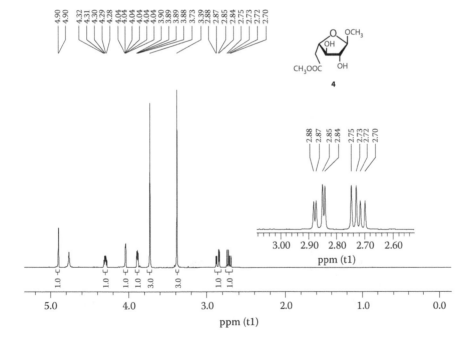

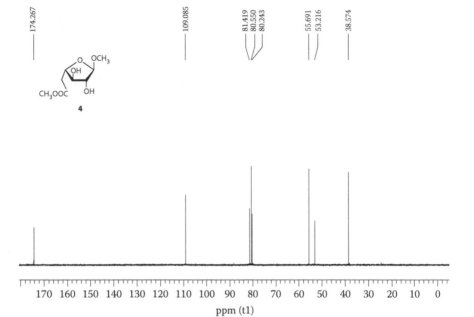

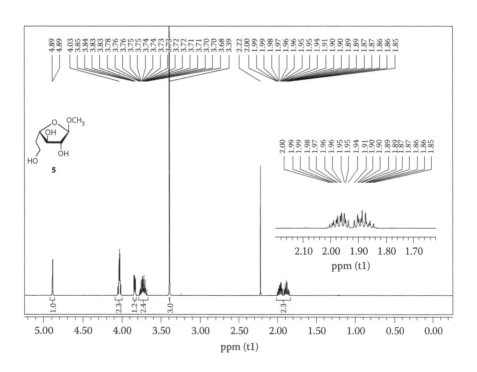

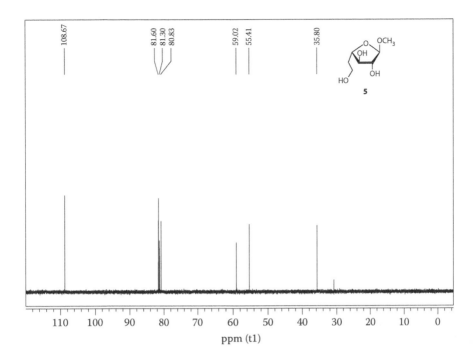

REFERENCES

1. Lederkremer, R. M.; Marino, C. *Adv. Carbohydr. Chem. Biochem.* **2007**, *61*, 143–216.
2. Ye, J.; Bhatt, R. K.; Falk, J. R. *Tetrahedron Lett.* **1993**, *34*, 8007–8010.
3. Matsuura, D.; Takabe, K.; Yoda, H. *Tetrahedron Lett.* **2006**, *47*, 1371–1374.
4. Barton, D. H. R.; McCombie, S. W. *J. Chem. Soc. Perkin Trans. 1*, **1975**, *16*, 1574–1585.
5. (a) Spiegel, D. A.; Wiberg, K. B.; Schacherer, L. N.; Medeiros, M. R.; Wood, J. L. *J. Am. Chem. Soc.* **2005**, *127*, 12513–12515. (b) Zhang, L.; Koreeda, M. *J. Am. Chem. Soc.* **2004**, *126*, 13190–13191.
6. Bordoni, A.; Lederkremer, R. M.; Marino, C. *Bioorg. Med. Chem.* **2010**, *18*, 5339–5345.
7. (a) Park, M.; Rizzo, C. J. *J. Org. Chem.* **1996**, *61*, 6092–6093; (b) Prudhomme, D. R.; Wang, Z.; Rizzo, C. J. *J. Org. Chem.* **1997**, *62*, 8257–8260; (c) Wang, Z.; Prudhomme, D. R.; Buck, J. R.; Park, M.; Rizzo, C. J. *J. Org. Chem.* **2000**, *65*, 5969–5985.
8. Chiocconi, A.; Marino, C.; Otal, E.; Lederkremer, R. M. *Carbohydr. Res.* **2002**, *337*, 2119–2126.
9. Bordoni, A.; Lederkremer, R. M.; Marino, C. *Carbohydr. Res.* **2006**, *341*, 1788–1795.
10. Bordoni, A.; Lederkremer, R. M.; Marino, C. *Tetrahedron* **2008**, *64*, 1703–1710.
11. Bordoni, A.; Lima, C.; Mariño, K.; Lederkremer, R. M.; Marino, C. *Carbohydr. Res.* **2008**, *343*, 1863–1869.
12. Bhanu Prasad, S.; Bhaskar Kanth, J. V.; Periasamy, M. *Tetrahedron* **1992**, *48*, 4623–4628.
13. (a) Wolfrom, M. L.; Matsuda, K.; Komitsky, Jr., F.; Whiteley, T. E. *J. Org. Chem.* **1963**, *28*, 3551–3553. (b) Szarek, W. A.; Ritchie, R. G. S.; Vyas, D. M. *Carbohydr. Res.* **1978**, *62*, 89–103.
14. Matsuhiro, B.; Zanlungo, A. B.; Dutton, G. G. S. *Carbohydr. Res.* **1981**, *97*, 11–18.

8 Glycal Transformation into 2-Deoxy Glycosides

Catarina Dias, Alice Martins, M. Soledade Santos, Amélia P. Rauter,* and Michał Malik †

CONTENTS

In this chapter, we illustrate the synthesis of 2-deoxy glycosides via reaction of glycals with alcohols in the presence of triphenylphosphane hydrobromide (TPHB). A special emphasis is given to the reaction of glycals with long-chain alcohols, which leads to materials with a variety of potential applications, in particular as antimicrobial agents.[1–3] The reaction is easy to perform and stereoselective, and when long-chain alcohols are used as synthons, the bioactivity of the formed 2-deoxy glycosides can be tuned by changing the glycal starting material structure, for example, 6-deoxygenation, and configuration.[2] The acid-catalyzed addition of an alcohol to an acetylated glycal occurs usually with concomitant Ferrier allylic rearrangement, giving the 2,3-unsaturated derivative as major reaction product. However, when TPHB is used, as first described by Bolitt et al.[4] to prepare 2-deoxy glycosides from glycals, formation of 2,3-unsaturated glycosides is minimal. Among the variety of glycals of D- and L-series already subjected to this procedure,[1–3] 3,4,6-tri-*O*-acetyl-1,5-anhydro-2-deoxy-D-*arabino*-hex-1-enitol (**1**) was selected to exemplify now the TPHB-mediated glycosylation of octanol and of dodecanol, which occurs with α-stereoselectivity, resulting from the anomeric effect. Glycosides **2–5** were obtained by reaction of glycal **1** in dichloromethane, using 2 equiv. alcohol at room temperature for 12 h. The 2-deoxy α-anomers were isolated in 69% yield (**2**) and 83% yield (**4**), and the

---

* Corresponding author; e-mail: aprauter@fc.ul.pt.
† Checker, Michał Malik, under supervision of Sławomir Jarosz; e-mail: slawomir.jarosz@icho.edu.pl.

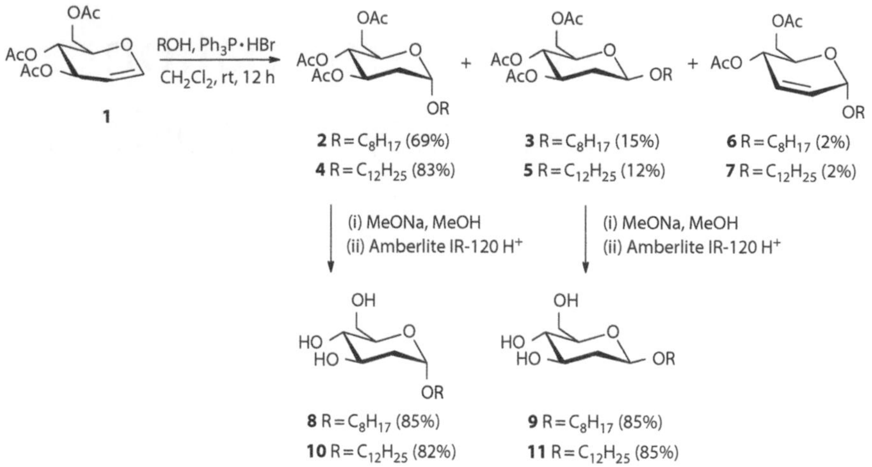

SCHEME 8.1 Synthesis of octyl and dodecyl 2-deoxy-D-*arabino*-hexopyranosides.

2-deoxy β-glycosides were isolated in 15% (**3**) and 12% yield (**5**). The α-anomer of the octyl or dodecyl 2,3-unsaturated glycosides (**6** and **7**, respectively) was isolated in 2% yield, when the reaction was run at room temperature and in 6% (**6**) and 4% (**7**) when the reaction was run under reflux. In order to simplify glycoside and nucleophile separation, the reaction was carried out with 1.05 equiv. alcohol at room temperature. Reaction time increased to 24 h, no Ferrier products were detected, and selectivity for the α-anomer was also improved as shown by the reaction yields obtained for compounds **2**, **3**, **4**, and **5**, isolated in 71%, 5%, 77%, and 7%, respectively. Zemplén deprotection of compounds **2–5** gave glycosides **8–11** in high yield (Scheme 8.1).

This two-step methodology to prepare **8** in 59% overall yield is an efficient alternative to the four-step pathway described in the literature, based on iodoacetoxylation of glycal **1**, followed by reaction with octanol in the presence of trimethylsilyl trifluoromethanesulfonate, deiodination with tributyltin hydride, and Zemplén deacetylation to furnish **8** in 48% overall yield.[5]

EXPERIMENTAL METHODS

GENERAL METHODS

Starting materials and reagents were purchased from Sigma-Aldrich, Fluka, and Acros. Solvents were dried prior to use with molecular sieves 4 Å, with the exception of methanol, dried over 3 Å molecular sieves. Thin layer cromatography (TLC) was carried out on aluminum sheets (20 × 20 cm) coated with silica gel 60 F-254, 0.2 mm thick (Merck) with detection by charring with 10% H_2SO_4 in ethanol. Column chromatography (CC) was performed using silica gel 230–400 mesh (Merck). Melting points were measured by differential scanning calorimetry, using a Setaram TG-DSC111 apparatus. Temperature was calibrated with LGC standard reference materials (Hg, In, Sn, Pb), and energy calibrated by both Joule effect and standard reference material (Sapphire NIST SRM 720). Samples of compounds **8–11** were weighed into aluminum crucibles, which were sealed under air, without protective atmosphere, and Differential Scanning Calorimetry (DSC) scans were performed at a rate of 1°C/min, from room temperature to 140°C. The stability of the samples and accuracy of the determinations were ensured performing successive heating and cooling scans that showed pronounced thermal hysteresis, giving reproducible results on successive heating and cooling cycles. No pretransitions or *double melting points* were observed demonstrating the absence of solvent loss or formation of liquid crystalline phases. Melting point of compounds **8–11** was also measured with a SMP3 melting point apparatus, Stuart Scientific, Bibby. Elemental analyses were performed at the Service of Microanalyses of Instituto Superior Técnico, Universidade Técnica de Lisboa. Optical rotations were measured with a PerkinElmer 343 polarimeter. Nuclear Magnetic Resonace (NMR) experiments were recorded on a Bruker Avance 400 spectrometer at 298 K, operating at 100.62 MHz for ^{13}C and at 400.13 MHz for ^{1}H for solutions in CDCl$_3$ containing 0.03% TMS or in CD$_3$OD (Sigma-Aldrich).

GENERAL PROCEDURE FOR THE GLYCOSYLATION REACTION

The nucleophile (36.5 mmol) and a solution of Ph$_3$P•HBr (686 mg, 2.0 mmol) in dry dichloromethane (16.0 mL) were added to a solution of 3,4,6-tri-*O*-acetyl-1,5-anhydro-2-deoxy-D-*arabino*-hex-1-enitol (**1**, 5.04 g, 18.5 mmol) in the same solvent (16.0 mL). The mixture was stirred at room temperature for 12 h and washed with a satd. NaHCO$_3$ (30 mL), the organic phase was concentrated, and CC (1:10 EtOAc cyclohexane) afforded the α-anomer of the corresponding 2-deoxy glycosides as major product with both the β-anomer and the Ferrier compound isolated as minor products. Isolated compounds are listed in the order of elution.

Octyl 3,4,6-tri-*O*-acetyl-2-deoxy-α-D-*arabino*-hexopyranoside (2), octyl 3,4, 6-tri-*O*-acetyl-2-deoxy-β-D-*arabino*-hexopyranoside (3), and octyl 4,6-di-*O*-acetyl-2,3-dideoxy-α-D-*erythro*-hex-2-enopyranoside (6). Reaction of glycal **1** with octanol (3.9 mL, 36.5 mmol) gave compounds **2**, **3**, and **6**.

Compound **2**: Syrup (5.11 g, 69%); [α]$_D^{20}$ +77 (*c* 1, CH$_2$Cl$_2$); R$_f$ 0.42 (1:3 EtOAc–PE). IR (neat): 1748 cm^{-1} (C=O). ^{1}H NMR (CDCl$_3$) δ: 5.33 (ddd, 1H, $J_{2a,3}$ 12.2 Hz, $J_{2e,3}$ 5.6 Hz, $J_{3,4}$ 10.0 Hz, H-3), 4.99 (t, 1H, $J_{4,5}$ 10.0 Hz, H-4), 4.94 (d, 1H, $J_{1,2a}$ 3.1 Hz, H-1), 4.31 (dd, 1H, $J_{5,6a}$ 4.7 Hz, H-6a), 4.05 (dd, 1H, $J_{6b,6a}$ 12.3 Hz, H-6b), 3.96 (ddd, 1H, $J_{5,6b}$ 1.6 Hz, H-5), 3.62 (dt, 1H, $J_{1'a,1'b}$ 13.5 Hz, $J_{1'a,2'a,b}$ 6.8 Hz, H-1'a), 3.38 (dt, 1H, H-1'b), 2.23 (dd, 1H, $J_{2e,2a}$ 12.2 Hz, $J_{2e,3}$ 5.6 Hz, H-2e), 2.09 (s, 3H, CH$_3$-Ac), 2.04 (s, 3H, CH$_3$-Ac), 2.01 (s, 3H, CH$_3$-Ac), 1.82 (dt, 1H, $J_{2a,3}$ 12.2 Hz, H-2a), 1.63–1.53 (m, 2H, H-2'a,b), 1.39–1.21 (m, 10H, H-3'a,b - H-7'a,b), 0.89 (t, 3H, $J_{7',8'}$ 7.2 Hz, H-8'). ^{13}C NMR (CDCl$_3$) δ: 170.6 (C=O), 170.1 (C=O), 169.9 (C=O), 96.8 (C-1), 69.4 (C-4), 69.1 (C-3), 67.8 (C-1'), 67.7 (C-5), 62.4 (C-6), 35.0 (C-2), 31.8, 29.3, 29.2, 29.1, 26.1, 22.6 (C2'-C-7'), 20.9, 20.7 (CH$_3$-Ac), 14.0 (C-8'). Anal. calcd for C$_{20}$H$_{34}$O$_8$: C, 59.68; H, 8.51. Found: C, 59.70; H, 8.80.

Compound **3**: Syrup (1.11 g, 15%); [α]$_D^{20}$ –18 (*c* 1, CH$_2$Cl$_2$); R$_f$ 0.40 (1:3 EtOAc–PE). IR (neat): 1749 cm^{-1} (C=O). ^{1}H NMR (CDCl$_3$) δ: 5.07–4.95 (m, 2H, H-3, H-4), 4.56 (dd, 1H, $J_{1,2a}$ 9.7 Hz, $J_{1,2e}$ 1.9 Hz, H-1), 4.30 (dd, 1H, $J_{6a,5}$ 4.9 Hz, $J_{6a,6b}$ 12.2 Hz, H-6a), 4.11 (dd, 1H, $J_{6b,5}$ 2.4 Hz, H-6b), 3.87 (dt, 1H, $J_{1'a,1'b}$ 9.4 Hz, $J_{1'a,2'a,b}$ 6.6 Hz, H-1'a), 3.60 (ddd, 1H, $J_{5,4}$ 9.3 Hz, H-5), 3.46 (dt, 1H, H-1'b), 2.32 (ddd, 1H, $J_{2e,3}$ 4.8 Hz, $J_{2e,2a}$ 12.6 Hz,

H-2e), 2.09 (s, 3H, CH_3-Ac), 2.04 (s, 3H, CH_3-Ac), 2.03 (s, 3H, CH_3-Ac), 1.75 (ddd, 1H, $J_{2a,3}$ 12.30 Hz, H-2a), 1.66–1.52 (m, 2H, H-2′a,b), 1.38–1.18 (m, 10H, H-3′a,b-H-7′a,b), 0.88 (t, 3H, $J_{7',8'}$ 7.0 Hz, H-8′). ^{13}C NMR ($CDCl_3$) δ: 170.9 (C=O), 170.4 (C=O), 169.8 (C=O), 99.6 (C-1), 71.9 (C-5), 70.8 (C-4), 70.0 (C-1′), 69.1 (C-3), 62.5 (C-6), 36.2 (C-2), 31.8, 29.7, 29.5, 29.3, 29.2, 26.0, 22.7 (C-2′-C-7′), 20.9, 20.8, 20.7 (CH_3-Ac), 14.0 (C-8′). Anal. calcd for $C_{20}H_{34}O_8$: C, 59.68; H, 8.51. Found: C, 59.60; H, 8.70.

Compound **6**: Syrup (0.11 g, 2%); $[\alpha]_D^{20}$ +38 (*c* 1, CH_2Cl_2); R_f 0.67 (1:3 EtOAc–PE). IR (neat): 1757 cm^{-1} (C=O). ^1H NMR ($CDCl_3$) δ: 5.92–5.81 (m, 2H, H-2, H-3), 5.32 (br d, 1H, $J_{4,5}$=9.7 Hz, H-4), 5.03 (br s, 1H, H-1), 4.28, 4.27, 4.25, 4.24 (Part AX of ABX system, 1H, $J_{6a,6b}$=11.9 Hz, $J_{6a,5}$=5.4 Hz, H-6a), 4.19, 4.16 (Part B of ABX system, 1H, H-6b), 4.11 (ddd, 1H, $J_{5,6b}$=1.7 Hz, H-5), 3.77 (dt, 1H, $J_{1'a,1'b}$=9.4 Hz, $J_{1'a,2'a,b}$=6.9 Hz, H-1′a), 3.50 (dt, 1H, $J_{1'b,2'a,b}$=6.7 Hz, H-1′b), 2.11 (s, 3H, CH_3-Ac), 2.09 (s, 3H, CH_3-Ac), 1.66–1.53 (m, 2H, H-2′a,b), 1.41–1.20 (m, 10H, H-3′a,b to H-7′a,b), 0.88 (t, 3H, $J_{7',8'}$=7.0 Hz, H-8′). ^{13}C NMR ($CDCl_3$) δ: 171.1 (C=O), 170.6 (C=O), 129.1, 128.1 (C-2, C-3), 94.5 (C-1), 69.1 (C-1′), 67.0 (C-5), 65.4 (C-4), 63.1 (C-6), 29.8 (C-2′), 31.9, 29.5, 29.4, 26.4, 22.8 (C-3′-C-7′), 21.1, 20.9 (CH_3-Ac), 14.2 (C-8′). Anal. calcd for $C_{18}H_{30}O_6$: C, 63.14; H, 8.83. Found: C, 62.90; H, 9.10.

Dodecyl 3,4,6-tri-*O*-acetyl-2-deoxy-α-D-*arabino*-hexopyranoside (4), dodecyl 3,4,6-tri-*O*-acetyl-2-deoxy-β-D-*arabino*-hexopyranoside (5), and **dodecyl 4,6-di-*O*-acetyl-2,3-dideoxy-α-D-*erythro*-hex-2-enopyranoside (7).** Reaction of **1** with dodecanol (5.6 mL, 36.5 mmol) gave compounds **4**, **5**, and **7**.

Compound **4**: Syrup (7.05 g, 83%); $[\alpha]_D^{20}$ +66 (*c* 1, CH_2Cl_2); R_f 0.48 (1:3 EtOAc–PE). IR (neat): 1748 cm^{-1} (C=O). ^1H NMR ($CDCl_3$) δ: 5.33 (ddd, 1H, $J_{3,2a}$ 11.9 Hz, $J_{3,2e}$ 5.5 Hz, $J_{3,4}$ 9.8 Hz, H-3), 5.00 (t, 1H, $J_{4,5}$ 10.0 Hz, H-4), 4.94 (d, 1H, $J_{1,2a}$ 3.1 Hz, H-1), 4.32 (dd, 1H, $J_{6a,5}$ 4.7 Hz, $J_{6a,6b}$ 12.3 Hz, H-6a), 4.06 (dd, 1H, $J_{6b,5}$ 2.2 Hz, H-6b), 3.97 (ddd, 1H, $J_{5,4}$=10.0 Hz, H-5), 3.62 (dt, 1H, $J_{1'a,1'b}$ 9.4 Hz, $J_{1'a,2'}$ 6.2 Hz, H-1′a), 3.38 (dt, 1H, $J_{1'b,2'a,b}$ 6.2 Hz, H-1′b), 2.24 (dd, 1H, $J_{2e,2a}$ 12.9 Hz, $J_{2e,3}$ 5.7 Hz, H-2e), 2.10 (s, 3H, CH_3-Ac), 2.05 (s, 3H, CH_3-Ac), 2.02 (s, 3H, CH_3-Ac), 1.83 (dt, 1H, H-2a), 1.63–1.53 (m, 2H, H- 2′a,b), 1.37–1.22 (m, 18H, H-3′- H-11′), 0.89 (t, 3H, $J_{12',11'}$ 7.1 Hz, H-12′). ^{13}C NMR ($CDCl_3$) δ: 170.8 (C=O), 170.2 (C=O), 170.0 (C=O), 96.9 (C-1), 69.5 (C-4), 69.2 (C-3), 67.9 (C-1′), 67.7 (C-5), 62.4 (C-6), 35.1 (C-2), 31.9, 29.7, 29.6, 29.5, 29.4, 29.3 26.2, 22.7 (C-3′-C-11′), 21.0, 20.8, 20.7 (CH_3-Ac), 14.1 (C-12′). Anal. calcd for $C_{24}H_{42}O_8$: C, 62.86; H, 9.23. Found: C, 63.20; H, 9.50.

Compound **5**: Syrup (1.01 g, 12%); $[\alpha]_D^{20}$ −19 (*c* 1, CH_2Cl_2); R_f 0.42 (1:3 EtOAc–PE). IR (neat): 1748 cm^{-1} (C=O). ^1H NMR ($CDCl_3$) δ: 5.03–4.99 (m, 2H, H-3, H-4), 4.56 (dd, 1H, $J_{1,2a}$ 9.7 Hz, $J_{1,2e}$ 2.0 Hz, H-1), 4.30 (dd, 1H, $J_{6a,6b}$ 12.0 Hz, $J_{5,6a}$ 5.0 Hz, H-6a), 4.11 (dd, 1H, $J_{6b,5}$ 2.5 Hz, H-6b), 3.87 (dt, 1H, $J_{1'a,1'b}$ 9.6 Hz, $J_{1'a,2'a,b}$ 6.3 Hz, H-1′a), 3.60 (ddd, 1H, $J_{4,5}$=9.4 Hz, H-5), 3.46 (ddd, 1H, H-1′b), 2.32 (ddd, 1H, $J_{2e,3}$ 4.4 Hz, $J_{2a,2e}$ 12.4 Hz, H-2e), 2.10 (s, 3H, CH_3-Ac), 2.05 (s, 3H,CH_3-Ac), 2.02 (s, 3H, CH_3-Ac), 1.75 (ddd, 1H, $J_{2a,3}$ 10.0 Hz, H-2a), 1.63–1.53 (m, 2H, H-2′a,b), 1.33–1.22 (m, 18H, H-3′a,b-H-11′a,b), 0.84 (t, 3H, $J_{11',12'}$ 6.4 Hz, H-12′). ^{13}C NMR ($CDCl_3$) δ: 99.6 (C-1), 71.9 (C-5), 71.0 (C-3), 70.8 (C-1′), 70.0 (C-4), 62.5 (C-6), 36.3 (C-2), 31.9, 29.7, 29.6, 29.5, 29.4, 26.0, 22.7 (C-3′- C-11′), 20.9, 20.8 (CH_3-Ac), 14.1 (C-12′). Anal. calcd for $C_{24}H_{42}O_8$: C, 62.86; H, 9.23. Found: C, 63.20; H, 9.40.

Compound **7**: Syrup (0.13 g, 2%); $[\alpha]_D^{20}$ +48 (*c* 1, CH_2Cl_2); R_f 0.61 (1:3 EtOAc–PE). IR (neat): 1757 cm^{-1} (C=O). ^1H NMR ($CDCl_3$) δ: 5.90–5.79 (m, 2H, H-2, H-3), 5.30

(br d, 1H, $J_{4,5}$=9.7 Hz, H-4), 4.95 (br s, 1H, H-1), 4.28, 4.27, 4.25, 4.24 (Part AX of ABX system, 1H, $J_{6a,6b}$ 12.1 Hz, $J_{6a,5}$ 5.3 Hz, H-6a), 4.19, 4.16 (Part B of ABX system, 1H, $J_{5,6b}$ 1.8 Hz, H-6b), 4.11 (ddd, 1H, H-5), 3.77 (dt, 1H, $J_{1'a,1'b}$ 9.4 Hz, $J_{1'a,2'a,b}$ 6.8 Hz, H-1'a), 3.50 (dt, 1H, $J_{1'b,2'a,b}$ 6.6 Hz, H-1'b), 2.10 (s, 3H, CH_3-Ac), 2.09 (s, 3H, CH_3-Ac), 1.66–1.54 (m, 2H, H-2'a,b), 1.35–1.24 (m, 18H, H-3'a,b to H-11'a,b), 0.88 (t, 3H, $J_{11',12'}$ 7.1 Hz, H-12'). ^{13}C NMR ($CDCl_3$) δ: 170.7 (C=O), 170.2 (C=O), 128.9, 127.9 (C-2, C-3), 94.3 (C-1), 68.9 (C-1'), 66.8 (C-5), 65.2 (C-4), 63.0 (C-6), 29.7 (C-20), 31.8, 29.6, 29.6, 29.5, 29.4, 29.3, 26.2, 22.6 (C-3'-C-11'), 20.9, 20.7 (CH_3-Ac), 14.1 (C-12'). Anal. calcd for $C_{22}H_{38}O_6$: C, 66.30; H, 9.61. Found: C, 66.00; H, 9.90.

GENERAL PROCEDURE FOR THE ZEMPLÉN DEACETYLATION

To a solution of a substrate (1 mmol) in dry methanol (10 mL) was added a solution of sodium methoxide in methanol (0.25 mL, *ca.* 1 M, prepared by dissolving sodium in dry methanol directly before use). The reaction mixture was stirred for 1.5 h and then neutralized with Amberlite (IR-120, H^+ form). The mixture was concentrated and filtered through silica gel (5 g, 230–400 mesh, eluted with AcOEt).

Octyl 2-deoxy-α-D-*arabino*-hexopyranoside (8)

Deacetylation of **2** gave **8** (0.49 g, 85%); mp. 107.8–109.6 (EtOAc–*n*-hexane); mp. by DSC=108.7°C; $[\alpha]_D^{20}$ +68 (*c* 1, MeOH); R_f 0.56 (EtOAc). IR (neat): 3371 cm^{-1} (C-OH). ^1H NMR (CD_3OD) δ: 4.90 (br d, 1H, $J_{1,2a}$ 2.7 Hz, H-1), 3.90–3.81 (m, 2H, H-6a, H-3), 3.74–3.67 (m, 2H, H-6b, H-1'a), 3.54 (ddd, 1H, H-5), 3.38 (dt, 1H, $J_{1'a,1'b}$ 9.8 Hz, $J_{1'a,2'a}$=$J_{1'a,2'b}$ 6.3 Hz, H-1'b), 3.26 (t, 1H, $J_{3,4}$=$J_{4,5}$ 9.2 Hz, H-4), 2.07 (dd, 1H, $J_{2e,3}$ 4.5 Hz, $J_{2e,2a}$ 12.8 Hz, H-2e), 1.67–1.56 (m, 3H, H-2a, H-2'a, H-2'b), 1.46–1.26 (m, 10H, H-3'a,b—H-7'a,b), 0.93 (t, 3H, $J_{7',8'}$ 6.4 Hz, H-8'). ^{13}C NMR (CD_3OD) δ: 99.4 (C-1), 74.8 (C-5), 74.2 (C-4), 70.8 (C-3), 69.1 (C-1'), 63.7 (C-6), 39.8 (C-2), 33.9, 31.5, 31.4, 31.3, 28.3, 24.6 (C-2'-C-7'), 15.3 (C-8'). Anal. calcd for $C_{14}H_{28}O_5$: C, 60.84; H, 10.21. Found: C, 60.50; H, 10.50.

Octyl 2-deoxy-β-D-*arabino*-hexopyranoside (9)

Deacetylation of **3** gave **9** (0.50 g, 85%); mp. 83.5°C–85.1°C (EtOAc–*n*-hexane); mp. by DSC=84.3°C; $[\alpha]_D^{20}$ −13 (*c* 1, MeOH); R_f 0.54 (EtOAc). IR (neat): 3465 cm^{-1} (C-OH). ^1H NMR (CD_3OD) δ: 4.48 (d, 1H, $J_{1,2a}$ 9.8 Hz, H-1), 3.89–3.77 (m, 2H, H-6a, H-1'a), 3.63 (dd, 1H, $J_{6b,5}$ 5.0 Hz, $J_{6b,6a}$ 11.6 Hz, H-6b), 3.49 (ddd, 1H, $J_{3,2e}$ 4.5 Hz, $J_{3,2a}$ 11.7 Hz, $J_{3,4}$ 12.0 Hz, H-3), 3.42 (dt, 1H, $J_{1'b,2'a}$=$J_{1'b,2'b}$ 6.8 Hz; $J_{1'b,1'a}$ 13.6 Hz, H-1'b), 3.18–3.07 (m, 2H, H-4, H-5), 2.04 (dd, 1H, $J_{2e,2a}$ 12.3 Hz, $J_{2e,3}$ 4.5 Hz, H-2e), 1.58–1.47 (m, 2H, H-2'a,b), 1.42 (ddd, 1H, $J_{2a,2e}$=$J_{2a,3}$=$J_{2a,1}$ 11.7 Hz, H-2a), 1.35–1.16 (m, 10H, H-3'a,b-H-7'a,b), 0.85 (t, 3H, $J_{7',8'}$ 7.2 Hz, H-8'). ^{13}C NMR (CD_3OD) δ: 101.9 (C-1), 78.8 (C-4), 73.9 (C-5), 73.3 (C-3), 71.1 (C-1'), 63.7 (C-6), 41.2 (C-2), 33.8, 31.5, 31.3, 31.2, 27.9, 24.5 (C-2'-C-7'), 15.2 (C-8'). Anal. calcd for $C_{14}H_{28}O_5$: C, 60.84; H, 10.21. Found: C, 60.60; H, 10.40.

Dodecyl 2-deoxy-α-D-*arabino*-hexopyranoside (10)

Deacetylation of **4** gave **10** (0.57 g, 82%); mp. 113.9°C–115.5°C (EtOAc/*n*-hexane); mp. by DSC=114.8°C; $[\alpha]_D^{20}$ +64 (*c* 1, MeOH); R_f 0.48 (EtOAc). IR (neat): 3354 cm^{-1} (C–OH). ^1H NMR (CD_3OD) δ: 4.90 (d, 1H, $J_{1,2a}$ 2.8 Hz, H-1), 3.90–3.81 (m, 2H, H-3,

H-6a), 3.75–3.67 (m, 2H, H-6b, H-1′a), 3.55 (ddd, 1H, $J_{5,4}$ 9.2 Hz, $J_{5,6a}$ 2.0 Hz, $J_{5,6b}$ 5.3 Hz, H-5), 3.38 (dt, 1H, $J_{1′b,2′a}=J_{1′b,2′b}$ 6.3 Hz, $J_{1′b,1′a}$ 9.8 Hz, H-1′b), 3.26 (t, 1H, $J_{4,3}=J_{4,5}=$9.2 Hz, H-4), 2.07 (dd, 1H, $J_{2e,2a}$ 12.8 Hz, $J_{2e,3}$ 5.2 Hz, H-2e), 1.67–1.54 (m, 3H, H-2a, H-2′a, H-2′b), 1.45–1.26 (m, 18H, H-3′a,b-H-11′a,b), 0.93 (t, 3H, $J_{11′,12′}$ 6.5 Hz, H-12′). ^{13}C NMR (CD$_3$OD) δ: 99.4 (C-1), 74.8 (C-4), 74.2 (C-5), 70.9 (C-3), 69.1 (C-1′), 63.7 (C-6), 39.8 (C-2), 34.0, 31.7, 31.6, 31.6, 31.5, 31.4, 28.3, 24.6 (C-2′-C-11′), 15.3 (C-12′). Anal. calcd for C$_{14}$H$_{36}$O$_5$: C, 65.03; H, 10.91. Found: C, 65.10; H, 11.20.

Dodecyl 2-deoxy-β-ᴅ-*arabino*-hexopyranoside (11)

Deacetylation of **5** gave **11** (0.59 g, 85%); mp. 101.9°C–104.9°C (EtOAc/*n*-hexane); mp. by DSC = 103.4°C; $[\alpha]_D^{20}$ −1.6 (*c* 1, MeOH); R$_f$ 0.45 (EtOAc). IR (neat): 3466 cm^{-1} (C-OH). ^{1}H NMR (CD$_3$OD) δ: 4.56 (dd, 1H, $J_{1,2e}$ 1.7 Hz, $J_{1,2a}$ 9.8 Hz, H-1), 3.97–3.87 (m, 2H, H-1′a, H-6a), 3.72 (dd, 1H, $J_{6b,5}$ 5.2 Hz, $J_{6b,6a}$ 11.8 Hz, H-6b), 3.58 (ddd, 1H, $J_{3,2e}$ 5.1 Hz, $J_{3,2a}$ 12.3 Hz, $J_{3,4}$ 12.4 Hz, H-3), 3.50 (dt, 1H, $J_{1′b,2′a}=J_{1′b,2′b}$ 6.7 Hz, $J_{1′b,1′a}$ 9.6 Hz, H-1′b), 3.25–3.16 (m, 2H, H-5, H-4), 2.13 (ddd, 1H, $J_{2e,2a}$ 12.3 Hz, H-2e), 1.65–1.58 (m, 2H, H-2′a, H-2′b), 1.51 (ddd, 1H, $J_{2a,3}=J_{2a,2e}=J_{2a,1}$ 12.3 Hz, H-2a), 1.44–1.26 (m, 18H, H-3′a,b-H11′a,b), 0.93 (t, 3H, $J_{11′,12′}$ 7.4 Hz, H-12′). ^{13}C NMR (CD$_3$OD) δ: 102.0 (C-1), 78.9 (C-4), 74.0 (C-5), 73.4 (C-3), 71.2 (C-1′), 63.8 (C-6), 41.3 (C-2), 33.9, 31.7, 31.6, 31.6, 31.4, 31.3, 28.1, 24.6 (C-2′-C-11′), 15.3 (C-12′). Anal. calcd for C$_{14}$H$_{36}$O$_5$: C, 65.03; H, 10.91. Found: C, 65.30; H, 11.10.

ACKNOWLEDGMENTS

The authors thank QREN–COMPETE program for the support of FACIB project (QREN–SI I&DT Co-Promotion Project nr. 21547). Fundação para a Ciência e a Tecnologia is gratefully acknowledged for the research grant SFRH/BDE/51998/2012 and for financial support of CQB Strategic Project PEst-OE/QUI/UI0612/2013.

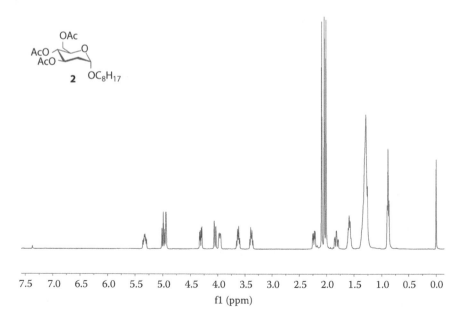

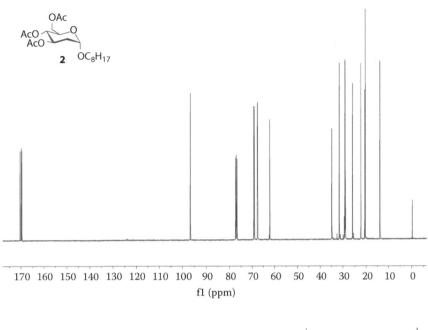

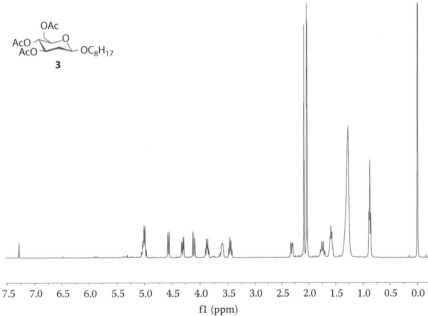

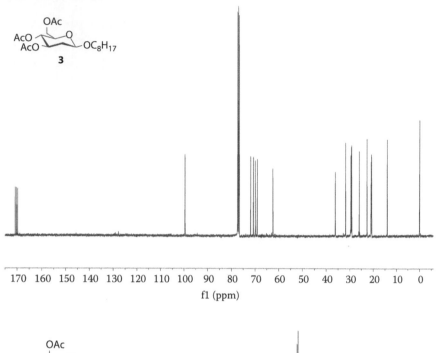

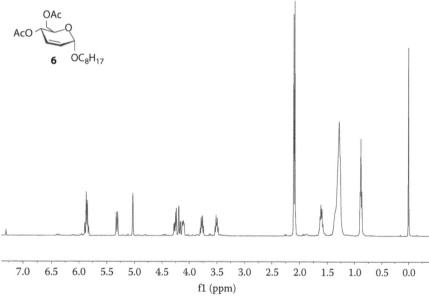

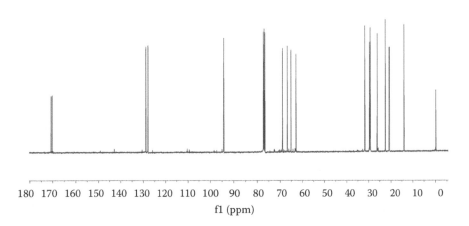

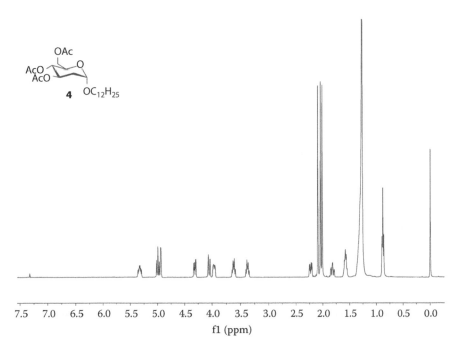

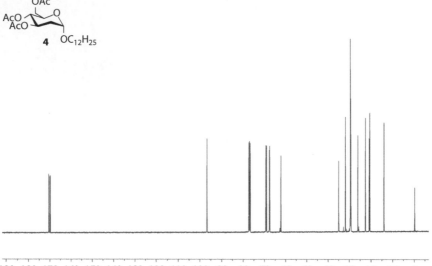

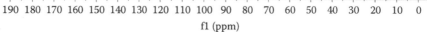

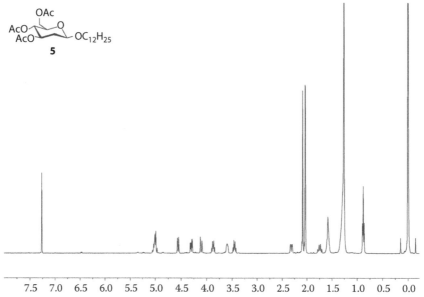

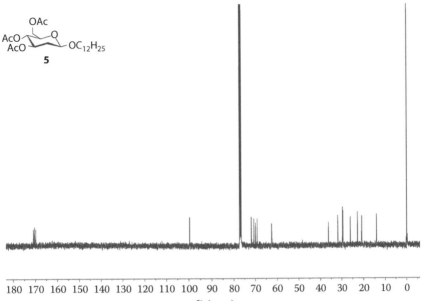

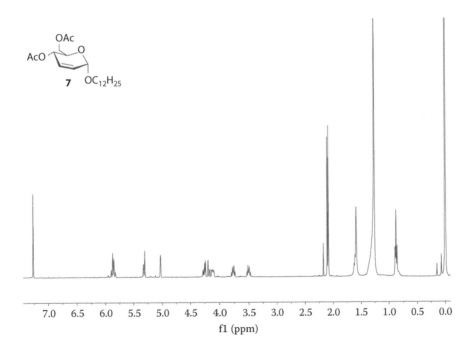

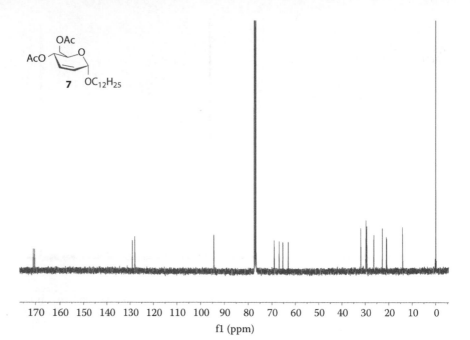

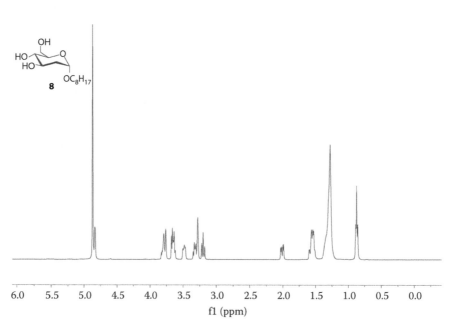

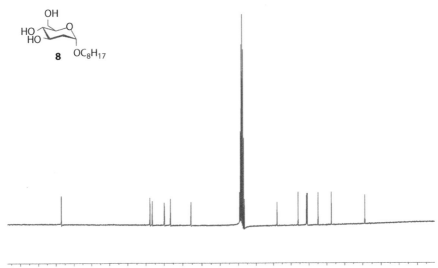

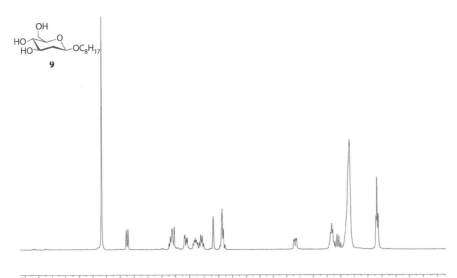

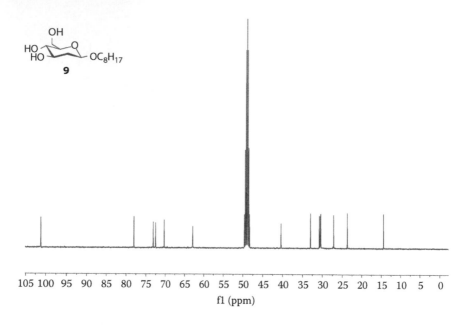

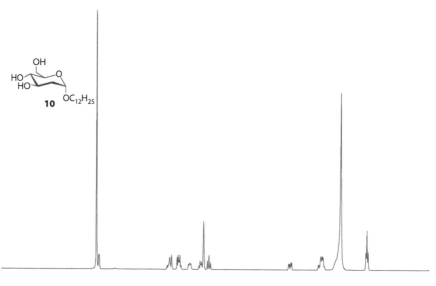

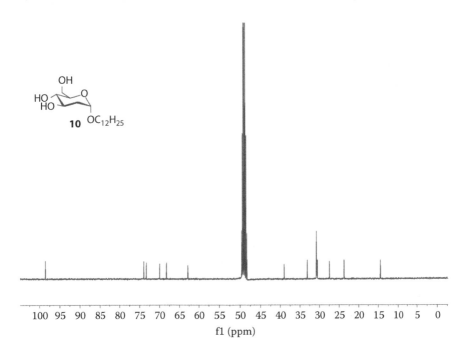

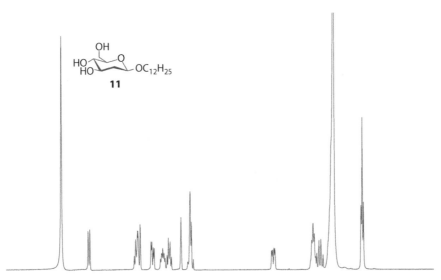

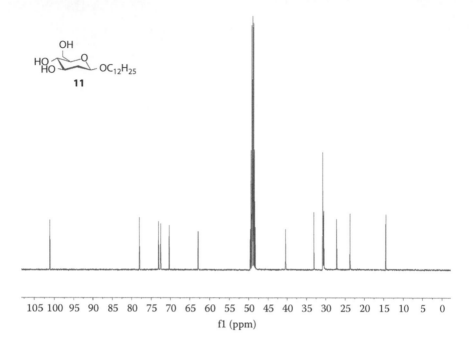

REFERENCES

1. F. V. M. Silva, M. Goulart, J. Justino, A. Neves, F. Santos, J. Caio, S. Lucas, A. Newton, D. Sacoto, E. Barbosa, M.S. Santos, A. P. Rauter, *Bioorg. Med. Chem.*, **2008**, *16*, 4083–4092.
2. A. Martins, M. S. Santos, C. Dias, P. Serra, V. Cachatra, J. Pais, J. Caio, V. H. Teixeira, M. Machuqueiro, M. S. Silva, A. Pelerito, J. Justino, M. Goulart, F. V. Silva, A. P. Rauter, *Eur. J. Org. Chem.*, **2013**, *8*, 1458–1459.
3. A. P. Rauter, S. Lucas, T. Almeida, D. Sacoto, V. Ribeiro, J. Justino, A. Neves, F. V. M. Silva, M. C. Oliveira, M. J.; Ferreira, M. S. Santos, E. Barbosa, *Carbohydr. Res.*, **2005**, *340*, 191–201.
4. V. Bolitt, C. Mioskowski, S.-G. Lee, J. R. Falck, *J. Org. Chem.*, **1990**, *55*, 5812–5813.
5. D. Lafont, P. Boullanger, M. Rosenzweig, *J. Carbohydr. Chem.*, **1998**, *17*, 1377–1393.

9 Regioselective Preparation of 4-Deoxy-*erythro*-hex-4-enopyranoside Enol Ethers through Acetone Elimination

Giorgio Catelani, Felicia D'Andrea,
Tiziana Gragnani, Irene Agnolin,[†]
and Lorenzo Guazzelli*

CONTENTS

* Corresponding author; e-mail: giorgio.catelani@farm.unipi.it.
† Checker, under supervision of Alberto Marra; e-mail: alberto.marra@univ-montp2.fr.

Reagents and conditions: **i**: t-BuOK, THF, reflux (only for **1**); **ii**: (a) t-BuOK, THF, reflux; (b) BnBr, wet THF, KOH, 18-crown-6 (only for **2**).

Pyranoside enol ethers possessing a double bond between C-1 and C-2 (glycals) have been known since 1913,[1] and their popularity reached a great height[2] due to their easy preparation and their potential use as intermediates in synthetic carbohydrate chemistry.[3] Other types of pyranoside enol ethers are much less studied. In this area, we reported the synthesis of hex-2- and hex-3-enopyranoside enol ethers promoted by NaH and N,N'-sulfuryldiimidazole (Im$_2$SO$_2$) in N,N-dimethylformamide (DMF).[4]

Furthermore, in 1989, we obtained a 4-deoxy-hex-4-L-*threo*-enopyranoside[5] as the main reaction product of the allyl isomerization[6] (t-BuOK, DMF, 80°C) in methyl 2-*O*-allyl-3,4-*O*-isopropylidene-6-*O*-(1-methoxy-1-methylethyl)-β-D-galactopyranoside. This serendipitous result prompted us to explore the scope of the reaction using these conditions on analogous galactopyranoside derivatives bearing base stable protecting groups in position 2.[7]

The reaction is formally similar to the base-promoted acetone elimination from 3,4-*O*-isopropylidene-D-galacturonates investigated several years ago by Kováč.[8] In the case of galacturonates, the acidic nature of H-5 represents the driving force for the reaction. This leads to a conjugate heterodiene being formed not only by elimination of acetone from C-3 and C-4 but also by elimination of C-4 substituents characterized by low nucleofugicity such as methoxy, hydroxy, and glycosyloxy groups.[8] However, for the galactopyranoside series, the sole factor favoring the acetone elimination is the antiperiplanar orientation between H-5 and the C(4)-O bond and the release of the strain of the *cis*-fused dioxolane ring.

Recently, we extended the acetone elimination reaction to the *talo* series[9] and we now report here two new examples (Table 9.1). Initially, the reaction was carried out by adding t-BuOK to a solution of the talopyranoside in DMF at 80°C followed, upon completion, by direct evaporation of the solvent. These conditions, used for galactopyranoside derivatives, were too harsh and only degradation products were observed. These problems were overcome by running the reaction in refluxing Tetrahydrofuran (THF) (66°C) and carrying out an aqueous workup, involving a saturated aqueous NaHCO$_3$ wash, followed by a CH$_2$Cl$_2$ extraction. A fast reaction took place leading to the formation of the elimination product after 15–20 min.

The optimized experimental protocol gave excellent results for methyl 2,6-di-*O*-benzyl 3,4-*O*-isopropylidene-β-D-talopyranoside[10a] (**1**, 90% yield) and good results for 4-*O*-(2-acetamido-6-*O*-benzyl-2-deoxy-3,4-*O*-isopropylidene-β-D-talopyranosyl)-2,3:5,6-di-*O*-isopropylidene-*aldehydo*-D-glucose dimethyl acetal[10b] (**2**). In the latter

TABLE 9.1

Regioselective Acetone Elimination from 3,4-*O*-Isopropylidene-β-D-talopyranosides

Entry	Talopyranoside	Product (Yield%)

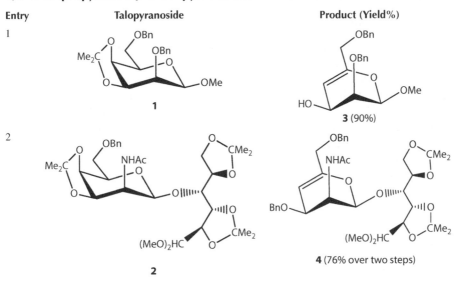

| 1 | **1** | **3** (90%) |
| 2 | **2** | **4** (76% over two steps) |

case, the enol ether and the starting material have similar R_f values. The crude intermediate was therefore subjected to benzylation with BnBr/KOH/18-crown-6 in wet THF, in order to facilitate the purification, to afford **4** in 76% overall yield. These reaction conditions avoid the *N*-benzylation of the acetamido group. In our experience, this side reaction takes place for the *talo* series in substantial amount using the common NaH/DMF conditions.

The higher reactivity of the *talo* compared to the *galacto* series is related to the different configuration of position 2. The change of configuration has a dual effect: on the one hand, it gives an unfavorable *syn* interaction between the axial group and the 3,4-*O*-isopropylidene ring, which provides a higher strain release with the acetone elimination, and on the other hand, freeing the alpha face facilitates access to the H-5 for the bulky *t*-BuOK.

The importance of steric factors has been previously reported when the reaction was attempted with methyl-3,4-*O*-isopropylidene-α-D-galactopyranosides.[5] No elimination products were formed even after increasing the temperature and the reaction time.

In conclusion, the regioselective acetone elimination reaction represents an efficient way to prepare 4-deoxy-hex-4-enopyranoside enol ethers with either *threo* or *erythro* configuration starting from fully protected 3,4-*O*-isopropylidene-galacto- or talopyranosides, respectively. The reaction works well on mono- and disaccharides; the presence of an *N*-acetyl substituent slightly affects the observed yield.

Limitations to the reaction arise with the use of base labile protecting groups and by its sensitivity to steric effects.

EXPERIMENTAL METHODS

GENERAL METHODS

Melting points were determined with a Kofler hot stage and are uncorrected. Optical rotations were measured on a PerkinElmer 241 polarimeter at 20°C ± 2°C. ^{1}H and ^{13}C NMR spectra were recorded with a Bruker Avance II 250 instrument operating at 250.13 MHz (^{1}H) and 62.9 MHz (^{13}C) in the reported solvent (internal standard Me$_4$Si), and the assignments were made, when possible, with DEPT, HETCOR, and COSY experiments. All reactions were monitored by thin-layer chromatography (TLC) on Kieselgel 60 F$_{254}$ with detection by UV light and/or by charring with 10% sulfuric acid in ethanol. Kieselgel 60 (Merck, 230–400 mesh) was used for flash chromatography. Solvents were dried by distillation according to standard procedures[11] and were stored over activated molecular sieves. Solutions in organic solvents were dried with MgSO$_4$. All reactions were performed under argon. t-BuOK was purchased from Aldrich and used as received.* Compounds **1** and **2** have been prepared according to the previously described procedures.[10]

Methyl 2,6-Di-*O*-benzyl-4-deoxy-α-L-*erythro*-hex-4-enopyranoside (3)

A solution of **1** (1.05 g, 2.54 mmol) in dry THF (30 mL) was gently warmed to reflux and treated with 3.10 g of solid t-BuOK (25.4 mmol). After 15 min, TLC analysis (3:7 hexane–EtOAc) revealed the disappearance of the starting material (**1**, R_f 0.66) and formation of a product (R_f 0.58). The mixture was cooled to 0°C and saturated aqueous NaHCO$_3$ solution (30 mL) was added. The aqueous phase was extracted with CH$_2$Cl$_2$ (3×60 mL); the organic extracts were collected, dried for a short time (5 min), filtered, and concentrated under diminished pressure. The crude residue was subjected to flash chromatography on silica gel (6:4 hexane–EtOAc,† 0.1% Et$_3$N) to give **3** as a colorless syrup (815.5 mg, 90%); R_f 0.40 (1:1 hexane–EtOAc); [α]$_D$ +6.2 (*c* 1.6, CHCl$_3$). ^{1}H NMR (250.13 MHz, CD$_3$CN) δ: 7.44–7.27 (m, 10H, Ar-*H*), 5.15 (dt, 1H, $J_{3,4}$ 5.1 Hz, $J_{4,6a}$=$J_{4,6a}$ 0.7 Hz, H-4), 5.05 (dd, 1H, $J_{1,2}$ 2.3 Hz, $J_{1,3}$ 1.1 Hz, H-1), 4.75, 4.61 (AB system, 2H, $J_{A,B}$ 11.7 Hz, C*H*$_2$Ph), 4.54, 4.48 (AB system, 2H, $J_{A,B}$ 11.8 Hz, C*H*$_2$Ph), 4.16 (m, 1H, H-3), 3.95, 3.87 (2dt, each 1H, $J_{6a,6b}$12.7 Hz, $J_{3,6a}$=$J_{3,6b}$ 0.9 Hz, H-6a, H-6b), 3.70 (dd, 1H, $J_{2,3}$ 5.1 Hz, H-2); 3.42 (s, 3H, OCH$_3$), 2.99 (d, 1H, $J_{3,OH}$ 10.8 Hz, OH-3). ^{13}C NMR (62.9 MHz, CD$_3$CN) δ: 148.6 (*C*-5), 139.4, 139.3 (2×Ar-*C*), 129.3–128.5 (Ar-*C*H), 103.9 (C-4), 100.3 (C-1), 75.5 (C-2), 72.6, 71.8 (*C*H$_2$Ph), 69.9 (C-6), 61.7 (C-3), 56.8 (O*C*H$_3$). Anal. calcd for C$_{21}$H$_{24}$O$_5$: C, 70.77; H, 6.79. Found: C, 70.81; H, 6.81.

---

* From our experience, older batches of t-BuOK resulted in reaction failure.
† The flash column chromatography can also be performed using a 8:2 cyclohexane–EtOAc mixture (containing 0.1% of Et$_3$N).

4-*O*-(2-Acetamido-3,6-di-*O*-benzyl-2,4-dideoxy-α-L-*erythro*-hex-4-enopyranosyl)-2,3:5,6-di-*O*-isopropylidene-*aldehydo*-D-glucose Dimethyl Acetal (4)

A solution of **2** (5.00 g, 7.81 mmol) in dry THF (74 mL) was gently warmed to reflux and treated with solid *t*-BuOK (9.60 g, 78.4 mmol). After 20 min, TLC analysis (EtOAc) revealed the disappearance of the starting material **2** (R_f 0.29) and formation of a major component (R_f 0.26).* The mixture was cooled to 0°C, saturated aqueous NaHCO$_3$ solution (90 mL) was added, and the aqueous phase was extracted with CH$_2$Cl$_2$ (3 × 100 mL). The organic extracts were collected, dried, filtered, and concentrated under diminished pressure to give a residue (4.80 g) containing mainly (NMR) the elimination product. 18-Crown-6 (330 mg, 1.2 mmol) was added to a solution of the crude product in wet THF (100 mL + 0.5% of water). The mixture was cooled to 0°C and pulverized KOH (2.36 g, 42.14 mmol) was added. After 30 min of stirring at 0°C, the suspension was treated with benzyl bromide (2.24 mL, 18.86 mmol), and stirring was continued until the starting material (TLC, EtOAc, R_f 0.26) was completely consumed (4 h)† and a new spot was observed (R_f 0.51, EtOAc). CH$_3$OH (15 mL) was added, the mixture was stirred for 30 min and concentrated under diminished pressure, and the residue was partitioned between CH$_2$Cl$_2$ (50 mL) and H$_2$O (30 mL). The aqueous phase was extracted with CH$_2$Cl$_2$ (3 × 50 mL) and the organic extracts were collected, dried, filtered, and concentrated under diminished pressure. The crude product (6.48 g, yellow oil) was chromatographed (3:7 hexane–EtOAc + 0.1% Et$_3$N)‡ to give pure **4** (3.99 g, 76% yield) as a colorless syrup, R_f 0.51 (EtOAc), $[\alpha]_D$ +6.29 (c 1.2, CHCl$_3$). ^{1}H NMR (250.13 MHz, CD$_3$CN) δ: 7.38–7.26 (m, 10H, Ar-*H*), 6.46 (d, 1H, $J_{2',NH}$ 9.3 Hz, NH), 5.41 (d, 1H, $J_{1',2'}$ 2.4 Hz, H-1'), 5.04 (d, 1H, $J_{3',4'}$ 3.8, H-4'), 4.53, 4.46 (AB system, 2H, $J_{A,B}$ 11.9 Hz, C*H$_2$*Ph), 4.52 (s, 2H, C*H$_2$*Ph), 4.45 (m, 1H, H-2'), 4.31 (m, 1H, H-5), 4.28 (dd, 1H, $J_{1,2}$ 6.7, $J_{2,3}$ 8.0, H-2), 4.25 (d, 1H, H-1), 4.10–3.99 (m, 2H, H-4, H-6b), 4.02 (dd, 1H, $J_{2',3'}$ 6.7, H-3'), 3.97–3.92 (m, 4H, H-3, H-6a, H-6'a, H-6'b), 3.35, 3.34 (2s, each 3H, 2 × OC*H$_3$*), 1.89 (s, 3H, C*H$_3$*CO), 1.36, 1.25 (2s, each 3H, C(C*H$_3$*)$_2$), 1.28 (s, 6H, C(C*H$_3$*)$_2$). ^{13}C NMR (62.9 MHz, CD$_3$CN) δ: 170.6 (*C*O), 150.7 (C-5'), 139.8, 139.5 (2 × Ar-*C*), 129.3–128.5 (Ar-*C*H), 110.7, 108.6 [2 × *C*(CH$_3$)$_2$], 106.7 (C-1), 99.4 (C-4'), 98.1 (C-1'), 79.3 (C-3), 78.8 (C-2), 76.3 (C-5), 74.9 (C-4), 73.1, 71.9 (2 × *C*H$_2$Ph), 69.7 (C-6'), 69.0 (C-3'), 65.4 (C-6), 56.6, 54.4 (2 × O*C*H$_3$), 48.2 (C-2'), 27.6, 26.6, 26.3, 24.4 (2 × C(*C*H$_3$)$_2$), 23.2 (*C*H$_3$CO). Anal. calcd for C$_{36}$H$_{49}$NO$_{11}$ (671.77): C, 64.36; H, 7.35; N, 2.09. Found: C, 64.38; H, 7.34; N, 2.10.

---

* Owing to the very similar R_f values of the starting compound and the elimination product, it is necessary to carefully check the reaction progress. The two compounds have a different behavior on TLC when heated after treatment with ethanolic sulfuric acid: the elimination enol ether appears first and is characterized by a brown color.

† In some attempts, a substantial drop of the reaction rate was observed and additional KOH/18-crown-6 was required to achieve complete conversion.

‡ The flash column chromatography can also be performed using a 3:7 cyclohexane–EtOAc mixture (containing 0.1% of Et$_3$N).

Carbohydrate Chemistry

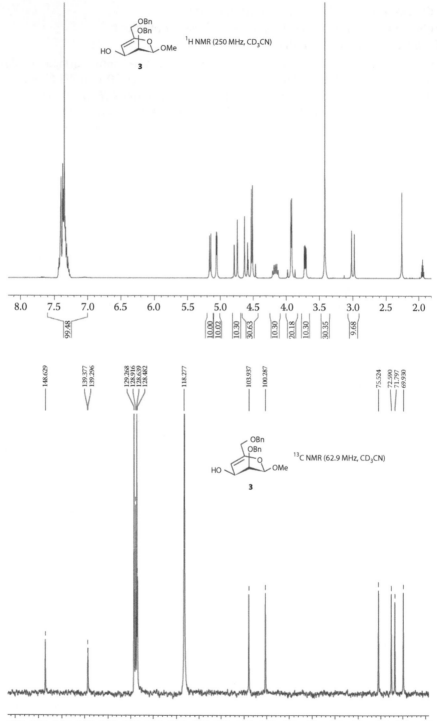

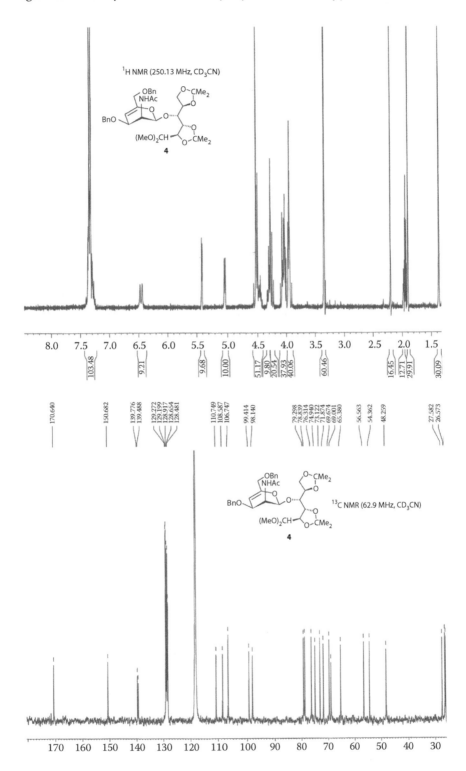

REFERENCES

1. Fischer, E.; Zach, K. *Sitzber Kgl Preuss Akad Wiss* **1913**, *16*, 311–317.
2. (a) Ferrier, R. J. Unsaturated sugars. In *The Carbohydrates, Chemistry and Biochemistry*, Pigman, W.; Horton, D., Eds.; Academic Press: San Diego, CA, 1980, Vol. 1B, pp. 843–879; (b) Ferrier, R. J.; Hoberg, J. O. *Adv. Carbohydr. Chem. Biochem.* **2003**, *58*, 55–119.
3. (a) Danishefsky, S. J.; Bilodeau, M. T. *Angew. Chem. Int. Ed. Engl.* **1996**, *35*, 1380–1419; (b) Fraser-Reid, B. *Acc. Chem. Res.* **1996**, *29*, 57–66; (c) Priebe, W.; Grynkiewicz, G. Reaction and formation of glycal derivatives. In *Glycoscience: Chemistry and Chemical Biology*, Fraser-Reid, B.; Tatsuda, K.; Thiem, J.; Eds. Springer-Verlag: Berlin, Germany, 2001, Vol. 1, pp. 749–781.
4. Attolino, E.; Catelani, G.; D'Andrea, F.; Guazzelli, L.; Scherrmann, M.-C. *Carbohydrate Chemistry: Proven Synthetic Methods*, Taylor & Francis Group, Boca Raton, FL, 2012, Vol. 1, pp. 11–26.
5. Barili, P. L.; Berti, G.; Catelani, G.; Colonna, F.; D'Andrea, F. *Carbohydr. Res.* **1989**, *190*, 13–21.
6. Gigg, J.; Gigg, R. *J. Chem. Soc. C* **1996**, 82–86.
7. The following are some references for the elimination reaction to give α-L-*threo*-4-deoxy-hex-4-enopyranosides. (a) Barili, P. L.; Berti, G.; Catelani, G.; D'Andrea, F. *Gazz. Chim. Ital.* **1992**, *122*, 135–142; (b) Barili, P. L.; Berti, G.; Catelani, G.; D'Andrea, F.; Gaudiosi, A. *Gazz. Chim. Ital.* **1994**, *124*, 57–63; (c) Catelani, G.; Corsaro, A.; D'Andrea, F.; Mariani, M.; Pistarà, V.; Vittorino, E. *Carbohydr. Res.* **2003**, *338*, 2349–2358; (d) Kováč, P.; Hirsch, J.; Kováčik, V. *Carbohydr. Res.* **1977**, *58*, 327–336; (e) Hirsch, J.; Kováč, P.; Kováčik, V. *J. Carbohydr. Nucleosides Nucleotides* **1974**, *1*, 431–448.
8. Kováč, P.; Kováčik, V. *Carbohydr. Res.* **1974**, *32*, 360–365 and references cited therein.
9. (a) Guazzelli, L.; Catelani, G.; D'Andrea, F. *Carbohydr. Res.* **2010**, *345*, 369–376; (b) Guazzelli, L.; Catelani, G.; D'Andrea, F.; Gragnani, T. *Carbohydr. Res.* **2014**, *388*, 44–49.
10. (a) For compound **1**: Attolino, E.; Catelani, G.; D'Andrea, F. *Eur. J. Org. Chem.* **2006**, 5279–5292; (b) for compound **2**: Attolino, E.; Catelani, G.; D'Andrea, F.; Nicolardi, M. *J. Carbohydr. Chem.* **2004**, *23*, 179–190.
11. Perrin, D. D.; Armarego, W. L. F.; Perrin, D. F. *Purification of Laboratory Chemicals*, 2nd edn., Pergamon Press, Oxford, U.K., 1980.

10 Stereoselective Reduction Using Sodium Triacetoxyborodeuteride
Synthesis of Methyl 2,3-Di-O-benzyl-α-D-(4-²H)-glucopyranoside

*Hani Mobarak, Olof Engström,
Martina Lahmann,† and Göran Widmalm**

CONTENTS

Reagents and conditions: (a) Bu_2SnO, 3 Å MS, toluene, reflux, 5 h; then dibromantin, DCM, r.t., 30 min. (b) $NaBD(OAc)_3$, AcOD/MeCN, r.t., 2 h.

* Corresponding author; e-mail: gw@organ.su.se.
† Checker; e-mail: m.lahmann@bangor.ac.uk.

INTRODUCTION

Replacing a specific atom with an isotope has wide applications in chemistry, where it plays an important role in the elucidation of reaction mechanisms[1] as well as tracing biological processes and unraveling biosynthetic routes in biochemistry.[2,3] In these studies, mass spectrometry (MS) and nuclear magnetic resonance (NMR) spectroscopy serve as important techniques in monitoring or revealing the information sought. Specifically, isotope-labeled nucleic acids have been synthesized for use in MS,[4] and specific labeling has been carried out to reduce spectral overlap, reveal specific nuclear spin–spin interactions, enhance the signal-to-noise ratio, or enable triple resonance experiments in NMR spectroscopy applications, resulting in compounds containing ^{2}H,[5–7] ^{13}C,[8,9] ^{15}N,[10] and ^{17}O.[11]

Sodium triacetoxyborohydride NaBH(OAc)$_3$ is a mild selective reducing reagent for aldehydes in the presence of ketones,[12] and it has been used for the reduction of axially oriented α- and β-hydroxy ketones to give anti-diols through hydroxy-mediated hydride delivery.[13–15] The natural abundance compound methyl 2,3-di-O-benzyl-α-D-glucopyranoside has been utilized as an intermediate in synthesis of oligosaccharides,[16,17] and the site specifically deuterium-labeled isotopologue was used in the synthesis of C4-deuterated methyl α-cellobioside,[18] employing the deuterated version of the reducing agent, namely, NaBD(OAc)$_3$, to facilitate conformational studies based on NMR spectroscopy.[19]

The present synthesis starts with activation of 2,3-di-O-benzyl-α-D-galactopyranoside (**1**) by dibutyltin oxide. The crude stannylene derivative is then oxidized selectively at position 4 by 1,3-dibromo-5,5-dimethylhydantoin[20] (dibromantin) to obtain methyl 2,3-di-O-benzyl-α-D-$xylo$-hexopyranosid-4-ulose (**2**). Subsequent reduction is effected with NaBD(OAc)$_3$, prepared by mixing sodium borodeuteride and acetic acid-d (see section "Experimental"). In the intramolecular reaction, the hydroxymethyl group in **2** is used as part of the complex, thereby delivering the deuteride in a highly stereoselective way resulting in methyl 2,3-di-O-benzyl-α-D-(4-^{2}H)-glucopyranoside (**3**). It should be noted that carrying out the synthesis on a scale smaller than ~1 g may result in poorer yields. The presence and absence of H4 or D4 in the three compounds are readily evident both from the ^{1}H and ^{2}H NMR spectra as well as the ^{13}C NMR spectrum of **3**, where the C4 resonance appears as a triplet due to the scalar coupling to D4.

GENERAL METHODS

All reagents were used as received. Column chromatography was performed with a Biotage Isolera flash chromatography system with KP-Sil SNAP silica gel cartridges. Thin-layer chromatography (TLC) was carried out on silica gel 60 F254 (20 × 20 cm, 0.2 mm thickness) and monitored with UV light 254–360 nm or by ceric ammonium sulfate staining[21] using 40% 2M sulfuric acid instead of phosphoric acid. NMR spectra were recorded at 25°C on a spectrometer operating at frequencies of 600 MHz for ^{1}H, 150 MHz for ^{13}C, and 92 MHz for ^{2}H, using CDCl$_3$ as a solvent; for ^{2}H NMR, CHCl$_3$ was used as solvent with a trace amount of CDCl$_3$. The NMR chemical shifts are reported in ppm and referenced to the residual CHCl$_3$ solvent peak at 7.26 and 77.16 ppm for ^{1}H and ^{13}C, respectively. Mass spectra were recorded on a Bruker Daltonics micrOTOF spectrometer in the positive mode. Optical rotation was measured with an AUTOPOL IV polarimeter at 589 nm and 30°C in a quartz cuvette of 10 cm length. 1,3-Dibromo-5,5-dimethylhydantoin was purchased from Sigma-Aldrich.

EXPERIMENTAL METHODS

METHYL 2,3-DI-O-BENZYL-α-D-XYLO-HEXOPYRANOSID-4-ULOSE (2)

Methyl 2,3-di-O-benzyl-α-D-galactopyranoside **1** (1.0 g, 2,67 mmol), dibutyltin oxide (0.90 g, 3.62 mmol), and molecular sieves (3 Å) were mixed in dry toluene (10 mL) and kept under reflux for 4–5 h (oil bath 130°C). The solvent was evaporated and the crude material was dried under vacuum for 1 h.* The material was dissolved in dry CH_2Cl_2 (10 mL) under nitrogen at r.t. before 1,3-dibromo-5,5-dimethylhydantoin (0.40 g, 1.40 mmol) was added producing a yellow-orange reaction mixture that was stirred until the coloration had disappeared (30 min–2 h, TLC toluene/EtOAc, 1:3, R_f **1** ~0.3, R_f **2** ~0.6). The mixture was filtered through Celite and the filtrate was concentrated. The residue was taken up in EtOAc (30 mL), washed with sat. aqueous $Na_2S_2O_3$ (2×20 mL) and brine (20 mL), and dried with sodium sulfate, and the organic layer was concentrated under reduced pressure followed by column chromatography (toluene/EtOAc 9:1 to 3:1) over silica gel to obtain the product as a colorless syrup (0.89 g, 2.39 mmol, 89%). ^{1}H NMR (CDCl$_3$) δ: 2.44 (broad, 1H, OH-6), 3.48 (s, 3H, OMe), 3.80 (dd, $J_{H1,H2}$ 3.5 Hz, $J_{H2,H3}$ 10.0 Hz, 1H, H-2), 3.88 (dd, $J_{H6a,H6b}$ −12.1 Hz, $J_{H5,H6a}$ 5.0 Hz, 1H, H-6a), 3.89 (dd, $J_{H6a,H6b}$ −12.1 Hz, $J_{H5,H6b}$ 4.8 Hz, 1H, H-6b), 4.13 (dd, $J_{H5,H6a}$ 5.0 Hz, $J_{H5,H6b}$ 4.8 Hz, 1H, H-5), 4.46 (d, $J_{H2,H3}$ 10.0 Hz, 1H, H-3), 4.67 (d, $J_{HBn2a,HBn2b}$ −12.1 Hz, 1H, H-Bn2a), 4.70 (d, $J_{HBn3a,HBn3b}$ −11.3 Hz, 1H, H-Bn3a), 4.79 (d, $J_{H1,H2}$ 3.5 Hz, 1H, H-1), 4.86 (d, $J_{HBn2a,HBn2b}$ −12.1 Hz, 1H, H-Bn2b), 4.96 (d, $J_{HBn3a,HBn3b}$ −11.3 Hz, 1H, H-Bn3b), 7.30–7.44 (m, 10 H, H-Ar). ^{13}C NMR (CDCl$_3$) δ: 56.2 (OMe), 60.7 (C-6), 72.9 (C-5), 74.0 (PhCH$_2$-2), 74.6 (PhCH$_2$-3), 80.1 (C-2), 82.6 (C-3), 98.6 (C-1), 128.0–128.6 (8 C-Ar), 137.7 (C-*ipso* OBn-3), 137.8 (C-*ipso* OBn-2), 204.0 (C-4). ESIMS [M+Na]$^+$ m/z calcd for $C_{21}H_{24}O_6Na$: 395.1465. Found: 395.1451.

METHYL 2,3-DI-O-BENZYL-α-D-(4-^{2}H)-GLUCOPYRANOSIDE (3)

Sodium borodeuteride (0.35 g, 8.63 mmol) was added to acetic acid-d (10 mL) at a temperature between 10°C and 20°C and stirred for 30 min. The mixture was placed in an ice bath and a solution of **2** (0.89 g, 2.39 mmol) in acetonitrile (3 mL) was added. After stirring for 2 h (TLC toluene/EtOAc, 1:3, R_f **3** ~0.3, R_f **2** ~0.6), the solvent was evaporated and column chromatography (toluene/EtOAc 4:1 to 3:2) provided **3** as a colorless syrup (0.80 g, 2.13 mmol, 89%). The product was crystallized from 2-propanol and n-pentane at 4°C; $[\alpha]_D$ = +19.7° (c 1.0, CHCl$_3$). ^{1}H NMR (CDCl$_3$) δ: 2.19 (dd, $J_{H6a,OH6}$ 7.2 Hz, $J_{H6b,OH6}$ 5.3 Hz, 1H, OH-6), 2.62 (s, 1H, OH-4), 3.37

---

* The material can be kept under vacuum for at least 12 h without degradation.

(s, 3 H, OMe), 3.49 (dd, $J_{H1,H2}$ 3.5 Hz, $J_{H2,H3}$ 9.5 Hz, 1H, H-2), 3.60 (dd, $J_{H5,H6a}$ 4.5 Hz, $J_{H5,H6b}$ 3.4 Hz, 1H, H-5), 3.74 (ddd, $J_{H6a,H6b}$ −11.8 Hz, $J_{H5,H6a}$ 4.5 Hz, $J_{H6a,OH6}$ 7.2 Hz, 1H, H-6a), 3.77 (ddd, $J_{H6a,H6b}$ −11.8 Hz, $J_{H5,H6b}$ 3.4 Hz, $J_{H6b,OH6}$ 5.3 Hz, 1H, H-6b), 3.79 (d, $J_{H2,H3}$ 9.5 Hz, 1H, H-3), 4.60 (d, $J_{H1,H2}$ 3.5 Hz, 1H, H-1), 4.65 (d, $J_{HBn2a,HBn2b}$ −12.1 Hz, 1H, H-Bn2a), 4.72 (d, $J_{HBn3a,HBn3b}$ −11.5 Hz, 1H, H-Bn3a), 4.76 (d, $J_{HBn2a,HBn2b}$ −12.1 Hz, 1H, H-Bn2b), 5.01 (d, $J_{HBn3a,HBn3b}$ −11.5 Hz, 1H, H-Bn3b), 7.30–7.37 (m, 10 H, H-Ar). ^{13}C NMR (CDCl$_3$) δ: 55.3 (OMe), 62.3 (C-6), 69.9 (t, J 21.6 Hz, C-4), 70.8 (C-5), 73.2 (PhCH$_2$-2), 75.5 (PhCH$_2$-3), 79.9 (C-2), 81.4 (C-3), 98.3 (C-1), 127.9–128.6 (8 C-Ar), 138.1 (C-*ipso* OBn-2), 138.8 (C-*ipso* OBn-3). ESIMS [M+Na]$^+$ m/z calcd for C$_{21}$H$_{25}$DO$_6$Na: 398.1684. Found: 398.1697. Anal. calcd for C$_{21}$H$_{25}$DO$_6$ (375.18): C 67.18, H(D) 7.25. Found: C 67.13, H(D) 6.87.

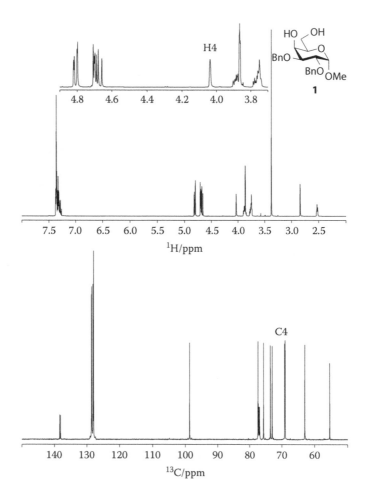

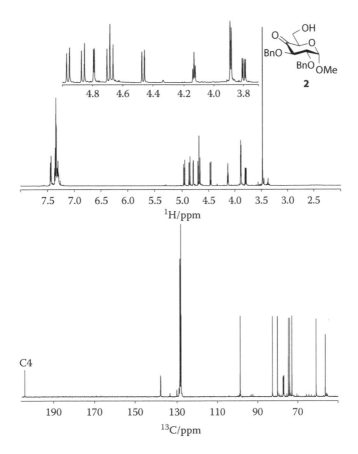

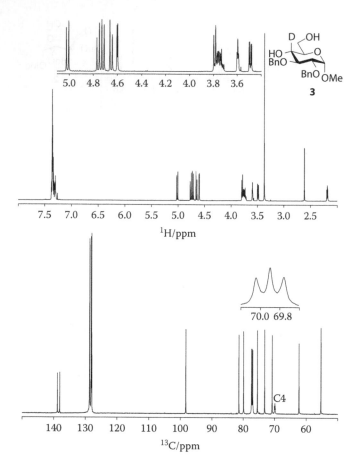

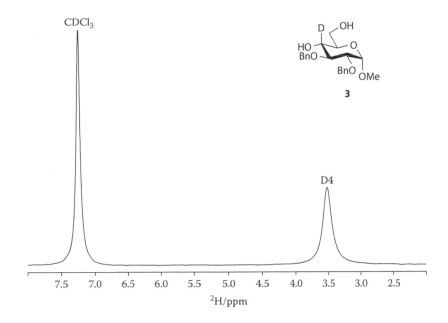

REFERENCES

1. Hanson, J. R. *The Organic Chemistry of Isotopic Labelling.* Royal Society of Chemistry, London, U.K., 2011.
2. Running, J. A.; Burlingame, R. P.; Berry, A. *J. Exp. Bot.* **2003**, *54*, 1841–1849.
3. Hancock, R. D.; Viola, R. *CRC Crit. Rev. Plant Sci.* **2005**, *24*, 167–188.
4. Burdzy, A. *Nucleic Acids Res.* **2002**, *30*, 4068–4074.
5. Hällgren, C.; Widmalm, G. *J. Carbohydr. Chem.* **1993**, *12*, 309–333.
6. Söderman, P.; Oscarson, S.; Widmalm, G. *Carbohydr. Res.* **1998**, *312*, 233–237.
7. Lycknert, K.; Helander, A.; Oscarson, S.; Kenne, L.; Widmalm, G. *Carbohydr. Res.* **2004**, *339*, 1331–1338.
8. Jonsson, K. H. M.; Pendrill, R.; Widmalm, G. *Magn. Reson. Chem.* **2011**, *49*, 117–124.
9. Jonsson, K. H. M.; Säwén, E.; Widmalm, G. *Org. Biomol. Chem.* **2012**, *10*, 2453–2463.
10. Coxon, B. *Carbohydr. Res.* **2007**, *342*, 1044–1054.
11. D'Souza, F. W.; Lowary, T. L. *J. Org. Chem.* **1998**, *63*, 3166–3167.
12. Gribble, G. W.; Ferguson, D. C. *J. Chem. Soc. Chem. Commun.* **1975**, 535–536.
13. Senda, Y.; Kikuchi, N.; Inui, A.; Itoh, H. *Bull. Chem. Soc. Jpn.* **2000**, *73*, 237–242.
14. Evans, D. A.; Chapman, K. T.; Carreira, E. M. *J. Am. Chem. Soc.* **1988**, *110*, 3560–3578.
15. McDevitt, R. E.; Fraser-Reid, B. *J. Org. Chem.* **1994**, *59*, 3250–3252.
16. Tiwari, P.; Misra, A. K. *J. Carbohydr. Chem.* **2007**, *26*, 239–248.
17. Tiwari, V. K.; Kumar, A.; Schmidt, R. R. *Eur. J. Org. Chem.* **2012**, *2012*, 2945–2956.
18. Söderman, P.; Widmalm, G. *J. Org. Chem.* **1999**, *64*, 4199–4200.
19. Larsson, E. A.; Staaf, M.; Söderman, P.; Höög, C.; Widmalm, G. *J. Phys. Chem. A* **2004**, *108*, 3932–3937.
20. Söderman, P.; Widmalm, G. *Carbohydr. Res.* **1999**, *316*, 184–186.
21. Pirrung, M. C. Appendix 3: Recipes for TLC stains, in *The Synthetic Organic Chemist's Companion.* John Wiley & Sons, Inc., Hoboken, NJ, 2007.

11 Selective Anomeric S-Deacetylation Using Aqueous Sodium Methanethiolate

Jun Rao, Gaolan Zhang,[†] Xiaojun Zeng, and Xiangming Zhu*

CONTENTS

Glycosyl thiols (sometimes called 1-thiosugars) are often used as key building blocks to construct thiooligosaccharides and other S-glycoconjugates.[1] Unlike sugar hemiacetals, which mutarotate easily under many conditions, glycosyl thiols are quite stable in terms of configuration, and their mutarotation does not occur readily. Mutarotation of glycosyl thiols is highly restricted and even blocked under basic conditions.[2] Due to this stability, glycosyl thiols have found wide application in thioglycoside chemistry. One approach to the preparation of glycosyl thiols is to convert peracetylated thiosugars to thiols, which requires selective anomeric S-deacetylation.[3] It is worth mentioning that sulfhydryl group can also be introduced directly onto anomeric carbon to obtain glycosyl thiols,[4] but conversion from anomeric thioacetates to glycosyl

---

* Corresponding author; e-mail: xiangming_zhu@hotmail.com.
[†] Checker, under supervision of Xin-Shan Ye; e-mail: xinshan@bjmu.edu.cn.

thiols is required sometimes to complete the synthesis. A number of methods for anomeric S-deacetylation have been reported, including the use of sodium methane-thiolate (NaSMe).[5] However, most of these methods require dry reaction conditions and low temperature,[6] which is not ideal for large-scale preparation. For example, anomeric S-acetyl group could be selectively removed with NaOMe in MeOH at −25°C to give a free thiol in high yield.[6a] Fully acetylated thiosugars were also treated with bromine at −78°C to achieve selective anomeric S-deacetylation.[6b]

Here, we wish to report a mild, experimentally simple and high-yielding method for selective anomeric S-deacetylation of a series of peracetylated thiosugars using commercially available and cheap aqueous NaSMe. Initially, fully acetylated thio-glucose 1[7] was chosen as the substrate and treated with 20% aqueous NaSMe in a mixture of dichloromethane and methanol (2:1 v/v). The reaction proceeded smoothly as indicated by thin-layer chromatography (TLC), and the desired glucosyl thiol 4[8] was isolated by flash column chromatography in 88% yield. Product 4 was characterized by NMR spectroscopy, which exhibited one doublet ($J_{1,SH}$ ca. 10 Hz, δ 2.31) for the sulfhydryl proton. Similarly, thiogalactose 2[9] could also be converted into the galactosyl thiol 5[8] in a highly regioselective way and isolated in excellent yield. The ready formation of 4 and 5 indicated that the present procedure may provide a general and convenient means to anomeric thiol. Indeed, the reaction yields with all investigated substrates were invariably high, as shown in Scheme 11.1, and the simplicity of the reaction conditions makes this approach a very attractive way of selective anomeric S-deacetylation. The maltose-derived thioacetate 3[10] was also subjected to the aforementioned conditions and gave rise to maltosyl thiol 6[11] in 93% yield. All the glycosyl thiols prepared earlier were stable to silica-gel chromatography. Disulfide formation was not observed so special precaution was not required during the purification.

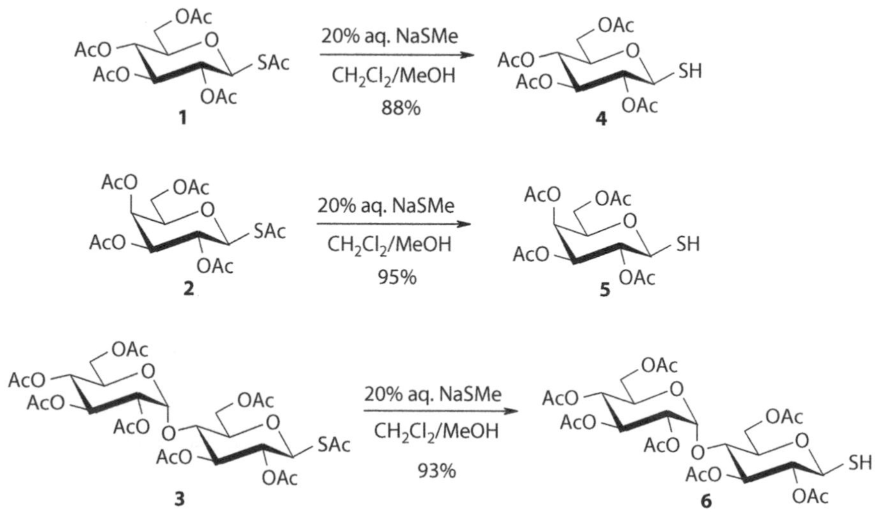

SCHEME 11.1 Selective anomeric S-deacetylation of peracetylated sugars.

EXPERIMENTAL METHODS

GENERAL METHODS

Solvents were evaporated under reduced pressure at <40°C. All reactions were monitored by TLC on silica gel 60 F_{254}–coated aluminum sheets and the spots were visualized by charring with 8% H_2SO_4 in methanol. Flash column chromatography was
performed using 46–60 μm silica gel. Optical rotations were measured at 20°C with
a PerkinElmer 343 polarimeter (1 dm cell). ^{1}H NMR (400 or 600 MHz) spectra were
measured for solutions in $CDCl_3$ with tetramethylsilane as internal standard. ^{13}C
NMR spectra were recorded at 100 or 150 MHz using $CDCl_3$ as solvent. Yields refer
to pure products isolated by chromatography.

GENERAL PROCEDURE FOR SELECTIVE ANOMERIC S-DEACETYLATION

To a stirred solution of the appropriate 1-thio-per-O-acetate (1.0 mmol) in a mixture of
CH_2Cl_2 (6 mL) and MeOH (3 mL) was added at 0°C 20% aqueous NaSMe (0.28 mL,
1.0 mmol). The solution was stirred for 5–20 min, after which time TLC indicated
that the reaction was complete. It is important to note that the R_f (TLC) values of
some anomeric thiols and their thioacetates were nearly identical. The reaction was
quenched with 10% aqueous HCl (1.0 mL) and extracted with CH_2Cl_2. The organic
layer was washed sequentially with H_2O and brine, dried over Na_2SO_4, filtered, and
concentrated under reduced pressure to give the crude product. Chromatography
(petroleum ether/ethyl acetate) gave the corresponding anomeric thiols.

2,3,4,6-Tetra-O-acetyl-1-thio-β-D-glucopyranose (4)

Purified by flash column chromatography (petroleum ether/EtOAc, 3:1) to give
the title compound, m.p. 117°C–118°C (from EtOAc–petroleum ether) (lit. [8]
69°C–70°C); $[\alpha]_D$ +4.4 (c 4.3, $CHCl_3$), [lit. [8] $[\alpha]_D$ +5.5 (c 1.0, $CHCl_3$)]; in view of the
value of $[\alpha]_D$ found here and that in the literature, the only explanation we can offer
for the considerable difference in m.p. is that the compound may be dimorphous. ^{1}H
NMR (400 MHz, $CDCl_3$) δ: 5.19 (t, J=9.4 Hz, 1H, H-3), 5.10 (t, J=9.7 Hz, 1H, H-4),
4.97 (t, J=9.5 Hz, 1H, H-2), 4.55 (t, J=9.9 Hz, 1H, H-1), 4.25 (dd, J=12.5, 4.8 Hz,
1H, H-6a), 4.12 (dd, J=12.5, 2.1 Hz, 1H, H-6b), 3.72 (ddd, J=9.9, 4.8, 2.2 Hz, 1H,
H-5), 2.31 (d, J=10.0 Hz, 1H, SH), 2.10 (s, 3H, Ac), 2.08 (s, 3H, Ac), 2.02 (s, 3H, Ac),
2.01 (s, 3H, Ac). ^{13}C NMR (150 MHz, $CDCl_3$) δ: 170.6, 170.1, 169.6, 169.3 (4MeCO),
78.7 (C-1), 76.3 (C-5), 73.6 (C-3), 73.55 (C-2), 68.1 (C-4), 62.0 (C-6), 20.7, 20.7, 20.6,
20.5 (4CH₃CO). ESI-HRMS calcd for $C_{14}H_{20}NaO_9S$ [M+Na]$^+$: 387.0726. Found:
387.0733. The NMR spectra data were consistent with literature.[8]

2,3,4,6-Tetra-O-acetyl-1-thio-β-D-galactopyranose (5)

Purified by flash column chromatography (petroleum ether/EtOAc, 3:1) to give the
title compound as a white amorphous solid: m.p. 121°C–122°C (recrystallized from
petroleum ether–EtOAc) (lit. [8] 86.5°C–88°C); $[\alpha]_D$ +25.6 (c 1.6, $CHCl_3$); [lit. [8]
$[\alpha]_D$ +26.4 (c 1.0, $CHCl_3$)]; in view of the value of $[\alpha]_D$ found here and that in the
literature, the only explanation we can offer for the considerable difference in m.p.

is that the compound may be dimorphous. ^{1}H NMR (600 MHz, CDCl$_3$) δ: 5.44 (d, J=3.2 Hz, 1H, H-4), 5.19 (t, J=9.9 Hz, 1H, H-2), 5.03 (dd, J=10.1, 3.3 Hz, 1H, H-3), 4.54 (t, J=9.8 Hz, 1H, H-1), 4.13 (d, J=6.5 Hz, 2H, H-6a, H-6b), 3.96 (t, J=6.5 Hz, 1H, H-5), 2.39 (d, J=10.0 Hz, 1H, SH), 2.17 (s, 3H, Ac), 2.10 (s, 3H, Ac), 2.06 (s, 3H, Ac), 1.99 (s, 3H, Ac). ^{13}C NMR (150 MHz, CDCl3) δ: 170.4, 170.2, 170.0, 169.9 (4MeCO), 79.2 (C-1), 74.95 (C-5), 71.6 (C-3), 70.8 (C-2), 67.2 (C-4), 61.5 (C-6), 20.9, 20.7, 20.7, 20.6 (4CH_3CO). ESI-HRMS calcd for C$_{14}$H$_{20}$NaO$_9$S [M+Na]$^+$: 387.0726. Found: 387.0724. The NMR spectra data were consistent with literature.[8]

2,3,4,6-Tetra-O-acetyl-α-ᴅ-glucopyranosyl-(1→4)-2,3,6-tri-O-acetyl-1-thio-β-ᴅ-glucopyranose (6)

Purified by flash column chromatography (petroleum ether/EtOAc, 2:1) to give the title compound as a white amorphous solid, which failed to crystallize from a range of solvents: [α]$_D$ +76.3 (c 2.6, CHCl$_3$) [lit. [11] [α]$_D$ +78.5 (c 1.0, CHCl$_3$)]. ^{1}H NMR (600 MHz, CDCl$_3$) δ: 5.43 (d, J=4.0 Hz, 1H, H-1′), 5.39–5.34 (m, 1H, H-3′), 5.27 (t, J=9.1 Hz, 1H, H-3), 5.08 (t, J=9.9 Hz, 1H, H-4′), 4.88 (dd, J=10.6, 4.0 Hz, 1H, H-2′), 4.83 (t, J=9.5 Hz, 1H, H-2), 4.61 (t, J=9.8 Hz, 1H, H-1), 4.47 (dd, J=12.2, 2.4 Hz, 1H, H-6a), 4.27 (dd, J=12.5, 3.8 Hz, 1H, H-6a′), 4.23 (dd, J=12.3, 4.6 Hz, 1H, H-6b), 4.06 (dd, J=12.0, 2.5 Hz, 1H, H-6b′), 4.03 (t, J=7.6 Hz, 1H, H-4), 3.96 (d, J=9.5 Hz, 1H, H-5′), 3.73 (ddd, J=9.7, 4.3, 2.6 Hz, 1H, H-5), 2.28 (d, J=9.7 Hz, 1H, SH), 2.18 (s, 3H, Ac), 2.12 (s, 3H, Ac), 2.08 (s, 3H, Ac), 2.07 (s, 3H, Ac), 2.04 (s, 3H, Ac), 2.03 (s, 3H, Ac), 2.02 (s, 3H, Ac). ^{13}C NMR (150MHz, CDCl$_3$) δ: 170.6, 170.55, 170.5, 170.1, 170.0, 169.9, 169.5 (7CH_3CO), 95.7 (C-1′), 78.2 (C-1), 76.5 (C-5), 76.1 (C-3), 74.3 (C-2), 72.6 (C-4), 69.9 (C-2′), 69.3 (C-3′), 68.6 (C-5′), 67.9 (C-4′), 63.0 (C-6), 61.5 (C-6′), 20.92, 20.89, 20.72, 20.63, 20.62 (7CH_3CO). ESI-HRMS calcd for C$_{26}$H$_{36}$NaO$_{17}$S [M+Na]$^+$: 675.1571. Found: 675.1574. The NMR spectra data were consistent with literature.[11]

ACKNOWLEDGMENTS

This work was supported by the Natural Science Foundation of Zhejiang Province (R4110195) and the Science and Technology Department of Zhejiang Province (2013C24004).

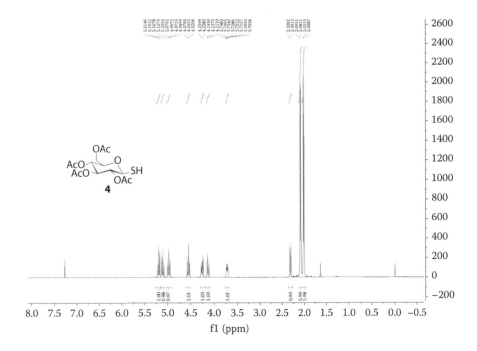

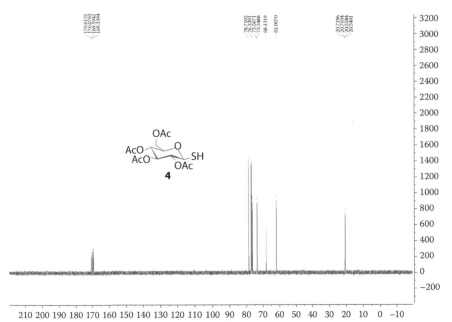

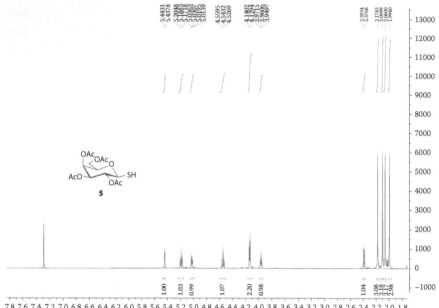

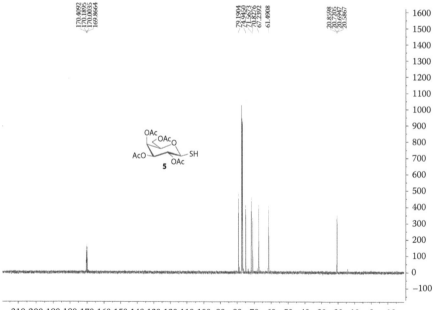

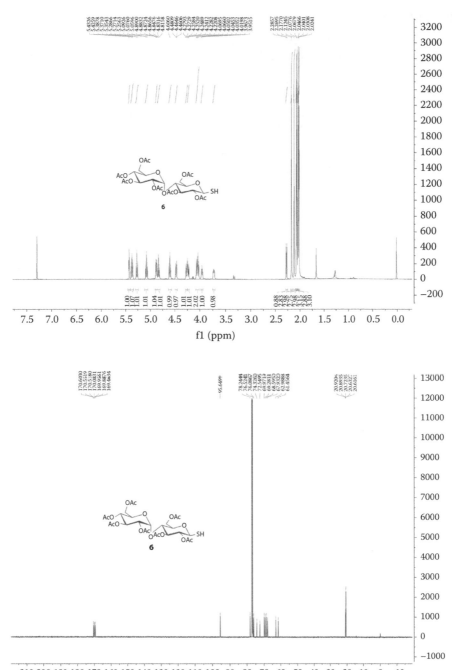

REFERENCES

1. Pachamuthu, K.; Schmidt, R. R. *Chem. Rev.* **2006**, *106*, 160–187.
2. Caraballo, R.; Deng, L.; Amorim, L.; Brinck, T.; Ramström, O. *J. Org. Chem.* **2010**, *75*, 6115–6121.
3. Driguez, H. *ChemBioChem* **2001**, *2*, 311–318.
4. Zhu, X.; Dere, R. T.; Jiang, J.; Zhang, L.; Wang, X. *J. Org. Chem.* **2011**, *76*, 10187–10197.
5. MacDougall, J. M.; Zhang, X. D.; Polgar, W. E.; Khroyan, T. V.; Toll, L.; Cashman, J. R. *J. Med. Chem.* **2004**, *47*, 5809–5815.
6. (a) Yamamoto, K.; Watanabe, N.; Matsuda, H.; Oohara, K.; Araya, T.; Hashimoto, M.; Miyairi, K. et al. *Bioorg. Med. Chem. Lett.* **2005**, *15*, 4932–4935. (b) Knapp, S.; Darout, E.; Amorelli, B. *J. Org. Chem.* **2006**, *71*, 1380–1389. (c) Zhang, P.; Zuccolo, A. J.; Li, W.; Zheng, R. B.; Ling, C.-C. *Chem. Commun.* **2009**, 4233–4235.
7. Czifrák, K.; Somsák, L. *Carbohydr. Res.* **2009**, *344*, 269–277.
8. Floyd, N.; Vijayakrishnan, B.; Koeppe, J. R.; Davies, B. G. *Angew. Chem. Int. Ed.* **2009**, *48*, 7798–7802.
9. Martín-Santamaría, S.; André, S.; Buzamet, E.; Caraballo, R.; Fernández-Cureses, G.; Morando, M.; Ribeiro, J. P. et al. *Org. Biomol. Chem.* **2011**, *9*, 5445–5455.
10. Huang, K. T.; Gorska, K.; Alvarez, S.; Barluenga, S.; Winssinger, N. *ChemBioChem* **2011**, *12*, 56–60.
11. Fujihira, T.; Chida, M.; Kamijo, H.; Takido, T.; Seno, M. *J. Carbohydr. Chem.* **2002**, *21*, 287–292.

12 Glycosylation of Phenolic Acceptors Using Benzoylated Glycosyl Trichloroacetimidate Donors

Jani Rahkila, Anup Kumar Misra,
Lorenzo Guazzelli,[†] and Reko Leino*

CONTENTS

Phenolic *O*-glycosides are common in nature, often possessing significant biological properties.[1] Their isolation from natural sources in sufficient quantities and purities is, however, often challenging. Thus, preparative approaches based on chemical synthesis methods are needed. In fact, the first reported chemical glycosylation reaction was the glycosylation of potassium phenolate with 2,3,4,6-tetra-*O*-acetyl-α-D-glucopyranosyl chloride.[2]

---

* Corresponding author; e-mail: reko.leino@abo.fi.
† Checker, under the supervision of Stefan Oscarson; e-mail: stefan.oscarson@ucd.ie.

Phenols are generally less nucleophilic than aliphatic alcohols, which makes the formation of O-glycosidic bonds difficult, especially in cases where the aromatic ring contains electron-withdrawing substituents. Consequently, phenolic glycosylations often require highly reactive glycosyl donors.[3] Donors commonly used for these types of reactions include glycosyl acetates, glycosyl halides, thioglycosides, and glycosyl trichloroacetimidates, of which glycosyl trichloroacetimidates generally give the best yields.[4]

Both ether and ester protecting groups are used in the glycosyl trichloroacetimidate donors.[4] The disadvantage of using acetylated glycosyl trichloroacetimidates compared to their benzoylated counterparts lies in the fact that the acetyl group at position 2 in the donor can be transferred to the acceptor molecule[1a,5] and that acetylated donors tend to form stable orthoesters.[6] Therefore, O-benzoylated glycosyl derivatives were selected as donors in the present work. Herein, we demonstrate the efficiency of benzoylated trichloroacetimidate glycosyl donors for performing phenolic glycosylations. As model compounds, benzoylated rhamnose donor was used for glycosylation of p-methoxyphenol (Scheme 12.1) and benzoylated lactose donor for glycosylation of p-nitrophenol (Scheme 12.2).

Anomerically deprotected L-rhamnose and lactose, which were required for making imidates, were prepared from their fully benzoylated counterparts by selective anomeric debenzoylation with MeNH$_2$ in THF.[7] After the selective deprotection, the trichloroacetimidate derivatives were prepared conventionally.[8]

The most common promoters used for this type of donors are TMSOTf and BF$_3$·OEt$_2$.[8] In the case of the L-rhamnose donor, no significant difference could be seen between TMSOTf and BF$_3$·OEt$_2$, and the reaction was complete almost instantly with both reagents. In the case of lactose donor, however, the use of TMSOTf resulted in a mixture of α- and β-glycosides. When using BF$_3$·OEt$_2$ as the promotor, only the β-glycoside was obtained. The glycosylation of lactose donor with p-nitrophenol was significantly slower than the glycosylation of rhamnose

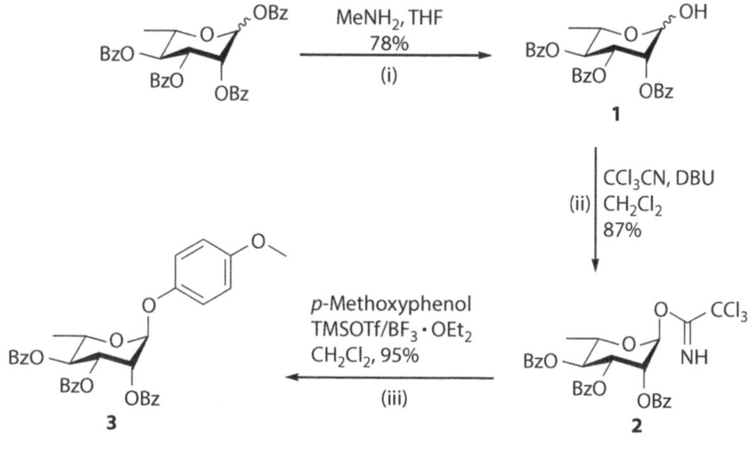

SCHEME 12.1   Reagents and conditions: (i) MeNH$_2$, THF, r.t., 5 h (78%); (ii) CCl$_3$CN, DBU, CH$_2$Cl$_2$, 0°C, 3 h (87%); (iii) p-methoxyphenol, TMSOTf/BF$_3$·OEt$_2$, −20°C, 5 min (95%).

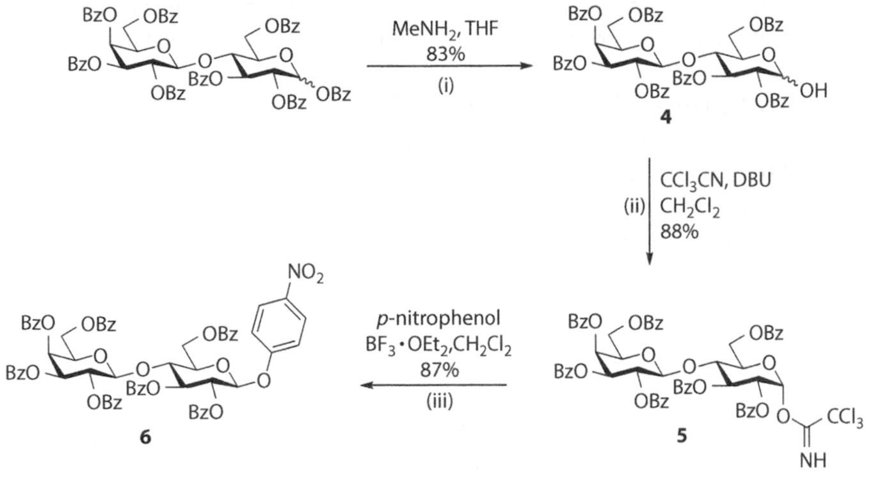

SCHEME 12.2 Reagents and conditions: (i) MeNH$_2$, THF, r.t., 15 h (83%); (ii) CCl$_3$CN, DBU, CH$_2$Cl$_2$, −5°C, 1.5 h (88%); (iii) p-nitrophenol, BF$_3$·OEt$_2$, CH$_2$Cl$_2$, −15°C, 1 h (87%).

with p-methoxyphenol. This is likely due to higher reactivity of rhamnose, due to the presence of the 6-CH$_3$ group donating electrons to the ring. Due to the electron-withdrawing effect of the nitro group, p-nitrophenol is less nucleophilic than p-methoxyphenol and therefore it reacts slower.

EXPERIMENTAL METHODS

GENERAL METHODS

All reagents, including 2 M solution of methylamine in THF, were purchased from Sigma-Aldrich or Fluka and used without further purification. Dichloromethane was distilled over CaH$_2$ under argon. The glassware used in moisture-sensitive reactions was flame dried prior to use and subsequently flushed with argon. The reactions were performed under argon atmosphere. Thin-layer chromatography (TLC) analysis was performed using aluminum sheets coated with silica gel 60 F$_{245}$ (Merck). The compounds were visualized with UV light and by charring with 20% H$_2$SO$_4$ in MeOH (v/v). Column chromatography was conducted using Merck silica gel 60, F$_{245}$ using mixtures of hexane and ethyl acetate as eluent. Optical rotations were measured on a PerkinElmer 241 polarimeter using the D-line of sodium at 589 nm. NMR spectra were recorded with a Bruker Avance spectrometer operating at a frequency of 600.13 MHz (^{1}H) and 150.9 MHz (^{13}C). The chemical shifts are referenced to an internal standard (tetramethylsilane δ=0.0 ppm) or the residual CDCl$_3$ signal (δ=7.26 ppm in ^{1}H and δ=77.16 ppm in ^{13}C). In addition to ^{1}H and ^{13}C NMR spectra, 2D NMR methods DQF-COSY, HSQC, and HMBC were used to fully assign the spectra. Accurate coupling constants were obtained by using NMR simulation software PERCH.[9] The anomeric configurations of the rhamnose units were determined by measuring the $^1J_{C-1,H-1}$ coupling constants. HRMS were recorded on a Bruker micrOToF-Q instrument with electrospray ionization operating in positive mode. Elemental analysis was performed on a FLASH 2000 organic elemental analyzer.

2,3,4-Tri-O-benzoyl-L-rhamnopyranose (1)[10]

To a solution of 1,2,3,4-tetra-O-benzoyl-L-rhamnopyranose[11] (1.3 g, 2.24 mmol, 1 eq.) in THF (10 mL) was added $MeNH_2$ (2.0 M in THF, 7 mL, 6 eq.). The mixture was stirred at room temperature for 5 h and then concentrated. Chromatography (4:1 hexane–EtOAc) afforded pure 1 as well as ~100 mg starting material (8%). Yield, 0.84 g (78%, isolated), α/β 6.7:1; white foam. R_f: 0.55 (1:1 hexane–EtOAc). α-anomer, ^1H NMR (600.13 MHz, $CDCl_3$, 25°C) δ: 8.15–7.20 (m, 15 H, aromatic H), 5.92 (dd, 1 H, $J_{2,3}$ 3.4 Hz, $J_{3,4}$ 10.2 Hz, H-3), 5.71 (dd, 1 H, $J_{1,2}$ 1.8 Hz, $J_{2,3}$ 3.4 Hz, H-2), 5.69 (dd, 1 H, $J_{3,4}$ 10.2 Hz, $J_{4,5}$ 9.8 Hz, H-4), 5.46 (dd, 1 H, $J_{1,2}$ 1.8 Hz, $J_{1,OH}$ 0.9 Hz, H-1), 4.46 (dq, 1 H, $J_{4,5}$ 9.8 Hz, $J_{5,6}$ 6.3 Hz, H-5), 3.16 (d, 1 H, $J_{1,OH}$ 0.9 Hz, OH), 1.37 (d, 3 H, $J_{5,6}$ 6.3 Hz, H-6). ^{13}C NMR (150.9 MHz, $CDCl_3$, 25°C) δ: 165.8, 165.7, 165.6 (COPh), 133.5, 133.3, 133.1 (3 Ph-p-C), 129.9, 129.7 (6 Ph-o-C), 129.4, 129.3, 129.2 (3 Ph-i-C), 128.6, 128.4, 128.3 (6 Ph-m-C), 92.3 (C-1), 71.9 (C-4), 71.2 (C-2), 69.6 (C-3), 66.8 (C-5), 17.8 (C-6); $^1J_{C-1,H-1}$ = 172.0 Hz.

β-anomer, ^1H NMR (600.13 MHz, $CDCl_3$, 25°C) δ: 8.15–7.20 (m, 15 H, aromatic H), 5.85 (dd, 1 H, $J_{1,2}$ 1.2 Hz, $J_{2,3}$ 3.3 Hz, H-2), 5.62 (dd, 1 H, $J_{3,4}$ 10.2 Hz, $J_{4,5}$ 9.6 Hz, H-4), 5.58 (dd, 1 H, $J_{2,3}$ 3.3 Hz, $J_{3,4}$ 10.2 Hz, H-3), 5.23 (dd, 1 H, $J_{1,2}$ 1.2 Hz, $J_{1,OH}$ 2.3 Hz, H-1), 3.91 (dq, 1 H, $J_{4,5}$ 9.6 Hz, $J_{5,6}$ 6.2 Hz, H-5), 3.72 (d, 1 H, $J_{1,OH}$ 2.3 Hz, OH), 1.44 (d, 3 H, $J_{5,6}$ 6.2 Hz, H-6). ^{13}C NMR (150.9 MHz, $CDCl_3$, 25°C) δ: 165.8, 165.7, 165.6 (3 COPh), 133.8, 133.4, 133.3 (3 Ph-p-C), 130.1, 129.7 (6 Ph-o-C), 129.4, 129.3, 129.2 (3 Ph-i-C), 128.7, 128.5, 128.3 (6 Ph-m-C), 93.2 (C-1), 71.9 (C-3), 71.6 (C-2), 71.2 (C-4), 71.0 (C-5), 17.7 (C-6); $^1J_{C-1,H-1}$ = 162.4 Hz. ESI-MS calcd for $C_{27}H_{24}NaO_8$ [M+Na]+: 499.1363. Found: 499.1321. Anal. calcd for $C_{27}H_{24}O_8$: C, 68.06; H, 5.08; O, 26.86. Found: C, 67.83; H, 5.08; O, 27.10.

1-O-(2,3,4-Tri-O-benzoyl-α-L-rhamnopyranosyl) Trichloroacetimidate (2)[12a]

A solution of 2,3,4-tri-O-benzoyl-L-rhamnopyranose[10] (1, 800 mg, 1.68 mmol, 1 eq.) in freshly distilled CH_2Cl_2 (10 mL) was kept under argon atmosphere and cooled on an ice bath. CCl_3CN (196 μL, 1.96 mmol, 1.17 eq.) was added, followed by DBU (32 μL, 0.21 mmol, 0.13 eq.), and the mixture was stirred in ice bath for 3 h. After concentration, chromatography (2:1 hexane–EtOAc + 0.1% Et_3N) afforded pure compound 2. Yield, 905 mg (87%, isolated); white foam. R_f: 0.77 (1:1 hexane–EtOAc); $[α]_D^{25}$ + 102° (c 1.0, CH_2Cl_2), lit. [12b] $[α]_D^{20}$ + 97.5 (c 1.0 $CHCl_3$). ^1H NMR (600.13 MHz, $CDCl_3$, 25°C) δ: 8.83 (s, 1 H, NH), 8.15–7.20 (m, 15 H, aromatic H), 6.50 (d, $J_{1,2}$ 1.9 Hz, H-1), 5.90 (each dd, each 1H, $J_{2,3}$ 3.5 Hz, $J_{3,4}$ 10.1 Hz, $J_{1,2}$ 1.9 Hz, $J_{2,3}$ 3.5 Hz, H-3, H-2), 5.78 (dd, $J_{3,4}$ 10.1 Hz, $J_{4,5}$ 9.8 Hz, H-4), 4.41 (dq, 1 H, $J_{4,5}$ 9.8 Hz, $J_{5,6}$ 6.2 Hz, H-5), 1.42 (d, 3 H, $J_{5,6}$ 6.2 Hz, H-6). ^{13}C NMR (150.9 MHz, $CDCl_3$, 25°C) δ: 165.7, 165.5, 165.3 (3 COPh), 160.1 (C=N), 133.7, 133.5, 133.2 (3 Ph-p-C), 130.0, 129.8, 129.7 (6 Ph-o-C), 129.1, 129.0 (3 Ph-i-C), 128.7, 128.5, 128.3 (6 Ph-m-C), 94.8 (C-1), 90.8 (CCl_3), 71.0 (C-4), 69.7 (C-5), 69.7 (C-3), 69.1 (C-2), 17.8 (C-6); $^1J_{C-1,H-1}$ = 179.8 Hz. ESI-MS calcd for $C_{29}H_{24}Cl_3NNaO_8$ [M + Na]+: 642.0460. Found: 642.0422.

p-Methoxyphenyl 2,3,4-Tri-O-benzoyl-α-L-rhamnopyranoside (3)

A mixture of 1-O-(2,3,4-tri-O-benzoyl-α-L-rhamnopyranosyl) trichloroacetimidate[12] (2, 300 mg, 0.48 mmol, 1 eq.) and 4-methoxyphenol (60 mg, 0.72 mmol 1.5 eq.) in freshly distilled CH_2Cl_2 (3 mL) was cooled under argon to −20°C. TMSOTf (10.5 μL, 0.058 mmol, 0.12 eq.) or $BF_3 \cdot OEt_2$ (7.2 μL, 0.058 mmol, 0.12 eq.) was added, and the reaction mixture was stirred at −20°C for 5 min, after which the reaction was quenched by adding a few drops of Et_3N and concentrated. Chromatography (3:1 hexane–EtOAc) afforded pure compound 3. Yield, 265 mg (95%); white foam. R_f: 0.33 (3:1 hexane–EtOAc); $[\alpha]_D^{25} + 62°$ (c 1.0, CH_2Cl_2). ^1H NMR (600.13 MHz, $CDCl_3$, 25°C) δ: 8.15–7.25 (m, 15 H, aromatic H), 7.14–7.10 (m, 2 H, 1-Ph-o-H), 6.89–6.85 (m, 2 H, 1-Ph-m-H), 6.04 (dd, 1 H, $J_{2,3}$ 3.5 Hz, $J_{3,4}$ 10.1 Hz, H-3), 5.85, (dd, 1 H, $J_{1,2}$ 1.9 Hz, $J_{2,3}$ 3.5 Hz, H-2), 5.76 (dd, 1 H, $J_{3,4}$ 10.1 Hz, $J_{4,5}$ 9.8 Hz, H-4), 5.63 (d, 1 H, $J_{1,2}$ 1.9 Hz, H-1), 4.35 (dq, 1 H, $J_{4,5}$ 9.8 Hz, $J_{5,6}$ 6.3 Hz, H-5), 3.79 (s, 3 H, OCH_3), 1.36 (d, 3 H, $J_{5,6}$ 6.3 Hz, H-6). ^{13}C NMR (150.9 MHz, $CDCl_3$, 25°C) δ: 165.8 (4-COPh), 165.6 (2-COPh, 3-COPh), 155.3 (1-Ph-p-C), 150.1 (1-Ph-i-C), 133.6, 133.4, 133.2 (3 Ph-p-C), 130.0, 129.8, 129.7 (6 Ph-o-C), 129.3, 129.2 (3 Ph-i-C), 128.6, 128.4, 128.3 (6 Ph-m-C), 117.7 (1-Ph-p-C), 114.7 (1-Ph-m-C), 96.6 (C-1), 71.7 (C-4), 70.7 (C-2), 69.8 (C-3), 67.3 (C-5), 55.7 (OCH_3), 17.7 (C-6); $^1J_{C-1,H-1} = 173.5$ Hz. ESI-MS calcd for $C_{34}H_{30}NaO_9$ [M+Na]⁺: 605.1782. Found: 605.1729. Anal. calcd for $C_{34}H_{30}O_9$: C, 70.09; H, 5.19; O, 24.72. Found: C, 70.10; H, 5.25; O, 24.88.

2,3,4,6-Tetra-O-benzoyl-β-D-galactopyranosyl-(1→4)-2,3,6-tri-O-benzoyl-D-glucopyranose (4)[14,15]

To a solution of lactose octabenzoate[13] (5 g, 4.25 mmol, 1 eq.) in THF (80 mL) was added, dropwise during 30 min, CH_3NH_2 in THF (2 M, 15 mL, 7 eq.), and the mixture was stirred at 20°C for 15 h. After concentration, chromatography (7:2 hexane–EtOAc) gave pure compound 4 as well as ~500 mg unreacted starting material (10%). Yield: 3.8 g (83%, based on consumed starting material) as amorphous α/β mixture (4.9:1). R_f: 0.2 (2:1 hexane–EtOAc); α-isomer, ^1H NMR (600.13 MHz, $CDCl_3$, 25°C) δ: 8.20–7.15 (m, 35 H, Ar-H), 6.19 (dd, 1 H, $J_{2,3}$ 10.2 Hz, $J_{3,4}$ 9.2 Hz, H-3ᴵ), 5.77 (dd, 1 H, $J_{3,4}$ 3.4 Hz, $J_{4,5}$ 1.1 Hz, H-4ᴵᴵ), 5.76 (dd, 1 H, $J_{1,2}$ 7.9 Hz, $J_{2,3}$ 10.3 Hz, H-2ᴵᴵ), 5.65 (dd, 1 H, $J_{1,2}$ 3.54 Hz, $J_{1,OH}$ 3.8 Hz, H-1ᴵ), 5.43 (dd, 1 H, $J_{2,3}$ 10.3 Hz, $J_{3,4}$ 3.4 Hz, H-3ᴵᴵ), 5.28 (dd, 1 H, $J_{1,2}$ 3.5 Hz, $J_{2,3}$ 10.2 Hz, H-2ᴵ), 4.96 (d, 1 H, $J_{1,2}$ 7.9 Hz, H-1ᴵᴵ), 4.56 (dd, 1 H, $J_{5,6a}$ 1.9 Hz, $J_{6a,6b}$ 12.2 Hz, H-6aᴵ), 4.49 (dd, 1 H, $J_{5,6b}$ 3.5 Hz, $J_{6a,6b}$ 12.2 Hz, H-6bᴵ), 4.37 (ddd, 1 H, $J_{4,5}$ 10.1 Hz, $J_{5,6a}$ 1.9 Hz, $J_{5,6b}$ 3.5 Hz, H-5ᴵ), 4.28 (dd, 1 H, $J_{3,4}$ 9.2 Hz, $J_{4,5}$ 10.1 Hz, H-4ᴵ), 4.08 (d, 1 H, $J_{1,OH}$ 3.8 Hz, OH), 3.93 (ddd, 1 H, $J_{4,5}$ 1.1 Hz, $J_{5,6a}$ 6.5 Hz, $J_{5,6b}$ 7.0 Hz, H-5ᴵᴵ), 3.85 (dd, 1 H, $J_{5,6a}$ 6.5 Hz, $J_{6a,6b}$ 11.4 Hz, H-6aᴵᴵ), 3.81 (dd, 1 H, $J_{5,6b}$ 6.9 Hz, $J_{6a,6b}$ 11.4 Hz, H-6bᴵᴵ). ^{13}C NMR (150.9 MHz, $CDCl_3$, 25°C) δ: 166.1, 166.0, 165.6, 165.5, 165.4, 165.3, 164.9 (7 COPh), 133.5–128.2 (Ar-C), 101.0 (C-1ᴵᴵ), 90.2 (C-1ᴵ), 76.0 (C-4ᴵ), 72.1 (C-2ᴵ), 71.9 (C-3ᴵᴵ), 71.3 (C-5ᴵᴵ), 70.2 (C-3ᴵ), 70.0 (C-2ᴵᴵ), 68.3 (C-5ᴵ), 67.5 (C-4ᴵᴵ), 62.4 (C-6ᴵ), 61.0 (C-6ᴵᴵ).

β-isomer, ^1H NMR (600.13 MHz, $CDCl_3$, 25°C) δ: 8.20–7.15 (m, 35 H, Ar-H), 5.86 (dd, 1 H, $J_{2,3}$ 9.9 Hz, $J_{3,4}$ 9.1 Hz, H-3ᴵ), 5.77 (dd, 1 H, $J_{3,4}$ 3.4 Hz, $J_{4,5}$ 0.9 Hz, H-4ᴵᴵ), 5.75 (dd, 1 H, $J_{1,2}$ 7.9 Hz, $J_{2,3}$ 10.4 Hz, H-2ᴵᴵ), 5.42 (dd, 1 H, $J_{2,3}$ 10.4 Hz, $J_{3,4}$ 3.4 Hz, H-3ᴵᴵ), 5.35 (dd, 1 H, $J_{1,2}$ 8.0 Hz, $J_{2,3}$ 9.9 Hz, H-2ᴵ), 4.95 (dd, 1 H, $J_{1,2}$ 8.0 Hz, $J_{1,OH}$

7.9 Hz, H-1^I), 4.92 (d, 1 H, $J_{1,2}$ 7.8 Hz, H-1^{II}), 4.57 (dd, 1 H, $J_{5,6a}$ 1.8 Hz, $J_{6a,6b}$ 11.6 Hz, H-6a^I), 4.51 (dd, 1 H, $J_{5,6b}$ 4.2 Hz, $J_{6a,6b}$ 11.6 Hz, H-6b^I), 4.43 (d, 1 H, $J_{1,OH}$ 7.9 Hz, OH), 4.28 (dd, 1 H, $J_{3,4}$ 9.1 Hz, $J_{4,5}$ 9.8 Hz, H-4^I), 3.93 (ddd, 1 H, $J_{4,5}$ 1.0 Hz, $J_{5,6a}$ 6.5 Hz, $J_{5,6b}$ 6.9 Hz, H-5^{II}), 3.83 (ddd, 1 H, $J_{4,5}$ 9.8 Hz, $J_{5,6a}$ 1.8 Hz, $J_{5,6b}$ 4.2 Hz, H-5^I), 3.78 (dd, 1 H, $J_{5,6a}$ 6.5 Hz, $J_{6a,6b}$ 11.3 Hz, H-6a^{II}), 3.81 (dd, 1 H, $J_{5,6b}$ 6.9 Hz, $J_{6a,6b}$ 11.3 Hz, H-6b^{II}). ^{13}C NMR (150.9 MHz, CDCl$_3$, 25°C) δ: 166.6, 166.0, 165.6, 165.5, 165.4, 165.3, 164.9 (7 COPh), 133.5–128.2 (Ar-C), 101.0 (C-1^{II}), 95.7 (C-1^I), 75.9 (C-4^I), 73.9 (C-2^I), 73.2 (C-5^I), 72.5 (C-3^I), 71.7 (C-2^{II}), 71.3 (C-5^{II}), 69.9 (C-3^{II}), 67.5 (C-4^{II}), 62.4 (C-6^I), 61.1 (C-6^{II}). ESI-MS calcd for C$_{61}$H$_{50}$NaO$_{18}$ [M+Na]$^+$: 1093.2889. Found: 1093.2837. Anal. calcd for C$_{61}$H$_{50}$O$_{18}$: C, 68.41; H, 4.71; O, 26.89. Found: C, 68.15; H, 5.04; O, 27.01.

1-O-[2,3,4,6-Tetra-O-benzoyl-β-D-galactopyranosyl-(1→4)-2,3,6-tri-O-benzoyl-α-D-glucopyranosyl] Trichloroacetimidate (5)^{15a}

To a solution of compound 4^{14,15} (2 g, 1.87 mmol, 1 eq.) in freshly distilled CH$_2$Cl$_2$ (25 mL) was added CCl$_3$CN (0.6 mL, 5.98 mmol, 3.2 eq.) and the reaction mixture was cooled to −5°C under argon. DBU (25 µL, 0.17 mmol, 0.09 eq.) was added, and the mixture was allowed to stir at the same temperature for 1.5 h. The solvents were removed under reduced pressure and the crude product was chromatographed (3:1 hexane–EtOAc + 0.1% Et$_3$N) to give pure compound 5. Yield, 2 g (88%). glass; R_f:0.4 (2:1 hexane–EtOAc); [α]$_D^{25}$ + 59° (c 1.0, CH$_2$Cl$_2$), lit. [15b] [α]$_D^{25}$ + 54.5° (c 1.0, CHCl$_3$). ^{1}H NMR (600.13 MHz, CDCl$_3$, 25°C) δ: 8.48 (br s, 1 H, NH), 7.95–7.07 (m, 35 H, Ar-H), 6.64 (d, 1 H, $J_{1,2}$ 3.8 Hz, H-1^I), 6.08 (dd, 1 H, $J_{2,3}$ 10.2 Hz, $J_{3,4}$ 9.1 Hz, H-3^I), 5.68 (dd, 1 H, $J_{3,4}$ 3.4 Hz, $J_{4,5}$ 1.0 Hz, H-4^{II}), 5.67 (dd, 1 H, $J_{1,2}$ 7.9 Hz, $J_{2,3}$ 10.3 Hz, H-2^{II}), 5.47 (dd, 1 H, $J_{1,2}$ 3.8 Hz, $J_{2,3}$ 10.2 Hz, H-2^I), 5.32 (dd, 1 H, $J_{2,3}$ 10.3 Hz, $J_{3,4}$ 3.4 Hz, H-3^{II}), 4.87 (d, 1 H, $J_{1,2}$ 7.9 Hz, H-1^{II}), 4.49 (dd, 1 H, $J_{5,6a}$ 1.7 Hz, $J_{6a,6b}$ 12.2 Hz, H-6a^I), 4.46 (dd, 1 H, $J_{5,6b}$ 3.7 Hz, $J_{6a,6b}$ 12.2 Hz, H-6b^I), 4.26 (dd, 1 H, $J_{3,4}$ 9.1 Hz, $J_{4,5}$ 10.1 Hz, H-4^I), 4.24 (ddd, 1 H, $J_{4,5}$ 10.2 Hz, $J_{5,6a}$ 1.7 Hz, $J_{5,6b}$ 3.7, Hz H-5^I), 3.82 (ddd, 1 H, $J_{4,5}$ 1.1 Hz, $J_{5,6a}$ 6.3 Hz, $J_{5,6b}$ 7.2 Hz, H-5^{II}), 3.75 (dd, 1 H, $J_{5,6a}$ 6.3 Hz, $J_{6a,6b}$ 11.3 Hz, H-6a^{II}), 3.65 (dd, 1 H, $J_{5,6b}$ 7.2 Hz, $J_{6a,6b}$ 11.3 Hz, H-6b^{II}). ^{13}C NMR (150.9 MHz, CDCl$_3$, 25°C) δ: 164.7, 164.5, 164.4, 164.2, 164.1, 163.8 (7 COPh), 159.6 (C=N), 132.5–124.2 (Ar-C), 100.3 (C-1^{II}), 92.0 (C-1^I), 89.6 (CCl$_3$), 74.6 (C-4^I), 70.9 (C-3^{II}), 70.3 (2 C, C-5^I, C-5^{II}), 69.5 (C-2^I), 69.2 (C-3^I), 68.9 (C-2^{II}), 66.4 (C-4^{II}), 60.8 (C-6^I), 59.9 (C-6^{II}). ESI-MS calcd for C$_{63}$H$_{50}$Cl$_3$NNaO$_8$ [M+Na]$^+$: 1236.1986. Found: 1236.1992.

p-Nitrophenyl 2,3,4,6-tetra-O-benzoyl-β-D-galactopyranosyl-(1→4)-2,3,6-tri-O-benzoyl-β-D-glucopyranoside (6)

A solution of 5^{15} (350 mg, 0.29 mmol, 1 eq.) and p-nitrophenol (60 mg, 0.43 mmol, 1.5 eq.) in anhydrous CH$_2$Cl$_2$ (5 mL) was cooled to −15°C under argon. BF$_3$·OEt$_2$ (10 µL, 0.08 mmol, 0.28 eq.) was added, and the mixture was stirred at the same temperature for 1 h [TLC; 3:2 hexane–EtOAc*]. The mixture was poured into

---

* The starting material and reaction product have almost identical R_f values in hexane:EtOAc eluents, which is why the use of a 9:1 mixture of toluene–EtOAc, where the difference is more clear, is advisable.

satd. $NaHCO_3$ and extracted with CH_2Cl_2 (25 mL). The organic layer was washed with water, dried (Na_2SO_4), and concentrated under reduced pressure to give the crude product. Chromatography (5:2 hexane–EtOAc) gave pure compound **6**. Yield: 300 mg (87%). R_f:0.4 (3:2 hexane–EtOAc); m.p. 124°C–126°C (EtOH); $[\alpha]_D^{25} + 33$ (c 1.0, CH_2Cl_2). ^{1}H NMR (600.13 MHz, $CDCl_3$, 25°C) δ: 8.03–7.20 (m, 37 H, Ar-H), 6.97 (d, 2 H, J 9.2 Hz, Ar-H), 5.92 (dd, 1 H, $J_{2,3}$ 9.0 Hz, $J_{3,4}$ 8.9 Hz, H-3^I), 5.79 (dd, 1 H, $J_{3,4}$ 3.4 Hz, $J_{4,5}$ 1.0 Hz, H-4^{II}), 5.77 (dd, 1 H, $J_{1,2}$ 7.9 Hz, $J_{2,3}$ 10.3 Hz, H-2^{II}), 5.75 (dd, 1 H, $J_{1,2}$ 7.3 Hz, $J_{2,3}$ 9.0 Hz, H-2^I), 5.46 (dd, 1 H, $J_{2,3}$ 10.4 Hz, $J_{3,4}$ 3.4 Hz, H-3^{II}), 5.41 (d, 1 H, $J_{1,2}$ 7.3 Hz, H-1^I), 4.95 (d, 1 H, $J_{1,2}$ 7.9 Hz, H-1^{II}), 4.68 (dd, 1 H, $J_{5,6a}$ 2.0 Hz, $J_{6a,6b}$ 12.0 Hz, H-6a^I), 4.52 (dd, 1 H, $J_{5,6b}$ 5.9 Hz, $J_{6a,6b}$ 12.0 Hz, H-6b^I), 4.35 (dd, 1 H, $J_{3,4}$ 8.9 Hz, $J_{4,5}$ 9.8 Hz, H-4^I), 4.12 (ddd, 1 H, $J_{4,5}$ 9.8 Hz, $J_{5,6a}$ 2.2 Hz, $J_{5,6b}$ 5.9 Hz, H-5^I), 4.00 (ddd, 1 H, $J_{4,5}$ 1.1 Hz, $J_{5,6a}$ 6.4 Hz, $J_{5,6b}$ 7.0 Hz, H-5^{II}), 3.83 (dd, 1 H, $J_{5,6a}$ 6.4 Hz, $J_{6a,6b}$ 11.3 Hz, H-6a^{II}), 3.76 (dd, 1 H, $J_{5,6b}$ 7.0 Hz, $J_{6a,6b}$ 11.3 Hz, H-6b^{II}). ^{13}C NMR (150.9 MHz, $CDCl_3$, 25°C) δ: 165.6, 165.5, 165.4, 165.3, 165.2, 165.0, 164.8 (7 COPh), 161.1–116.6 (Ar-C), 101.2 (C-1^{II}), 98.1 (C-1^I), 76.0 (C-4^I), 73.6 (C-5^I), 72.7 (C-3^I), 71.6 (C-3^{II}), 71.5 (C-5^{II}), 71.4 (C-2^I), 69.9 (C-2^{II}), 67.5 (C-4^{II}), 62.2 (C-6^I), 61.0 (C-6^{II}). ESI-MS calcd for $C_{67}H_{53}NNaO_{20}$ [M+Na]$^+$: 1214.3053. Found: 1214.3048. Anal. calcd for $C_{67}H_{53}NO_{20}$: C, 67.50; H, 4.48; O, 26.84; N, 1.17. Found: C, 67.46; H, 4.52; O, 26.66; N, 1.10.

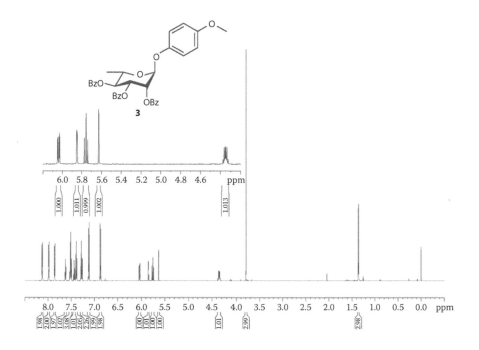

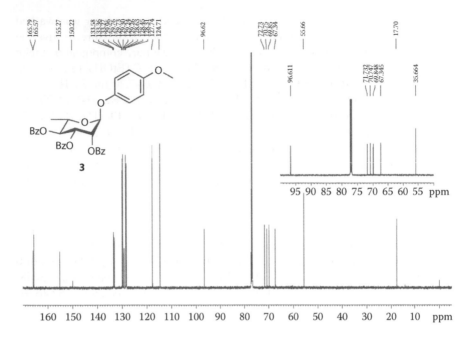

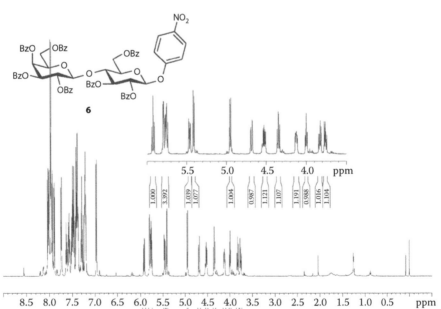

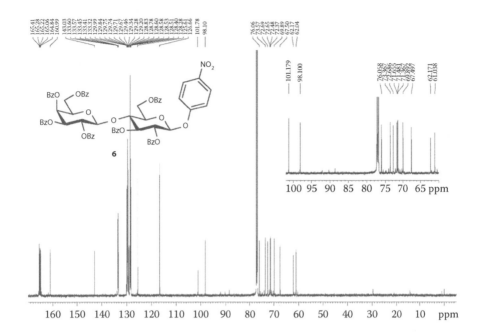

REFERENCES

1. (a) Ekholm, F. S.; Eklund, P.; Leino, R. *Carbohydr. Res.*, **2010**, *345*, 1963–1967; (b) Moon, B.-S.; Ryoo, I.-J.; Yun, B.-S.; Bae, K.-S.; Lee, K. D.; Yoo, I.-D.; Kim, J.-P. *J. Antibiot.*, **2006**, *59*, 735–739; (c) Gupta, P.; Sharma, U.; Gupta, P.; Siripurapu, K. B.; Maurya, R. *Bioorg. Med. Chem.*, **2013**, *21*, 1116–1122.
2. Michael, A. *Am. Chem. J.*, **1879**, *1*, 305–312.
3. Jensen, K. J. *J. Chem. Soc. Perkin Trans. 1*, **2002**, 2219–2233.
4. Schmidt, R. R.; Jung, K.-H. In: *Preparative Carbohydrate Chemistry*, Hanessian, S., Ed.; Marcel Dekker, Inc., New York, **1997**, pp. 283–312.
5. Ziegler, T.; Kováč, P.; Glaudemans, C. P. J. *Liebigs Ann. Chem.*, **1990**, 613–615.
6. Garegg, P. J.; Norberg, T. *Acta Chem. Scand.*, **1979**, *B33*, 116–118.
7. Egusa, K.; Kusumoto, S.; Fukase, K. *Eur. J. Org. Chem.*, **2003**, 3435–3445.
8. Schmidt, R. R.; Michel, J. *Angew. Chem. Int. Ed.*, **1980**, *19*, 731–732.
9. PERCH Solutions ltd, Kuopio, Finland, http://www.perchsolutions.com.
10. Cui, C.; Wen, X.; Cui, M.; Gao, J.; Sun, B.; Lou, H. *Drug Discov. Ther.*, **2012**, *6*, 9–17.
11. Kochetkov, N. K.; Byramova, N. E.; Tsvetkov, Yu. E.; Backinovskii, L. V. *Tetrahedron*, **1985**, *41*, 3363–3375.
12. (a) Vargas-Berenguel, A.; Meldal, M.; Paulsen, H.; Jensen, K. J.; Bock, K. *J. Chem. Soc. Perkin Trans. 1*, **1994**, 3287–3294; (b) Ziegler, T.; Bien, F.; Jurisch, C. *Tetrahedron: Asymmetry*, **1998**, *9*, 765–780.
13. Barrientos, Á. G.; de la Fuente, J. M.; Rojas, T. C.; Fernández, A.; Penadés, S. *Chem. Eur. J.*, **2003**, *9*, 1909–1921.
14. Douglas, S. P.; Whitfield, D. M.; Krepinsky, J. J. *J. Am. Chem. Soc.*, **1995**, *117*, 2116–2117.
15. (a) Boons, G. J. P. H.; Van Delft, F. L.; Van der Klein, P. A. M.; Van der Marel, G. A.; Van Boom, J. H. *Tetrahedron*, **1992**, *48*, 885–904; (b) Mei, X.; Heng, L.; Fu, M.; Li, Z.; Ning, J. *Carbohydr. Res.*, **2005**, *340*, 2345–2351.

13 Squaric Acid Diethyl Ester–Mediated Homodimerization of Unprotected Carbohydrates

Sarah Roy, Melissa Barrera Tomas,[†] and Denis Giguère*

CONTENTS

Squaric acid diesters (3,4-dialkyloxycyclobut-3-ene-1,2-dione) are versatile reagents used to couple two amino-functional groups.[1–3] The main feature of using squaric acid diester is its ability to successively react one of the ester groups selectively. In addition, it has limited reactivity with alcohols and phenols in comparison to amines. Squaric acid esters were first used as a coupling reagent for carbohydrates by the group

---

* Corresponding author; e-mail: denis.giguere@chm.ulaval.ca.
† Checker, under the supervision of René Roy; e-mail: roy.rene@uqam.ca.

of Tietze[4] and since then, they have been widely applied for linking carbohydrates to proteins.[1] Moreover, the combination of glycosides bearing an amine (2 equivalents or more) with squaric acid ester (1 equivalent) leads to divalent glycosides, which are valuable medically relevant synthetic tools.[5] Linker rigidity between two carbohydrates can modulate the cross-linking ability of various lectins.[6] It is thus important to construct efficiently a variety of rigid linkers to study the interactions between lectins and divalent sugar ligands.

We present herein an efficient synthesis of p-{N-[4-(4-aminophenyl α-D-glucopyranosyl)-2,3-dioxocyclobut-1-enyl]amino}phenyl   α-D-glucopyranoside **3a** and p-{N-[4-(4-aminophenyl β-D-glucopyranosyl)-2,3-dioxocyclobut-1-enyl] amino}phenyl β-D-glucopyranoside **3b** via the use of commercially available 3,4-diethoxy-3-cyclobutene-1,2-dione **2**.[7] The method is simple and purification involves only precipitation of products by treatment of the concentrated reaction mixture with MeOH–DCM (3:7), allowing dissolution of the unreacted starting materials **1** along with monomer p-[N-ethoxy-3,4-dioxocyclobut-1-enyl)amino]phenyl D-glucopyranoside.*

EXPERIMENTAL METHODS

GENERAL METHODS

Reactions in organic media were carried out under nitrogen atmosphere, and ACS-grade solvents were used without further purification. Reactions were monitored by thin-layer chromatography (TLC) using silica gel 60 F_{254}–coated plates (E. Merck). Optical rotations were measured with a JASCO DIP-360 digital polarimeter. NMR spectra were recorded on an Agilent DD2 500 MHz spectrometer. Proton and carbon chemical shifts (δ) are reported in ppm relative to the chemical shift of residual DMSO, which was set at 2.50 ppm (^{1}H) and 39.52 ppm (^{13}C). Coupling constants (J) are reported in hertz (Hz), and the following abbreviations are used: singlet (s), doublet (d), doublet of doublets (dd), triplet (t), multiplet (m), and broad (br). High-resolution mass spectra (HRMS) were measured with an Agilent 6210 LC time-of-flight mass spectrometer in electrospray mode.

p-{N-[4-(4-Aminophenyl α-D-glucopyranosyl)-2,3-dioxocyclobut-1-enyl]amino}phenyl α-D-glucopyranoside (3a)

4-Aminophenyl α-D-glucopyranoside (**1a**)[8] (140 mg, 0.517 mmol, 2.0 equiv.) was added in one portion to a solution of 3,4-diethoxy-3-cyclobutene-1,2-dione (44 mg, 0.259 mmol, 1.0 equiv.) in MeOH (5.0 mL). The resulting homogenous mixture was stirred at room temperature for 48 h. After this time, TLC analysis (3:7 MeOH–DCM) revealed an almost complete disappearance of the starting material (R_f 0.9) along with the monomer: p-[N-ethoxy-3,4-dioxocyclobut-1-enyl)amino]phenyl α-D-glucopyranoside (R_f 0.7). The mixture was concentrated under reduced pressure and the residue was stirred for 5 min with 10 mL of MeOH–DCM (3:7). The undissolved

* Products **3a** and **3b** are not soluble in the MeOH–DCM (3:7) mixture due to their high polarity.

solid was filtered and rinsed with 2 mL of MeOH–DCM (3:7).* Pure homodimer **3a** (108 mg) was isolated as an amorphous, off-white solid (67% yield). R_f=0.4 MeOH–DCM (1:1); $[\alpha]_D$ +55.0 (c 0.7, H_2O). ^1H NMR (DMSO-d_6) δ: 3.17 (t, J=9.2 Hz, 2H, 2H-6a), 3.35 (dd, J=3.6, 9.7 Hz, 2H, 2H-2), 3.48 (dt, J=5.7, 8.3 Hz, 4H, 2H-5, 2H-6b), 3.58 (d, J=10.0 Hz, 2H, 2H-4), 3.61 (t, J=9.2 Hz, 2H, 2H-3), 4.80 (br s, 8H, 8OH), 5.31 (d, J=3.6 Hz, 2H, 2H-1), 7.10 (d, J=8.5 Hz, 4H, Ar), 7.41 (d, J=8.7 Hz, 4H, Ar), 9.88 (br s, 2H, 2NH). ^{13}C NMR (DMSO-d_6) δ: 60.8 (2C), 70.03 (2C), 71.7 (2C), 73.11 (2C), 73.8 (2C), 98.6 (2C), 118.0 (4C), 119.9 (4C), 133.6 (2C), 153.3 (2C), 165.4 (2C), 181.6 (2C). HRMS [M+H]$^+$ calcd for $C_{28}H_{33}N_2O_{14}$: 621.1926. Found: 621.1935.

p-{*N*-[4-(4-Aminophenyl β-D-glucopyranosyl)-2,3-dioxocyclobut-1-enyl]amino}phenyl β-D-Glucopyranoside (3b)

4-Aminophenyl β-D-glucopyranoside (**1b**)[9] (150 mg, 0.553 mmol, 2.0 equiv.) was added in one portion to a solution of 3,4-diethoxy-3-cyclobutene-1,2-dione (47 mg, 0.276 mmol, 1.0 equiv.) in MeOH (5.0 mL). The resulting homogenous mixture was stirred at room temperature for 48 h. After this time, TLC analysis (3:7 MeOH–DCM) revealed an almost complete disappearance of the starting material (R_f 0.5) along with the monomer: *p*-[*N*-ethoxy-3,4-dioxocyclobut-1-enyl)amino]phenyl β-D-glucopyranoside (R_f 0.8). The mixture was concentrated under reduced pressure and the residue was stirred for 5 min with 10 mL of MeOH–DCM (3:7). The undissolved solid was filtered and rinsed with 2 mL of MeOH–DCM (3:7). Pure homodimer **3b** (112 mg) was isolated as a white, amorphous solid (65% yield). R_f=0.40 MeOH–DCM (1:1); $[\alpha]_D$ +24.3 (c 0.65, H_2O). ^1H NMR (DMSO-d_6) δ: 3.15 (t, J=8.3 Hz, 2H, 2H-3), 3.20–3.28 (m, 4H, 2H-2, 2H-4), 3.32 (ddd, J=2.2, 5.7, 9.7 Hz, 2H, 2H-6a), 3.46 (dd, J=5.8, 11.8 Hz, 2H, 2H-6b), 3.70 (dd, J=2.1, 11.9 Hz, 2H, 2H-5), 4.81 (d, J=7.4 Hz, 2H, 2H-1), 5.11 (br s, 8H, 8OH), 7.05 (d, J=9.0 Hz, 4H, Ar), 7.41 (d, J=9.0 Hz, 4H, Ar), 9.81 (br s, 2H, 2NH). ^{13}C NMR (DMSO-d_6) δ: 60.7 (2C), 69.7 (2C), 73.3 (2C), 76.6 (2C), 77.1 (2C), 100.9 (2C), 117.1 (4C), 119.8 (4C), 133.0 (2C), 153.7 (2C), 165.2 (2C), 181.3 (2C); HRMS [M+H]$^+$ calcd for $C_{28}H_{33}N_2O_{14}$: 621.1926. Found: 621.1929.

ACKNOWLEDGMENTS

We gratefully acknowledge the Université Laval for financial support of this work. We are also grateful to Professor Pierre Deslongchamps for generously sharing reagents. M.B.T. is thankful to T.C. Shiao for the helpful discussion.

---

* This allowed dissolution and removal of the remaining starting material and monomer.

^{1}H-NMR (500 MHz, DMSO-d_6) of compound **3a**.

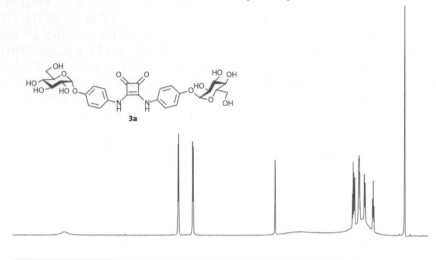

^{13}C-NMR (125 MHz, DMSO-d_6) of compound **3a**.

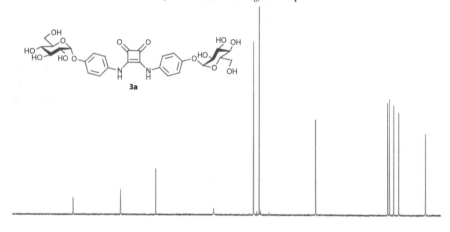

¹H-NMR (500 MHz, DMSO-d_6) of compound **3b**.

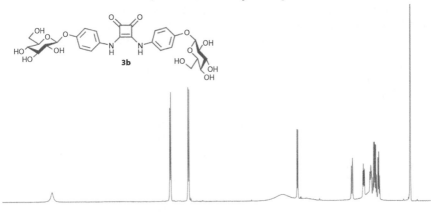

¹³C-NMR (125 MHz, DMSO-d_6) of compound **3b**.

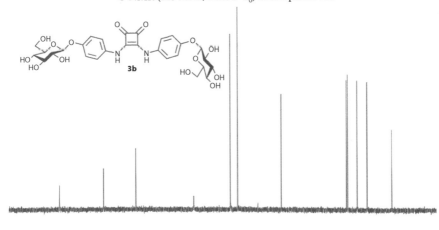

REFERENCES

1. Wurm, F. R.; Klok, H.-A. *Chem. Soc. Rev.* **2013**, *42*, 8220–8236.
2. Chernyak, A.; Karavanov, A.; Ogawa, Y.; Kováč, P. *Carbohydr. Res.* **2001**, *330*, 479–486.
3. Quinonero, D.; Frontera, A.; Ballester, P.; Deyà, P. M. *Tetrahedron Lett.* **2000**, *41*, 2001–2005.
4. Tietze, L. F.; Arlt, M.; Beller, M.; Glüsenkamp, K.-H.; Jähde, E.; Rajewsky, F. *Chem. Ber.* **1991**, *124*, 1215–1221.
5. Lindhorst, T. K.; Bruegge, K.; Fuchs, A.; Sperling, O. *Beilstein J. Org. Chem.* **2010**, *6*, 801–809.
6. (a) Bergeron-Brlek, M.; Giguère, D.; Shiao, T. C.; Saucier, C.; Roy, R. *J. Org. Chem.* **2012**, *77*, 2971–2977; (b) Roy, R.; Trono, M. C.; Giguère, D. *ACS Symp. Ser.* **2005**, *896*, 137–150; (c) Roy, R.; Das, S. K.; Santoyo-Gonzalez, F.; Hernandez-Mateo, F.; Dam, T. K.; Brewer, C. F. *Chem. Eur. J.* **2000**, *6*, 1757–1762; (d) Dam, T. K.; Oscarson, S.; Roy, R.; Das, S. K.; Pagé, D.; Macaluso, F.; Brewer, C. F. *J. Biol. Chem.* **2005**, *280*, 8640–8646; (e) Lameignere, E.; Shiao, T. C.; Roy, R.; Wimmerova, M.; Dubreuil, F.; Varrot, A.; Imberty, A. *Glycobiology* **2010**, *20*, 87–98.
7. (a) Tietze, L. F.; Schroeter, C.; Gabius, S.; Brinck, U.; Goerlach-Graw, A.; Gabius, H. J. *Bioconjug. Chem.* **1991**, *2*, 148–153; (b) Hou, S.-J.; Saksena, R.; Kováč, P. *Carbohydr. Res.* **2008**, *343*, 196–210; (c) Kamath, V. P.; Diedrich, P.; Hindsgaul, O. *Glycoconj. J.* **1996**, *13*, 315–319.
8. (a) Gary-Bobo, M.; Hocine, O.; Brevet, D.; Maynadier, M.; Raehm, L.; Richeter, S.; Charasson, V. et al. *Int. J. Parma.* **2012**, *423*, 509–515; (b) Hu, J.; Kuang, W.; Deng, K.; Zou, W.; Huang, Y.; Wei, Z.; Faul, C. F. J. *Adv. Funct. Mater.* **2012**, *22*, 4149–4158.
9. (a) Branco de Barros, A. L.; Cardoso, V. N.; das Graça Mota, L.; Leite, E. A.; de Oliveira, M. C.; Alves, R. J. *Bioorg. Med. Chem. Lett.* **2010**, *20*, 2478–2480; (b) Branco de Barros, A. L.; Cardoso, V. N.; das Gracas Mota, L.; Alves, R. J. *Bioorg. Med. Chem. Lett.* **2010**, *20*, 315–317.

14 Thiol–Yne Coupling between Propargyl 2,3,4,6-Tetra-O-acetyl-β-D-glucopyranoside and Thiol

Expedient Access to Neoglycolipids

*David Goyard, Tze Chieh Shiao,
Denis Giguère,[†] and René Roy**

CONTENTS

* Corresponding author; e-mail: roy.rene@uqam.ca.
† Checker; e-mail: denis.giguere@chm.ulaval.ca.

Glycolipids and glycosphingolipids consist of sugar residues that are attached to different lipid backbones.[1] Glycolipids are important components expressed at the cell surface where they play numerous roles in biological processes such as immunity and cell signaling, as well as structural roles.[2] They form the backbone of both prokaryotic and eukaryotic cell membranes. Novel biological activities involving glycolipids have been continuously discovered in recent years. Consequently, their synthesis as well as those of analogs,[3] including glycodendrimersomes,[4] has proved useful in understanding these latest functions. In addition, development of neoglycolipid microarrays has witnessed several recent applications toward deciphering their specific biological functions.[5]

Radical-initiated thiol–yne coupling has been first described for acetylenic hydrocarbons using thioacetic acid.[6] It was recently proved as a powerful tool for small molecules as well as polymer synthesis[7] and has drawn interest as it fulfills the criteria of a *click-chemistry* reaction. We present herein the first application of this strategy to efficient synthesis of a neoglycolipid. Hence, 3-(dodecanylthio)prop-2-enyl 2,3,4,6-tetra-*O*-acetyl-β-D-glucopyranoside **2** and 2,3-bis(dodecanylthio) propyl 2,3,4,6-tetra-*O*-acetyl-β-D-glucopyranoside **3** were prepared *via* a thiol–yne click reaction between dodecanethiol and propargyl 2,3,4,6-tetra-*O*-acetyl-β-D-glucopyranoside **1**.[8] Subsequent deacetylation of **3** leads to 2,3-bis(dodecanylthio) propyl β-D-glucopyranoside **4**, which can be considered as a glycolipid mimetic. Work in progress[9] is showing that they readily form self-assembling liposomes and that they constitute multivalent functional glycoconjugates by reacting with plant, bacterial, and animal lectins. In addition, the fact that the alkyne can react sequentially with a first thioether such as in **2** opens an unexplored entry to the construction of hybrid glycolipids constituted of two different lipophilic side chains.

EXPERIMENTAL METHODS

GENERAL METHODS

Reactions were carried out under argon using commercially available ACS-grade dioxane, which was stored over 4 Å molecular sieve. Commercially available dodecanethiol and azobisisobutyronitrile from Sigma-Aldrich Canada LTD were used without further purification. Progress of reactions was monitored by thin-layer chromatography (TLC) using silica gel 60 F_{254}–coated plates (E. Merck). NMR spectra were recorded on Varian Inova AS600 and Bruker Avance III HD 600 MHz spectrometer. Proton and carbon chemical shifts (δ) are reported in ppm relative to the chemical shift of residual CHCl$_3$, which was set at 7.26 ppm (^{1}H) and 77.16 ppm (^{13}C). Coupling constants (J) are reported in hertz (Hz), and the following abbreviations are used for peak multiplicities: singlet (s), doublet (d), doublet of doublets (dd), doublet of doublet with equal coupling constants (t_{ap}), triplet (t), and multiplet (m). The chiral center in **3** is denoted with an asterisk (C*). Analysis and assignments were made using COSY and HSQC experiments. High-resolution mass spectra (HRMS) were measured with a LC–MS-TOF spectrometer (Agilent Technologies) in positive and/or negative electrospray mode by the analytical platform of UQAM.

3-(Dodecanylthio)prop-2-enyl 2,3,4,6-Tetra-O-acetyl-β-D-glucopyranoside (2)

With the aid of a syringe, a commercial microwave vial was loaded with propargyl 2,3,4,6-tetra-O-acetyl-β-D-glucopyranoside[8] (100 mg, 0.26 mmol, 1.00 eq.), dodecanethiol (52 mg,* 1.00 mmol, 1.00 eq.), and 1 mL of dioxane. The solution was sonicated under a gentle stream of argon for 30 min,[†] then AIBN[‡] (5 mg, 0.03 mmol, 0.10 eq.) was added, and the mixture was heated under microwave conditions for 30 min at 100°C. TLC monitoring showed that the starting sugar was not completely converted, but the reaction could not be brought to completion by extension of the reaction time. Due to the fact that monoaddition was the goal for **2** and that further reaction to **3** needed to be avoided, the reaction was stopped at this level. After cooling to room temperature, the solution was transferred, with the aid of CH$_2$Cl$_2$ into a flask, and concentrated under reduced pressure. The pale yellow oil residue was chromatographed (15:85 → 20:80 EtOAc–hexane), to afford compound **2** as an amorphous solid (99 mg, 0.17 mmol, 65%, intractable 2:1 (E)/(Z) diastereomeric mixture. R_f=0.47, 7:3 hex–EtOAc). ^{1}H NMR (CDCl$_3$) δ: 6.20 (d, 1H, J = 15.2 Hz, Csp^2-H$_{(E)}$), 6.16 (d, 1H, J = 9.7 Hz, Csp^2-H$_{(Z)}$), 5.54 (ddd, 1H, J = 9.5, 7.1, 6.3 Hz, Csp^2-H$_{(Z)}$), 5.54 (dt_{ap}, 1H, J = 15.1, 6.4 Hz, Csp^2-H$_{(E)}$), 5.14 (t_{ap}, 2H, J = 9.5 Hz, 2×H-3), 5.03 (t_{ap}, 1H, J = 9.6 Hz, H-4$_{(Z)}$), 5.02 (t_{ap}, 1H, J = 9.7 Hz, H-4$_{(E)}$), 4.90–4.95 (m, 2H, 2×H-2), 4.50 (d, 1H, J = 8.1 Hz, H-1$_{(E)}$), 4.48 (d, 1H, J = 8.2 Hz, H-1$_{(Z)}$), 4.18–4.26 (m, 4H, 2×CH$_2$O, 2×H-6a), 4.03–4.09 (m, 4H, 2×CH$_2$O, 2×H-6b), 3.61–3.65 (m, 2H, 2×H-5), 2.59–2.63 (m, 4H, SCH$_2$-alk), 2.02, 1.98, 1.95, 1.93 (8s, 24H, acetyl), 1.52–1.59 (m, 4H, CH$_2$-alk), 1.28–1.34 (m, 4H, CH$_2$-alk), 1.19–1.23 (m, 32H, CH$_2$-alk), 0.81 (t, 6H, J = 6.9 Hz, 2×CH$_3$-alk). ^{13}C NMR (CDCl$_3$) δ: 170.6, 170.6, 170.2, 170.2, 169.3, 169.3, 169.2(8×C=O), 131.2 (Csp^2-H$_{(Z)}$), 130.0 (Csp^2-H$_{(E)}$), 123.0 (Csp^2-H$_{(Z)}$), 120.9 (Csp^2-H$_{(E)}$), 99.3 (C-1$_{(Z)}$), 99.1 (C-1$_{(E)}$), 72.9 (C-3$_{(Z)}$), 72.8 (C-3$_{(E)}$), 71.9 (C-5$_{(Z)}$), 71.8 (C-5$_{(E)}$), 71.2 (2×C-2), 69.6 (OCH$_{2(E)}$), 68.4 (C-4$_{(Z)}$), 68.4 (C-4$_{(E)}$), 65.7 (OCH$_{2(Z)}$), 61.9 (C-6$_{(E)}$), 61.9 (C-6$_{(Z)}$),

---

* Dodecanethiol was weighed for more accuracy.
[†] Degassing the solvent is essential. Experiments have shown that oxidation of sulfur atoms can occur during the reaction, resulting in decreased yield.
[‡] AIBN can be substituted by azobiscyanovaleric acid (ACVA) as initiator.

34.2 (SCH$_2$-alk), 32.0 (SCH$_2$-alk), 22.6–31.9 (20×CH$_2$-alk), 20.7, 20.7, 20.7, 20.6, 20.6, 20.6, 20.5, 20.5 (8×CH$_3$ acetyl), 14.1 (2×CH$_3$-alk). ESI-HRMS [M+COOH]⁻ calcd for C$_{30}$H$_{49}$O$_{12}$S: 633.2950. Found: 633.2938.

2,3-Bis(dodecanylthio)propyl 2,3,4,6-Tetra-O-acetyl-β-D-glucopyranoside (3)

Procedure A: From propargyl 2,3,4,6-tetra-O-acetyl-β-D-glucopyranoside, (1) with the aid of a syringe, a microwave vial was loaded with propargyl 2,3,4,6-tetra-O-acetyl-β-D-glucopyranoside[8] 1 (386 mg, 1.0 mmol, 1.0 eq.), dodecanethiol (959 μL, 4.0 mmol, 4.0 eq.), and 1.0 mL of dioxane. The solution was sonicated for 30 min under a gentle stream of argon.* AIBN† (16 mg, 0.1 mmol, 0.1 eq.) was added to the solution, and the mixture was heated at 100°C for 1 h, after which TLC showed complete disappearance of the starting material (R_f=0.18, 7:3 hexane–EtOAc). After cooling to room temperature, the solution was transferred with the aid of CH$_2$Cl$_2$ into a flask and concentrated under reduced pressure. The pale yellow oil residue was chromatographed (85:15 hexane–EtOAc), to afford compound 3 as white solid (751 mg, 0.95 mmol, 95%). Compound 3 was isolated as a 1:1 diastereomeric mixture.

Procedure B: From 3-(dodecanylthio)prop-2-enyl 2,3,4,6-tetra-O-acetyl-β-D-glucopyranoside, (2) with the aid of a syringe, a microwave vial was loaded with 2 (100 mg, 0.17 mmol, 1 eq.), dodecanethiol (82 μL, 0.34 mmol, 2.0 eq.), AIBN (3 mg, 0.02, 0.1 eq.), and 1 mL of dioxane. The solution was sonicated for 30 min, sealed, and heated at 100°C for 45 min. The vial was allowed to cool down to room temperature, the solution transferred in a flask, washed with CH$_2$Cl$_2$, and concentrated under reduced pressure. The crude mixture (pale yellow oil) was purified over silica gel chromatography (85:15 hex–EtOAc) to afford compound 3 as a white solid (118 mg, 0.15 mmol, 90%). R_f=0.56, 7:3 hex–EtOAc. ^{1}H NMR (CDCl$_3$) δ: 5.19 (2t_{ap}, 2H, *J*=9.5 Hz, 2×H-3), 5.06 (2t_{ap}, 2H, *J*=9.7 Hz, 2×H-4), 4.96, 5.00 (m, 2H, 2×H-2), 4.52 (d, 2H, *J*=8.0 Hz, 2×H-1), 4.25 (dd, 2H, *J*=12.4, 4.2 Hz, 2×H-6a), 4.1–4.13 (m, 3H, 2×H-6b, OC*H$_2$*C*H), 4.06 (dd, 1H, *J*=10.0, 5.8 Hz, OC*H$_2$*C*H), 3.67–3.72 (m, 3H, 2×H-5, OC*H$_2$*C*H), 3.60 (dd, 1H, *J*=9.7, 7.8 Hz, OC*H$_2$*CH*), 2.86–2.94 (m, 2H, 2×C*H), 2.69–2.82 (m, 4H, 2×SC*H$_2$*C*H), 2.50–2.57 (m, 4H, SC*H$_2$*-alk), 2.07, 2.04, 2.00, 1.98 (8s, 24H, acetyl), 1.52–1.57 (m, 4H, C*H$_2$*-alk), 1.24–1.34 (m, 34H, C*H$_2$*-alk), 0.86 (t, 6H, *J*=7.1 Hz, 2×C*H$_3$*-alk). ^{13}C NMR (CDCl$_3$) δ: 170.7, 170.7, 170.4, 170.3, 169.5, 169.5, 169.4, 169.3 (8×C=O), 101.4 (C-1), 101.0 (C-1), 72.9 (C-3), 72.8 (C-3), 71.9 (2×C-5), 71.9 (OC*H$_2$*C*H), 71.6 (OC*H$_2$*C*H), 71.3 (C-2), 71.2 (C-2), 68.5 (2×C-4), 62.0 (2×C-6), 45.8 (C*H), 45.3 (C*H), 34.6, 34.6 (2×SC*H$_2$*C*H), 33.5 (SC*H$_2$*-alk), 33.2 (SC*H$_2$*-alk), 32.0 (SC*H$_2$*-alk, 2×C*H$_2$*-alk),

---

* Special care must be taken in degassing the solvent. Experiments have shown that oxidation of sulfur can occur during the reaction, resulting in decreased yield.
† AIBN can be substituted by ACVA as initiator.

31.7 (SCH$_2$-alk), 30.0–29.0 (36×CH$_2$-alk), 22.8 (2×CH$_2$CH$_3$-alk), 20.9, 20.8, 20.7, 20.7 (8×CH$_3$ acetyl), 14.2 (2×CH$_3$-alk). ESI$^+$-HRMS [M+H]$^+$ calcd for C$_{41}$H$_{75}$O$_{10}$S$_2$: 791.4796. Found: 791.4789. Anal. calcd for C$_{41}$H$_{74}$O$_{10}$S$_2$: C, 62.24; H, 9.43. Found: C, 61.96; H, 9.40.

2,3-Bis(dodecanylthio)propyl β-D-Glucopyranoside (4)

Compound **3** (100 mg, 0.13 mmol) was dissolved in 1:1 MeOH–CH$_2$Cl$_2$ (4 mL), 1 M methanolic NaOMe (0.5 mL) was added, and the mixture was stirred at room temperature for 6 h. The mixture was neutralized with Amberlite H$^+$-resin, filtered, and concentrated, to afford analytically pure, deacetylated compound **4** (76 mg, 0.12 mmol, 96%). ^{1}H NMR (DMSO-d_6) δ: 4.96–4.90 (m, 3H, OH-2, OH-3, OH-4), 4.48–4.42 (m, 1H, OH-6), 4.13 (d, 1H, $^3J_{1,2}$=7.8 Hz, H-1), 3.92–3.90, 3.83–3.80, 3.71–3.65, 3.56–3.53, 3.46–3.42 (m, 4H, OCH$_2$, H-6a, H-6b), 3.31–3.02 (m, 3H, H-3, H-5, H-4), 2.97–2.90 (m, 3H, H-2, C*HCH$_2$S), 2.69–2.65 (m, 1H, C*H), 2.59–2.50 (m, 4H, 2×SCH$_2$), 1.49 (m, 4H, SCH$_2$CH$_2$), 1.33 (m, 4H, 2×S CH$_2$CH$_2$CH$_2$), 1.2 (s, 32H, CH$_2$-alk), 0.84 ppm (t, 6H, $J_{H,H}$=6.9 Hz, 2×CH$_3$-alk). ^{13}C NMR (DMSO-d_6) δ: 104.1 (C-1), 103.3 (C-1), 77.4 (C-5), 77.2 (C-5), 77.1 (2×C-3), 73.9 (2×C-2), 71.5 (OCH$_2$C*H), 70.8 (OCH$_2$C*H), 70.5 (2×C-4), 61.5 (2×C-6), 46.1 (C*H), 45.5 (C*H), 34.4 (C*HCH$_2$S), 34.3 (C*HCH$_2$S), 32.7 (SCH$_2$-alk), 32.5 (SCH$_2$-alk), 31.8 (CH$_2$-alk), 31.1 (2×CH$_2$-alk), 31.0 (CH$_2$-alk), 29.9-28.7 (16×CH$_2$-alk), 22.6 (2×CH$_2$CH$_3$-alk), 14.4 ppm (2×CH$_3$-alk). SI$^+$-HRMS [M+Na]$^+$ calcd for C$_{33}$H$_{66}$O$_6$S$_2$Na: 645.4193. Found: 645.4182. Anal. calcd for C$_{33}$H$_{66}$O$_6$S$_6$: C, 63.62; H, 10.68. Found: C, 62.78; H, 10.69.

ACKNOWLEDGMENTS

This work was supported from Natural Sciences and Engineering Research Council of Canada (NSERC) and a Canadian Research Chair in Therapeutic Chemistry to R.R.

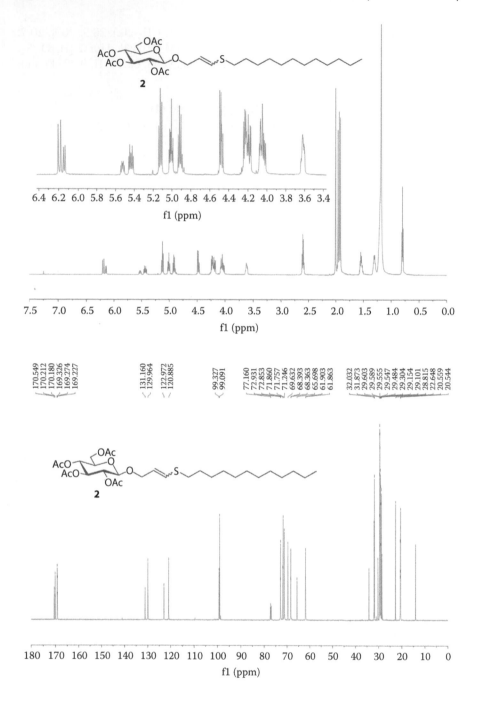

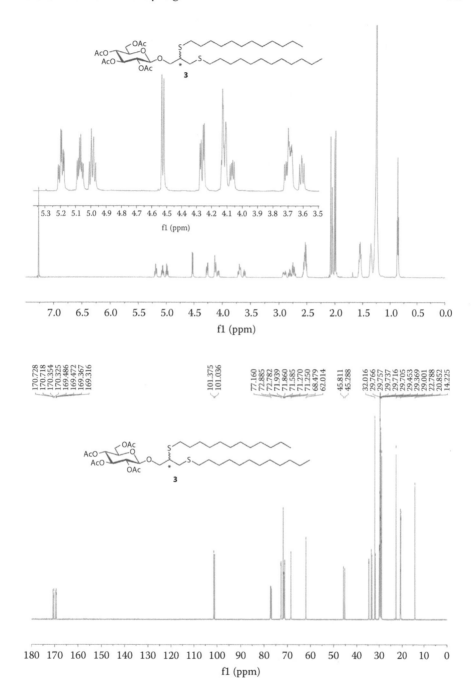

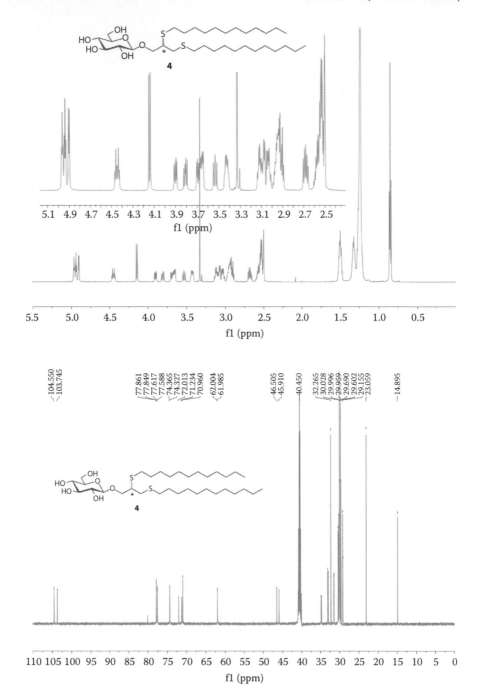

REFERENCES

1. Kolter, T.; Sandhoff, K. *Angew. Chem. Int. Ed.* **1999**, *38*, 1532–1568.
2. (a) Kazunori, M. *Trends Glycosci. Glycotechnol.* **2013**, *25*, 227–239; (b) Cortés-Sánchez, A. D. J.; Hernández-Sánchez, H.; Jaramillo-Flores, M. E. *Microbiol. Res.* **2013**, *168*, 22–32.
3. (a) Jayaraman, N.; Maiti, K.; Naresh, K. *Chem. Soc. Rev.* **2013**, *42*, 4640–4656; (b) Stocker, B. L.; Timmer, M. S. M. *ChemBioChem*, **2013**, *14*, 1164–1184; (c) Roy, R.; Letellier, M.; Fenske, D.; Jarrell, H. C. *J. Chem. Soc. Chem. Commun.* **1990**, 378–380.
4. (a) Percec, V.; Leowanawat, P.; Sun, H.-J.; Kulikov, O.; Nusbaum, C. D.; Tran, T. M.; Bertin, A. et al. *J. Am. Chem. Soc.* **2013**, *135*, 9055–9077; (b) Zhang, S.; Moussodia, R.-O.; Sun, H.-J.; Leowanawat, P.; Muncan, A.; Nusbaum, C. D.; Chelling, K. M. et al. *Angew. Chem Int. Ed.* 2014, *53*, 10899–10903.
5. (a) Palma, A. S.; Feizi, T.; Childs, R. A.; Chai, W.; Liu. Y. *Curr. Opin. Chem. Biol.* **2014**, *18*, 87–94; (b) Park, S.; Gildersleeve, J. C.; Blixt, O.; Shin, I. *Chem. Soc. Rev.* **2013**, *42*, 4310–4326.
6. Bader, H.; Cross, L. C.; Heilbron, I.; Jones, E. R. H. *J. Chem. Soc.* **1949**, 619–623.
7. Hoyle, C. E.; Lowe, A. B.; Bowman, C. N. *Chem. Soc. Rev.* **2010**, *39*, 1355–1387.
8. (a) Mereyala, H. B.; Gurrala, S. R. *Carbohydr. Res.* **1998**, *307*, 351–354; (b) Roy, R.; Das, S. K.; Santoyo-Gonzalez, F.; Hernandez-Mateo, F.; Dam, T. K.; Brewer, C. F. *Chem. Eur. J.* **2000**, *6*, 1757–1762.
9. Goyard, D.; Shiao, T. C.; Roy, R. Unpublished data.

REFERENCES

15 Lactose-Modified Triethoxysilane for the Surface Modification of Clay Nanocomposites

Tze Chieh Shiao, Radia Sennour,
Mohamed Touaibia,† Abdelkrim Azzouz,
and René Roy*

CONTENTS

Surface functionalized mesoporous materials have emerged as one of the most important research areas in the field of advanced functional materials.[1,2] They have found widespread applications in catalysis,[3–5] separation,[6] decontamination,[7] drug delivery,[8] and sensor design.[9] In our ongoing program[10] aimed at incorporating

---

* Corresponding author; e-mail: roy.rene@uqam.ca.
† Checker: Mohamed Touaibia; e-mail: mohamed.touaibia@umoncton.ca.

dendritic polyols into inorganic clay materials for the reversible adsorption of CO_2 gas, it was found that nonsugar polyols formed fragile carbonate bonds that could release CO_2 under much milder conditions than polyamines do. Since the polyols were noncovalently intercalated within the matrix interlamellar spaces, we became interested to compare the behavior and efficacy of covalently grafted polyols. For the later, we chose carbohydrates and particularly the readily available lactose. For the grafting chemistry, we decided to use the efficacious copper-catalyzed *click* chemistry for the incorporation of the siloxane moiety.[11] Hence, known β-propargylated lactoside **1**[12] was treated under standard click chemistry conditions with (3-azido-propyl)triethoxysilane **2**[13] to afford the desired reagent **3** ready for grafting onto montmorillonite.

EXPERIMENTAL METHODS

GENERAL METHODS

All reagents were used as supplied from Sigma-Aldrich unless otherwise stated. DMF was purified by refluxing with ninhydrin and distilled. Microwave reactions were conducted using Biotage Initiator Microwave Synthesizer under normal absorption level. After workup, the organic layers were dried over anhydrous Na_2SO_4. Reactions were monitored by analytical thin-layer chromatography using silica gel 60 F_{254}–precoated plates (E. Merck). The visualization of spots was performed by soaking in (1) charring solution (sulfuric acid/methanol/water: 5/45/45 v/v/v) and (2) oxidizing solution of molybdate (prepared from 25 g of ammonium molybdate and 10 g of ceric sulfate dissolved in 900 mL of water and 100 mL of concentrated sulfuric acid), followed by heating to 300°C. Preparative chromatography was performed using silica gel from Silicycle Inc. (Quebec, Canada) (60 Å, 40–63 µm) with the indicated eluent. The solvents employed for the chromatography were High Performance Liquid Chromatography (HPLC) quality. They were evaporated under reduced pressure (rotary evaporator under vacuum generated by a system of glass filter pump). Optical rotations were measured with a JASCO P-1010 polarimeter. NMR spectra were recorded for solutions in $CDCl_3$ with a Bruker Avance III HD 600 MHz spectrometer. Proton and carbon chemical shifts are reported in ppm (δ) relative to the signal of $CDCl_3$ (δ 7.27 and 77.23 ppm for ^1H- and ^{13}C, respectively). Coupling constants (J) are reported in hertz (Hz), and the following abbreviations are used for signal multiplicities: singlet (s), doublet (d), doublet of doublets (dd), and multiplet (m). Analysis and assignments were made by COSY and HSQC experiments. The superscripts (I and II) in the ^1H NMR assignments refer to the Glc and Gal residues, respectively. High-resolution mass spectra (HRMS) were measured with an LC–MS-TOF (liquid chromatography–mass spectrometry time of flight) (Agilent Technologies) in positive and/or negative electrospray mode by the analytical platform of UQAM.

4-(2,3,4,6-Tetra-*O*-acetyl-β-D-galactopyranosyl-(1→4)-2,3,6-tri-*O*-acetyl-β-D-glucopranosyloxymethyl)-1-[3-(triethoxysilyl)propyl]-1*H*-1,2,3-triazole (3)

Propargyl lactoside **1**[12] (200 mg, 0.30 mmol, 1.0 equiv.), (3-azidopropyl)triethoxysilane **2**[13] (111 mg, 0.45 mmol, 1.5 equiv.), sodium ascorbate (12 mg, 0.06 mmol,

0.2 equiv.), and $CuSO_4 \cdot 5H_2O$ (15 mg, 0.06 mmol, 0.2 equiv.) were dissolved in pure DMF (3 mL) in a sealable microwave tube. The mixture was stirred under microwave irradiation for 30 min at 70°C. The mixture was then diluted with EtOAc and a little amount of hexane (~5% by volume: to push DMF into the aqueous phase) was added, followed by aqueous washings (3×). The organic layer was treated with an aqueous solution of EDTA 5% (2×) and brine, dried over Na_2SO_4, filtered, and concentrated. The residue was chromatographed using DCM 100% → DCM/MeOH 100:3 to give **3** as a white foam (224 mg, 82%). $R_f = 0.25$, 97:3 DCM–MeOH; $[\alpha]_D -17.2$ (c 1, $CHCl_3$). ^1H NMR ($CDCl_3$) δ: 7.49 (s, 1H, H-triazole), 5.32 (dd, 1H, $J_{3,4} = 3.5$ Hz, $J_{4,5} = 1.2$ Hz, H-4^{II}), 5.15 (dd, 1H, $J_{3,4} = J_{4,5} = 9.3$ Hz, H-4^I), 5.08 (dd, 1H, $J_{2,3} = 10.4$ Hz, $J_{1,2} = 7.9$ Hz, H-2^{II}), 4.93 (dd, 1H, $J_{2,3} = 10.4$ Hz, $J_{3,4} = 3.5$ Hz, H-3^{II}), 4.89 (dd, 1H, $J_{1,2} = J_{2,3} = 7.9$ Hz, H-2^I), 4.88 (d, 1H, $J_{H,H} = 12.6$ Hz, OCH_2), 4.76 (d, 1H, $J_{H,H} = 12.6$ Hz, OCH_2), 4.61 (d, 1H, $J_{1,2} = 7.9$ Hz, H-1^I), 4.49 (dd, 1H, $J_{6a,6b} = 12.1$ Hz, $J_{6a,5} = 2.1$ Hz, H-6a^I), 4.46 (d, 1H, $J_{1,2} = 7.9$ Hz, H-1^{II}), 4.32 (t, 1H, $J_{H,H} = 7.3$ Hz, $NCH_2CH_2CH_2Si$), 4.08 (m, 3H, H-6b^I, H-6a^{II}, H-6b^{II}), 3.85 (ddd, 1H, $J_{6a,5} = 7.4$ Hz, $J_{6b,5} = 6.3$ Hz, $J_{4,5} = 1.2$ Hz, H-5^{II}), 3.79 (q, 6H, $J_{CH2,CH3} = 7.0$ Hz, OCH_2CH_3), 3.79 (dd, 1H, $J_{2,3} = 7.9$ Hz, $J_{3,4} = 9.3$ Hz, H-3^I), 3.62 (ddd, 1H, $J_{6a,5} = 9.9$ Hz, $J_{6b,5} = 5.1$ Hz, $J_{4,5} = 2.1$ Hz, H-5^I), 2.12, 2.11, 2.03, 2.02, 2.01, 1.96, 1.94 (7×s, 21H, 7×$COCH_3$), 1.19 (t, 9H, $J_{CH2,CH3}$ 7.0 Hz, OCH_2CH_3), and 0.58 ppm (m, 2H, CH_2Si). ^{13}C NMR ($CDCl_3$) δ: 170.2, 170.2, 170.0, 170.0, 169.6, 169.5, 168.9 (CO), 143.8 (C_q-triazole), 122.6 (CH-triazole), 100.9 (C-1^{II}), 99.5 (C-1^I), 76.0 (C-3^I), 72.7 (C-4^I and C-5^I), 71.5 (C-2^I), 70.9 (C-3^{II}), 70.6 (C-5^{II}), 69.0 (C-2^{II}), 66.6 (C-4^{II}), 62.8 (OCH_2), 61.8 (C-6^I), 60.7 (C-6^{II}), 58.4 (OCH_2CH_3), 52.4 ($NCH_2CH_2CH_2Si$), 24.1 ($NCH_2CH_2CH_2Si$), 20.8, 20.7, 20.6, 20.5, 20.5, 20.5, 20.4 ($COCH_3$), 18.2 (OCH_2CH_3), and 7.4 ppm (CH_2Si). ESI$^+$–HRMS [M+H]$^+$ calcd for $C_{38}H_{60}N_3O_{21}Si$: 922.3483. Found: 922.3527. Anal. calcd for $C_{38}H_{59}N_3O_{21}Si$: C, 49.50; H, 6.45; N, 4.56. Found: C, 49.24; H, 6.36; N, 4.32.

ACKNOWLEDGMENTS

This work was supported by Natural Sciences and Engineering Research Council of Canada (NSERC) and a Canadian Research Chair in Therapeutic Chemistry to R.R and a FQRNT Quebec grant to R.R and A.A.

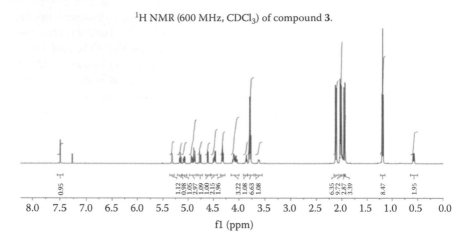

¹H NMR (600 MHz, CDCl₃) of compound **3**.

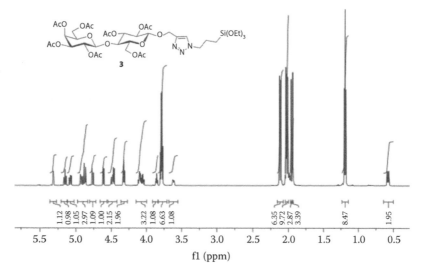

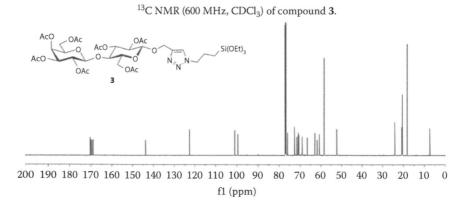

¹³C NMR (600 MHz, CDCl₃) of compound **3**.

REFERENCES

1. Trewyn, B. G.; Slowing, I. I.; Giri, S.; Chen, H.-T.; Lin, V. S. Y. *Acc. Chem. Res.*, **2007**, *40*, 846–853.
2. Wan, Y.; Zhao, D. Y. *Chem. Rev.*, **2007**, *107*, 2821–2860.
3. Kesanli, B.; Lin, W. *Chem. Commun.*, **2004**, 2284–2285.
4. Nozaki, C.; Lugmair, C. G.; Bell, A. T.; Tilley, T. D. *J. Am. Chem. Soc.*, **2002**, *124*, 13194–13203.
5. Terry, T. J.; Dubois, G.; Murphy, A.; Stack, T. D. P. *Angew. Chem. Int. Ed.*, **2007**, *46*, 945–947.
6. Schiel, J. E.; Mallik, R.; Soman, S.; Joseph, K. S.; Hage, D. S. *J. Sep. Sci.*, **2006**, *29*, 719–737.
7. Li, J.; Qi, T.; Wang, L.; Liu, C.; Zhang, Y. *Mater. Lett.*, **2007**, *61*, 3197–3200.
8. Lai, C. Y.; Trewyn, B. G.; Jeftinija, D. M.; Jeftinija, K.; Xu, S.; Jeftinija, S.; Lin, V. S. Y. *J. Am. Chem. Soc.*, **2003**, *125*, 4451–4459.
9. Nozawa, K.; Osono, C.; Sugawara, M. *Sens. Actuators B Chem.*, **2007**, *126*, 632–640.
10. (a) Azzouz, A.; Ursu, A.-V.; Nistor, D.; Sajin, T.; Assad, E.; Roy, R. *Thermochim. Acta*, **2009**, *496*, 45–49; (b) Azzouz, A.; Assad, E.; Ursu, A.-V.; Sajin, T.; Nistor, D.; Roy, R. *Appl. Clay Sci.*, **2010**, *48*, 133–137; (c) Azzouz, A.; Platon, N.; Nousir, S.; Ghomari, K.; Nistor, D.; Shiao, T. C.; Roy, R. *Sep. Purif. Technol.*, **2013**, *108*, 181–188; (d) Nousir, S.; Platon, N.; Ghomari, K.; Sergentu, A.-S.; Shiao, T. C.; Hersant, G.; Bergeron, J. Y.; Roy, R.; Azzouz, A. *J. Colloid Interface Sci.*, **2013**, *402*, 215–222; (e) Azzouz, A.; Nousir, S.; Platon, N.; Ghomari, K.; Shiao. T. C.; Hersant, G.; Bergeron, J. Y.; Roy, R. *Int. J. Greenhouse Gas Control*, **2013**, *17*, 140–147; (f) Azzouz, A.; Nousir, S.; Platon, N.; Ghomari, K.; Hersant, G.; Bergeron, J. Y.; Shiao, T. C.; Rej, R.; Roy, R. *Mater. Res. Bull.*, **2013**, *48*, 3466–3473; (g) Azzouz, A.; Aruş, V. A.; Platon, N.; Ghomari, K.; Nistor, D.; Shiao, T. C.; Roy, R. *Adsorption*, **2013**, *19*, 909–918.
11. (a) Kolb, A. C.; Finn, M. G.; Sharpless, K. B. *Angew. Chem. Int. Ed.*, **2001**, *40*, 2004–2021; (b) Hein, J. E.; Fokin, V. V. *Chem. Soc. Rev.*, 2010, *39*, 1302–1315; (c) Santoyo-González, F.; Hernández-Mateo, F. *Top. Heterocycl. Chem.*, **2007**, *7*, 133–177; (d) Meldal, M.; Tornoe C. W. *Chem. Rev.*, **2008**, *108*, 2952–3015; (e) Dondoni, A.; Marra, A. *Chem. Rev.*, **2010**, *110*, 4949–4977; (f) Dedola, S.; Nepogodiev, S. A.; Field, R. A. *Org. Biomol. Chem.*, 2007, *5*, 1006–1017; (g) Pieters, R. J.; Rijkers, D. T. S.; Liskamp, R. M. J. *QSAR Comb. Sci.*, **2007**, *26*, 1181–1190; (h) Kushwaha, D.; Dwivedi, P.; Kuanar, S. K.; Tiwari, V. K. *Curr. Org. Synth.*, **2013**, *10*, 90–135.
12. Yang, R.; Ding, H.; Song, Y.; Xiao, W.; Xiao, Q.; Wu, J. *Lett. Org. Chem.*, **2008**, *5*, 518–521.
13. Puglisi, A.; Benaglia, M.; Annunziata, R.; Chiroli, V.; Porta, R.; Gervasini, A. *J. Org. Chem.*, **2013**, *78*, 11326–11334.

16 Synthesis of 2-(Mannopyranosyloxy-methyl) Benzo[*b*] furan through Sonogashira Coupling and Intramolecular Dehydrocyclization

Mohamed Touaibia,[†] Tze Chieh Shiao, and René Roy*

CONTENTS

In recent years, benzo-fused cyclic ether skeletons have received increasing interest,[1] mainly due to their presence in a wide variety of natural products with promising

---

* Corresponding author; e-mail: roy.rene@uqam.ca.
† Checker; e-mail: mohamed.touaibia@umoncton.ca.

biological activities.[2] These include insecticidal, fungicidal, antimicrobial, antiproliferative, cytotoxic, and antioxidant properties.[3]

In our ongoing research efforts to improve and develop new mannosides with promising bioactivities,[4,5] we describe herein the synthesis of mannopyranoside **2** from prop-2-ynyl 2,3,4,6-tetra-*O*-acetyl-α-D-mannopyranoside **1**[6] and 2-iodophenol using a tandem Sonogashira cross coupling and intramolecular dehydrocyclization reactions.

In a previous work, Roy et al.[7] have demonstrated that couplings of propargyl glycosides and aryl halides could be successfully performed at 60°C in DMF using either Pd(0) or Pd(II) catalysts with Et_3N in the absence of CuI.[8] Initially, Sonogashira coupling and dehydrocyclization between **1** and 2-iodophenol was attempted following the preceding procedure. Surprisingly, these conditions did not provide **2** in any acceptable yields. Therefore, CuI addition and solvent effects were reinvestigated. It was ultimately found that $Pd(PPh_3)_2Cl_2$ in Et_3N and in the presence of CuI catalyst provided the best results.

EXPERIMENTAL METHODS

GENERAL METHODS

The reaction was carried out under nitrogen using freshly distilled solvent. After workup, the organic layer was dried over anhydrous $MgSO_4$ and concentrated at reduced pressure. Progress of the reaction was monitored by thin-layer chromatography using silica gel 60 F_{254} precoated plates (E. Merck). Optical rotation was measured with a JASCO P-1010 polarimeter. NMR spectra were recorded on a Varian Inova AS600 spectrometer. Proton and carbon chemical shifts are reported in ppm relative to the signal of $CDCl_3$. Coupling constants (J) are reported in Hz, and the following abbreviations are used: singlet (s), doublet (d), doublet of doublets (dd), triplet (t), multiplet (m), and broad (b). Analysis and assignments were made using COSY and HSQC experiments. High-resolution mass spectra (HRMSs) were measured with a liquid chromatography–mass spectrometry–time of flight (LC-MS-TOF) in (Agilent Technologies) positive and/or negative electrospray mode by the analytical platform of UQAM.

2-(2,3,4,6-TETRA-*O*-ACETYL-α-D-
MANNOPYRANOSYLOXYMETHYL)BENZO[*B*]FURAN (2)

To a solution of prop-2-ynyl 2,3,4,6-tetra-*O*-acetyl-α-D-mannopyranoside **1**[6] (386 mg, 1.0 mmol, 1.0 equiv.) in Et_3N (7.5 mL) was added $PdCl_2(PPh_3)_2$ (35 mg, 0.05 mmol, 0.05 equiv.). The mixture was flushed with argon for 5 min. CuI (9.5 mg, 0.05 mmol, 0.05 equiv.) and 2-iodophenol (330 mg, 1.5 mmol, 1.5 equiv.) were added. The mixture was flushed with argon for 10 min and stirred at reflux for 6 h. TLC* (9:1 DCM–EtOAc) indicated total consumption of starting material R_f=0.51

---

* When hexane–EtOAc (with or without toluene) is used for TLC, the starting alkyne and the final product have the same R_f.

and appearance of a major spot at R_f=0.65. The mixture was partitioned between dichloromethane and water, the organic layer was separated, and the aqueous layer was extracted with dichloromethane (2×). Combined organic layer was washed with saturated aqueous NH_4Cl, brine, and concentrated. The residue was chromatographed (7:3 hexane–EtOAc), to afford the benzofuran derivative **2** as a colorless oil (49 mg, 50%); R_f 0.20, 7:3 hexane–EtOAc; $[\alpha]_D^{20}$ = +39 (c = 1.0, $CHCl_3$). ^{1}H NMR ($CDCl_3$): δ = 7.57 (d, 1H, $J_{H,H}$ 7.9 Hz, H-arom), 7.50 (d, 1H, $J_{H,H}$ 7.9 Hz, H-arom), 7.32 (dd, 1H, $J_{H,H}$ 8.0 7.9 Hz, H-arom), 7.24 (dd, 1H, $J_{H,H}$ 7.9 7.4 Hz, H-arom), 6.75 (s, 1H. C=CH), 5.39 (dd, 1H, $J_{3,4}$ = 10.0 Hz, $J_{2,3}$ = 3.3 Hz, H-3), 5.32 (dd, 1H, $J_{2,3}$ = 3.3 Hz, $J_{1,2}$ = 1.6 Hz, H-2), 5.31 (dd, 1H, $J_{3,4}$ = 10.0 Hz, H-4), 5.00 (d, 1H, $J_{1,2}$ = 1.6 Hz, H-1), 4.76 (dd, 2H, $J_{H,H}$ = 34.0, 12.7 Hz, OCH_2), 4.30 (dd, 1H, $J_{6a,6b}$ = 12.3 Hz, $J_{6a,5}$ = 5.2 Hz, H-6a), 4.16–4.06 (m, 2H, H-5 and H-6b), 2.15, 2.10, 2.05, 1.99 ppm (4×s, 12H; $COCH_3$). ^{13}C NMR ($CDCl_3$): δ = 170.5, 169.8, 169.7, 169.6 (CO), 155.3, 152.4 (C_q-arom), 127.8 (CH_2–Cq), 124.8, 122.9, 121.2, 111.4 (CH-arom), 106.7 (C=CH), 96.9 (C-1), 69.5 (C-4), 69.0 (C-3), 68.9 (C-5), 66.1 (C-2), 62.4 (C-6), 62.0 (OCH_2), 20.8, 20.6, 20.6, 20.6 ppm (CH_3). ESI$^+$-HRMS [M+Na]$^+$ calcd for $C_{23}H_{26}O_{11}Na$: 501.1367. Found: 501.1369. Anal. calcd. for $C_{23}H_{26}O_{11}$: C, 57.74; H, 5.48. Found: C, 57.70; H, 5.55.

ACKNOWLEDGMENTS

This work was carried out with financial support from CORPAQ (Quebec, Canada) and the Natural Sciences and Engineering Research Council of Canada.

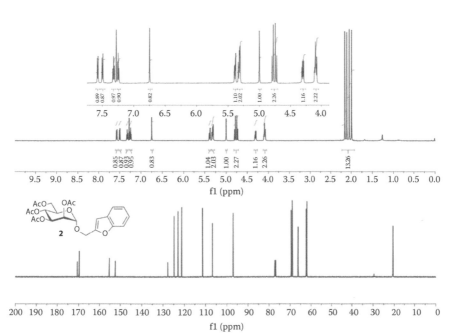

REFERENCES

1. Vyvyan, J. R.; Looper, R. E. *Tetrahedron Lett.* **2000**, *41*, 1151–1154.
2. (a) Schneiders, G. E.; Stevenson, R. *J. Org. Chem.* **1979**, *44*, 4710–4711; (b) Yang, Z.; Hon, P. M.; Chui, K. Y.; Xu, Z. L.; Chang, H. M.; Lee, C. M.; Cui, Y. X.; Wong, H. N. C.; Poon, C. D.; Fung, B. M. *Tetrahedron Lett.* **1991**, *32*, 2061–2064.
3. Ward, R. S. *Nat. Prod. Rep.* **1999**, *16*, 75–96.
4. Chabre, Y. M.; Giguère, D.; Blanchard, B.; Rodrigue, J.; Rocheleau, S.; Neault, M.; Rauthu, S. et al. *Chem. Eur. J.* **2011**, *17*, 6545–6562.
5. (a) Touaibia, M.; Roy, R. *Mini. Rev. Med. Chem.* **2007**, *7*, 1270–1283; (b) Touaibia, M.; Wellens, A.; Shiao, T. C.; Wang, Q.; Sirois, S.; Bouckaert, J.; Roy, R. *ChemMedChem.* **2007**, *2*, 1190–1201; (c) Touaibia, M.; Shiao, T. C.; Papadopoulos, A.; Vaucher, J.; Wang, Q.; Benhamioud, K.; Roy, R. *Chem. Commun.* **2007**, *4*, 380–382; (d) Fortier, S.; Touaibia, M.; Lord-Dufour, S.; Galipeau, J.; Roy, R.; Annabi, B. *Glycobiology* **2008**, *18*, 195–204; (e) Wellens, A.; Lahmann, M.; Touaibia, M.; De Greve, H.; Oscarson, S.; Roy, R.; Remaut, H.; Bouckaert, J. *Biochemistry* **2012**, *51*, 4790–4799; (f) Roos, G.; Wellens, A.; Touaibia, M.; Yamakawa, N.; Geerlings, P.; Roy, R.; Wyns, L.; Bouckaert, J. *ACS Med. Chem. Lett.* **2013**, *4*, 1085–1090.
6. Das, S. K.; Trono, M. C.; Roy, R. *Methods Enzymol.* **2003**, *362*, 3–17.
7. (a) Roy, R.; Das, S. K.; Dominique, R.; Trono, M. C.; Hernandez-Mateo, F.; Santoyo-Gonzalez, F. *Pure Appl. Chem.* **1999**, *71*, 565–571; (b) Roy, R.; Das, S. K.; Santoyo-Gonzalez, F.; Hernandez-Mateo, F.; Dam, T. K.; Brewer, C. F. *Chem. Eur. J.* **2000**, *6*, 1757–1762.
8. Liu, B.; Roy, R. J. *Chem. Soc. Perkin Trans.* **2001**, *1*, 773–779.

17 Stereocontrolled β- and α-Phosphorylations of D-Mannose

Tianlei Li, Abdellatif Tikad, Weidong Pan,
Yoan Brissonnet,[†] and Stéphane P. Vincent*

CONTENTS

As biochemical entities, glycosyl phosphates are ubiquitous in all living organisms. In particular, glycosyl-1-phosphates are of paramount importance since they are the direct precursors of nucleotide sugars, the substrates of the so-called Leloir enzymes.[1] Therefore, the stereoselective phosphorylation of carbohydrate at the anomeric position has naturally been a topic of intense research aiming at the preparation of enzyme's substrates or at the synthesis of analogues of natural metabolites.[2] The α- and β-stereoselective phosphorylations of *manno*-structures have been particularly studied due to their biochemical relevance in natural products,[3] including GDP-α-mannose,[1] the substrate of mannosyl transferases and β-D-*manno*-heptose 1-phosphate, a very important bacterial glycoside.[4] Indeed, the latter is the biosynthetic precursor of heptosides found in the lipopolysaccharide, a key component of the outer membrane of Gram-negative bacteria.

Both α- and β-stereoselective phosphorylations of D-mannosides can be achieved thanks to procedures derived from the pioneering work of Sabesan and collaborators in 1992.[5] The phosphorylation of mannoside **1** by the commercially available diphenyl chlorophosphate gives α-phosphate **2**[5,6] with a α/β selectivity of 35:1 when the reaction is conducted at −10°C (Scheme 17.1). If the reaction is performed at room temperature with an extremely slow addition of the same chlorophosphate, the kinetic β-phosphate **3**[5,7] is predominantly formed with a α/β selectivity of 1:4.

---

* Corresponding author; e-mail: stephane.vincent@fundp.ac.be.
† Checker, under the supervision of Sebastien Gouin; e-mail: sebastien.gouin@univ-nantes.fr.

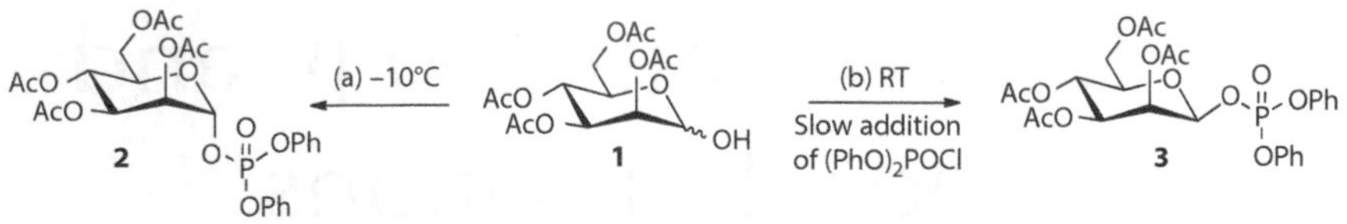

OAc
OAc
AcO
AcO
AcO
2
O
P
OPh
OPh

(a) −10°C

OAc
OAc
AcO
AcO
1
OH

(b) RT
Slow addition
of (PhO)₂POCl

OAc
OAc
AcO
AcO
3
O
P
OPh
OPh

SCHEME 17.1 *Reagents and conditions:* (a) (PhO)₂POCl, DMAP, CH₂Cl₂, α/β (35:1), −10°C (81%). (b) (PhO)₂POCl, DMAP, CH₂Cl₂, α/β (1:3.8), 22°C (69% yield for the β-phosphate).

Given the extreme difficulty to synthesize β-mannosides in general,[8] the latter anomeric stereoselectivity is significant. It should be mentioned that products **2** and **3** have limited stability when stored at room temperature. Moreover, the β-phosphate may partially decompose during the purification by silica gel chromatography. Long purification times should thus be avoided.

Zamyatina et al.[5] synthesized for the first time, according to a related procedure based on a slow addition of phosphorylating reagent, the challenging ADP-L-*glycero*-β-D-mannoheptose.[9] In our laboratory, the procedure could be performed on a multigram scale and applied, on the parent D- and L-*glycero*-β-D-mannoheptoses, for syntheses of substrates of all enzymes involved in the biosynthesis of bacterial heptose.[10] α-Phosphate **2** has also been used for the total synthesis of bleomycins,[6] and β-phosphate **3** was a key building block for the synthesis of glycolipid characteristic of *Mycobacterium tuberculosis*.[7]

EXPERIMENTAL SECTION

GENERAL METHODS

All chemicals were purchased from Sigma, Aldrich, Fluka, or Acros and were used without further purification. Dichloromethane was freshly distilled over CaH₂. NMR spectra (22°C) were recorded with a Bruker Avance 300 Ultrashield. Spectra were interpreted with the aid of ¹H–¹H and ¹H–¹³C correlation experiments. Column chromatography was performed on silica gel Kieselgel Si 60 (40–63 μm).

Diphenyl (2,3,4,6-Tetra-*O*-acetyl-α-D-mannopyranosyl)phosphate (2)

Solution A: In a 25 mL round bottom two-neck flask, a solution of 2,3,4,6-tetra-*O*-acetyl-D-mannopyranose (**1**)[5] (100 mg, 0.29 mmol, 1.0 eq) in anhydrous dichloromethane (7.5 mL) was stirred at −10°C under argon for 20 min of "cooling time."

Solution B: In a separate 25 mL round bottom flask, diphenyl chlorophosphate (0.3 mL, 1.45 mmol, 5.0 eq) was added dropwise at 0°C to a solution of DMAP (177 mg, 1.45 mmol, 5.0 eq) in anhydrous CH₂Cl₂ (1.25 mL).* The resulting mixture was diluted with anhydrous CH₂Cl₂ to reach a final volume of 2.5 mL (C=0.58 M for DMAP and diphenyl chlorophosphate).

---

* To insure a good reproducibility, solution B should be prepared during the 20 min of solution A's cooling time.

After the 20 min solution A's cooling time, solution B was added to solution A with the aid of a syringe pump (injection rate = 5 mL/h for a 5 mL syringe) at −10°C. The entire solution B was injected in 30 min, while the temperature was maintained at −10°C with a bath containing ice, brine, and dry ice.* When the addition of solution B was complete, the mixture was stirred at −10°C for 5 h, and thin-layer chromatography (TLC, 1:1 cyclohexane–EtOAc) showed disappearance of **1** and formation of **2** and **3**, both less polar than the starting material; the major product **2** showed R_f value of 0.5 and the minor β-phosphate **3** R_f 0.4). The solution was diluted with a saturated aq NaHCO$_3$ (12.5 mL) and extracted with CH$_2$Cl$_2$ (3×6 mL). The combined organic phases were dried over MgSO$_4$, filtered, and concentrated, to give a colorless oil. Chromatography (9:1→7:3 cyclohexane–EtOAc) afforded a colorless syrup[†] (138 mg, 81% yield), $[\alpha]^{20}_D = +40$ (*c* 0.3, CHCl$_3$).[‡] ¹H-NMR (300 MHz, CDCl$_3$): 7.42–7.17 (m, 10H, PO(OPh)$_2$), 5.87 (dd, J = 1.5, 6.6 Hz, 1H, H-1), 5.42–5.27 (m, 3H, H-2, H-3, and H-4), 4.19 (dd, J = 4.8, 12.4 Hz, 1H, H-6a), 4.11–4.03 (m, 1H, H-5), 3.92 (dd, J = 2.1, 12.4 Hz, 1H, H-6b), 2.16 (s, 3H, CH$_3$), 2.04 (s, 3H, CH$_3$), 2.00 (s, 3H, CH$_3$), 1.98 (s, 3H, CH$_3$). ¹³C-NMR (75 MHz, CDCl$_3$): 170.7 (1C, C=O), 169.9 (1C, C=O), 169.6 (s, 2C, C=O), 150.3 (d, 1C, J_{C-P} = 7.0 Hz, C-Ar), 150.2 (d, 1C, J_{C-P} = 7.2 Hz, C-Ar), 130.1 (s, 2C, CH-Ar), 130.1 (s, 2C, CH-Ar), 126.0 (1C, CH-Ar), 125.9 (1C, CH-Ar), 120.3 (d, 2C, CH-Ar), 120.2 (d, 2C, CH-Ar), 96.2 (d, J_{C-P} = 5.8 Hz, CH-1), 70.8 (CH-5), 68.8 (d, J_{C-P} = 11.5 Hz, CH-2), 68.3(CH-3), 65.2 (CH-4), 61.8 (CH$_2$-6), 20.8 (CH$_3$), 20.8 (CH$_3$), 20.7 (s, 2C, CH$_3$). ³¹P-NMR (75 MHz, CDCl$_3$): −14.2 ppm. HRMS: m/z: calcd for C$_{26}$H$_{29}$O$_{13}$P [M+Na]+ 634.0976, found: 634.0985.

Diphenyl (2,3,4,6-Tetra-*O*-acetyl-β-D-mannopyranosyl)phosphate (3)

In a 25 mL round bottom two-neck flash, DMAP (179 mg, 1.46 mmol, 5.0 eq.) was added under argon at room temperature to a solution of 2,3,4,6-tetra-*O*-acetyl-D-mannopyranose (**1**)[5] (100 mg, 0.29 mmol, 1.0 eq) (DMAP and mannopyranose were previously dried on the high vacuum pump for 1 h) in anhydrous dichloromethane (7.5 mL). After 15 min, a solution of diphenyl chlorophosphate (0.3 mL, 1.44 mmol, 5.0 eq.) in anhydrous dichloromethane (5 mL, C = 0.29 M) was added with the aid of a syringe pump (injection rate 0.397 mL/h with a 10 mL syringe) at room temperature (23°C). After 5 h, TLC analysis showed disappearance of **1** and the formation of **2** and **3** (see R_f values in the previous procedure) in a α/β ratio of 1:3.8 as shown by ³¹P-NMR. The solution was diluted with a saturated sodium bicarbonate solution (15 mL) and extracted with dichloromethane (3×10 mL). The combined organic phases were dried over MgSO$_4$, filtered, and concentrated to give a colorless solid. Chromatography (9:1→3:2 cyclohexane–EtOAc) afforded α-phosphate **2**

---

* The temperature must be rigorously controlled to give high and reproducible α/β ratio.
† The α-phosphate **2** slowly decomposes at room temperature. It should be stored at or below −23°C under argon.
‡ Literature data: $[\alpha]^{22}_D = +32.5 \pm 2.3$ (*c* = 0.87, CHCl$_3$, *Carbohydr. Res.*, **1992**, *223*, 169–185). 1. S. Sabesan and S. Neira, *Carbohydr. Res.*, 1992, **223**, 169–185, and $[\alpha]^{22}_D = +40$ (*c* = 0.5, CHCl$_3$, *J. Am. Chem. Soc.*, **1995**, *117*, 7344–7356).

(29 mg, 18%) and β-phosphate **3*** (115 mg, 69%) as colorless syrups, $[\alpha]^{20}{}_D = -0.5$ (*c* 0.2, CHCl$_3$). ^{1}H-NMR (400 MHz, CDCl$_3$): 7.40–7.10 (m, 10H, PO(OPh)$_2$, 5.59 (dd, *J* = 1.0 Hz, *J* = 7.1 Hz, 1H, H-1), 5.48 (dd, *J* = 1.0 Hz, 3.3 Hz, 1H, H-2), 5.25 (t, *J* = 9.6 Hz, 1H, H-4), 5.07 (dd, *J* = 3.3 Hz, 9.6 Hz, 1H, H-3), 4.26 (dd, *J* = 5.5 Hz, 12.3 Hz, 1H, H-6a), 4.12 (dd, *J* = 2.7 Hz, 12.3 Hz, 1H, H-6b), 3.82–3.75 (m, 1H, H-5), 2.10 (s, 3H, CH$_3$), 2.05 (s, 6H, 2xCH$_3$), 1.98 (s, 3H, CH$_3$). ^{13}C-NMR (100 MHz, CDCl$_3$): 170.7 (1C, C=O), 169.9 (1C, C=O), 169.8 (1C, C=O), 169.7 (1C, C=O), 150.5 (d, 1C, $J_{C\text{-}P}$ = 7.8 Hz, C-Ar), 150.1 (d, 1C, $J_{C\text{-}P}$ = 7.8 Hz, C-Ar), 130.0 (s, 2C, CH-Ar), 129.8 (s, 2C, CH-Ar), 126.0 (1C, CH-Ar), 125.8 (1C, CH-Ar), 120.6 (d, 2C, CH-Ar), 120.4 (d, 2C, CH-Ar), 94.9 (d, $J_{C\text{-}P}$ = 4.6 Hz, CH-1), 73.3 (CH-5), 70.3 (CH-3), 68.2 (d, $J_{C\text{-}P}$ = 8.9 Hz, CH-2), 65.6 (CH-4), 62.2 (CH-6), 20.8 (s, 2C, CH$_3$), 20.7 (CH$_3$), 20.6 (CH$_3$). ^{31}P-NMR (100 MHz, CDCl$_3$): −13.9 ppm. HRMS: m/z: calcd for C$_{26}$H$_{29}$O$_{13}$P [M+Na]+ 634.0976, found: 634.0984.

ACKNOWLEDGMENTS

This work was funded by FNRS (PDR grant T.0170.13). PhD thesis fellowship for T.L. was funded by the China Scholarship Council.

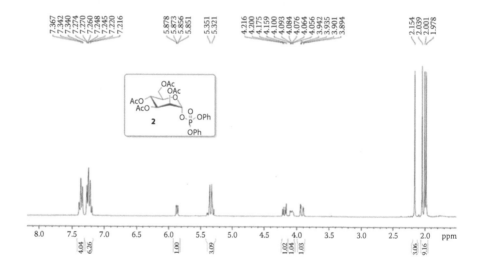

---

* The β-phosphate **3** is less stable than α-phosphate **2**. When stored at room temperature, it decomposed either by anomerization or by phosphate removal. During the silica gel chromatography, we also found that it could decompose if the purification time was too long (typically less than 3 h). The purified material was stored at −23°C under argon for weeks without noticeable decomposition.

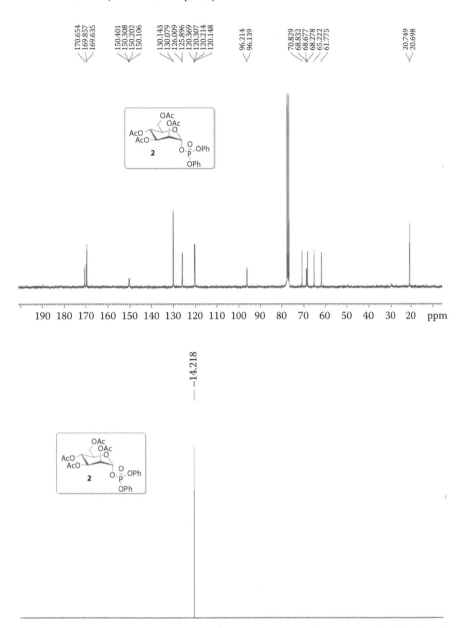

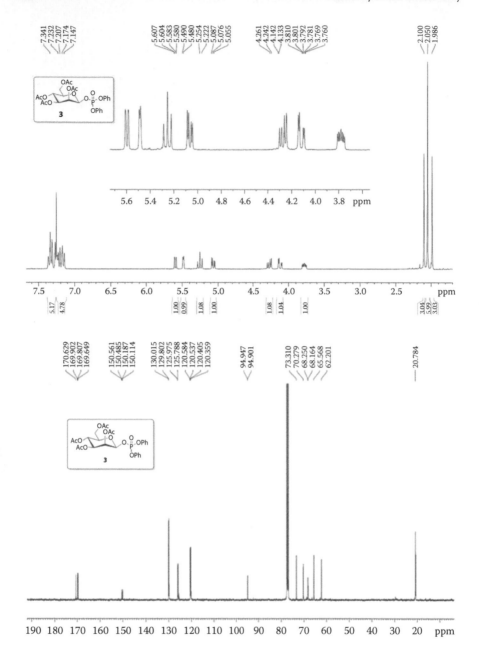

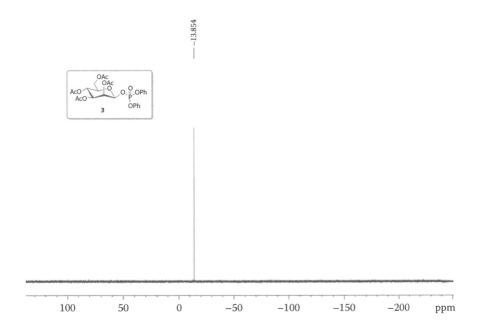

REFERENCES

1. G. K. Wagner, T. Pesnot, and R. A. Field, *Nat. Prod. Rep.* **2009**, *26*, 1172–1194.
2. A. V. Nikolaev, I. V. Botlinko, and A. J. Ross, *Carbohydr. Res.* **2007**, *342*, 297–344.
3. (a) D. B. Moody, T. Ulrichs, W. Muehlecker, D. C. Young, S. S. Gurcha, E. Grant, J.-P. Rosat et al., *Nature* **2000**, *404*, 884–888; (b) D. Crich and V. Dudkin, *J. Am. Chem. Soc.* **2002**, *124*, 2263–2266; (c) D. Majumdar, G. A. Elsayed, T. Buskas, and G.-J. Boons, *J. Org. Chem.* **2005**, *70*, 1691–1697; (d) D. Crich and S. Picard, *J. Org. Chem.* **2009**, *74*, 9576–9579; (e) A. Ravida, X. Liu, L. Kovacs, and P. H. Seeberger, *Org. Lett.* **2006**, *8*, 1815–1818; (f) S. Sanyal and A. K. Menon, *Proc. Natl. Acad. Sci. USA.* **2010**, *107*, 11289–11294.
4. A. Tikad and S. P. Vincent, Synthetic methodologies towards aldoheptoses and their applications to the synthesis of biochemical probes and LPS fragments, Eds.: S. Vidal and D. Werz, in *Modern Synthetic Methods in Carbohydrate Chemistry*, Wiley-VCH, Berlin, Germany, **2013**, chapter 2, pp: 29–65.
5. S. Sabesan and S. Neira, *Carbohydr. Res.* **1992**, *223*, 169–185.
6. D. L. Boger, S. Teramoto, and J. Zhou, *J. Am. Chem. Soc.* **1995**, *117*, 7344–7356.
7. R. P. van Summeren, D. B. Moody, B. L. Feringa, and A. J. Minnaard, *J. Am. Chem. Soc.* **2006**, *128*, 4546–4547.
8. D. Crich and S. Sun, *J. Am. Chem. Soc.* **1998**, *120*, 435–436.
9. A. Zamyatina, S. Gronow, C. Oertelt, M. Puchberger, H. Brade, and P. Kosma, *Angew. Chem. Int. Ed.* **2000**, *39*, 4150–4153; A. Zamyatina, S. Gronow, M. Puchberger, A. Graziani, A. Hofinger, and P. Kosma, *Carbohydr. Res.* **2003**, *338*, 2571–2789.
10. (a) H. Dohi, R. Périon, M. Durka, M. Bosco, Y. Roué, F. Moreau, S. Grizot, A. Ducruix, S. Escaich, and S. P. Vincent, *Chem. Eur. J.* **2008**, *14*, 9530–9539; (b) M. Durka, K. Buffet, J. Iehl, M. Holler, J.-F. Nierengarten, and S. P. Vincent, *Chem. Eur. J.* **2012**, *18*, 641–651; (c) M. Durka, A. Tikad, R. Périon, M. Bosco, M. Andaloussi, S. Floquet, E. Malacain et al., *Chem. Eur. J.* **2011**, *17*, 11305–11313.

Section II

Synthetic Intermediates

18 Improved Synthesis of Ethyl 1-Thio-α-D-galactofuranoside

Anushka B. Jayasuiya, Wenjie Peng,
Laure Guillotin,[†] and Todd L. Lowary*

CONTENTS

Galactose is an important constituent of many glycoconjugates, existing in the pyranose form (Galp) in mammalian cells and in both the pyranose and furanose (Galf) forms in bacteria, fungi, and protozoa.[1,2] Many of the later are pathogenic, and Galf residues can be found in, for example, the galactomannan of *Aspergillus fumigatus*; mucin-like proteins in *Trypanosoma cruzi*, the causative agent of Chagas disease; galactan I of *Klebsiella pneumoniae*; the O-antigen of *Escherichia coli* and *K. pneumonia* lipopolysaccharide; and the arabinogalactan of *Mycobacteria tuberculosis*, the causative agent of tuberculosis.

   The chemical synthesis of putative substrates and substrate analogs is essential to elucidating the function of the biosynthetic enzymes responsible for the assembly of Galf-containing polysaccharides, as well as identifying leads for drug development that inhibit these processes. In nature, Galf residues are biosynthesized at

---

* Corresponding author; e-mail: tlowary@ualberta.ca.
[†] Checker, under supervision of Richard Daniellou; e-mail: richard.daniellou@univ-orleans.fr.

the sugar nucleotide level by an enzymatic ring contraction of uridine diphospho-galactopyranose (UDP-Gal*p*) to UDP-Gal*f*.[1] A number of synthetic methods have been developed for accessing the thermodynamically less stable Gal*f* ring form from galactose, which exists predominantly in the pyranose ring form.[3-12] These include kinetically controlled Fischer glycosylations,[4,5] high-temperature perben-zoylation of galactose in pyridine,[6] reduction of D-galactonolactones with borane reagents,[7] reduction of furanoside derivatives of D-galacturonic acid with sodium borohydride and iodine,[8] and electrophile-induced cyclization of dithioacetals or *O,S*-mixed acetals.[9-12] We provide here an optimized procedure for the synthe-sis of a Gal*f* thioglycoside (**2**) *via* the kinetically favored 5-*exo*-trig ring closure of D-galactose diethyl dithioacetal (**1**), promoted by mercuric chloride, a method developed originally in the 1930s by Pacsu and Wilson[10] and then later refined by Wolfrom et al.[13] This procedure, which involved careful monitoring of the pH dur-ing the reaction, allowed the yield to be increased to 71% from the previous report of 48%.[13] Products obtained after appropriate protection of the hydroxyl groups can be used as donors for the synthesis of Gal*f*-containing glycoconjugates,[13-16] as well as more complex monosaccharides.[17]

EXPERIMENTAL METHODS

GENERAL METHODS

Reactions were carried out in oven-dried glassware.* All reagents used were pur-chased from commercial sources and were used without further purification. The reaction was monitored by TLC on Silica Gel 60 F_{254} (0.25 mm, E. Merck). Spots were detected by charring with acidified *p*-anisaldehyde solution in ethanol.† The ^1H NMR spectrum was recorded at 600 MHz and chemical shifts were referenced to $CDCl_3$ at 7.26 ppm. The ^1H data are reported as though they were first order. The ^{13}C NMR (APT) spectrum was recorded at 125 MHz, and ^{13}C chemical shifts were referenced to internal $CDCl_3$ (77.06, $CDCl_3$). Assignments of resonances in NMR spectra were done using $^1H-^1H$ COSY and HMQC experiments.

Ethyl 1-Thio-α-D-galactofuranoside (2)

To a solution of D-galactose diethyl dithioacetal (**1**)[18] (20.0 g, 0.0699 mol) in water (400 mL) was added a solution of mercury (II) chloride (24.7 g, 0.0909 mol) in water (290 mL),‡ and then 1.0 M NaOH was added portion-wise, to maintain a solution pH of 6–7.§ Once disappearance of the starting material was observed by TLC (~3 h), the excess mercury (II) chloride was filtered, and the filtrate was con-centrated under reduced pressure.¶ The resulting residue was purified by column

---

* The entire operation should be done in an efficient fume hood.
† Prepared by dissolving *p*-anisaldehyde (5 mL), glacial acetic acid (2 mL), and concentrated sulfuric acid (7 mL) in 95% ethanol (186 mL).
‡ If the reaction is done under more dilute conditions, hydrolysis, leading to the formation of D-galactose, is observed, leading to reduced yield.
§ Reduced yields are observed if the pH varies outside this range.
¶ 1 mL of octanol was added to prevent foaming.

chromatography (6:1 CH_2Cl_2–CH_3OH) to give ethyl 1-thio-α-ᴅ-galactofuranoside
(2) (11.2 g, 71%) as an oil, $[\alpha]_D^{21}$ +108.7 (c 1.20, H_2O), Reference $[\alpha]_D^{25}$ +124
(c 1.36, H_2O).[13] Attempts to further dry the compound did not result in better
agreement with the reported rotation. ^{1}H NMR (500 MHz, D_2O, δ_H), 5.40 (d,
1H, J = 5.4 Hz, H-1), 4.27 (app t, 1H, J = 5.4 Hz, H-2), 4.18 (dd, 1H, J = 5.4, 5.1
Hz, H-3), 3.85–3.87 (m, 1H, H-5), 3.80 (app t, 1H, J = 5.1 Hz, H-4), 3.71 (dd, 1H,
J = 4.2, 11.4 Hz, H-6a), 3.62 (dd, 1H, J = 7.2, 11.4 Hz, H-6b), 2.74 (q, 2H, J = 7.2
Hz, CH_3CH_2), 1.28 (t, 1 H, J = 7.2 Hz, CH_3CH_2). ^{13}C NMR (125 MHz, $CDCl_3$, δ_C):
87.5 (C-1), 83.7 (C-4), 77.6 (C-2), 76.8 (C-3), 71.7 (C-5), 62.6 (C-6), 25.0 (CH_2),
14.5 (CH_3) (Figures 18.1 and 18.2).

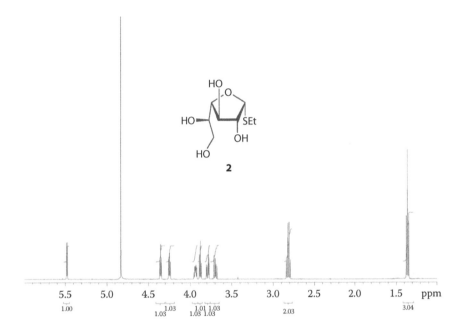

FIGURE 18.1 ^{1}H NMR spectrum (500 MHz, D_2O) of compound **2**.

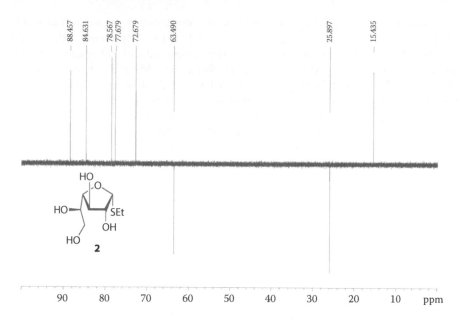

FIGURE 18.2   ¹³C NMR spectrum (125 MHz, D₂O) of compound **2**.

REFERENCES

1. M. R. Richards, T. L. Lowary, *ChemBioChem.* **2009**, *10*, 1920–1938.
2. C. Marino, C., C. Gallo-Rodriguez, R. M. de Lederkremer, Galactofuranosyl-containing glycans: Occurrence, synthesis and biochemistry, in *Glycans: Biochemistry, Characterization and Applications*, H. M. Mora-Montes, Ed. Nova Science Publishers: Hauppage, New York, **2012**; pp. 207–268.
3. A. Imamura, T. L. Lowary, *Trends Glycosci. Glycotechnol.* **2011**, *23*, 134–152.
4. A. Arasappan, B. Fraser Reid, *Tetrahedron Lett.* **1995**, *36*, 7967–7970.
5. R. Velty, T. Benvegnu, M. Gelin, E. Privat, D. Plusquellec, *Carbohydr. Res.* **1997**, *299*, 7–14.
6. N. B. D'Accorso, I. M. E. Thiel, M. Schuller, *Carbohydr. Res.* **1983**, 124, 177–184.
7. P. Kohn, R. H. Samaritano, L. M. Lerner, *J. Am. Chem. Soc.* **1965**, *87*, 5475–5480.
8. A. Bordoni, R. M. de Lederkremer, C. Marino, *Tetrahedron* **2008**, *64*, 1703–1710.
9. G. C. Completo, T. L. Lowary, *J. Org. Chem.* **2008**, *73*, 4513–4525.
10. E. Pacsu, E. J. Wilson, *J. Am. Chem. Soc.* **1938**, *60*, 2288–2289.
11. J. C. McAuliffe, O. Hindsgaul, *J. Org. Chem.* **1997**, *62*, 1234–1239.
12. S. K. Madhusudan, A. K. Misra, *Carbohydr. Res.* **2005**, *340*, 497–502.
13. M. L. Wolfrom, Z. Yosizawa, B. O. Juliano, *J. Org. Chem.* **1959**, *24*, 1529–1530.
14. P. K. Mandal, A. K. Misra, *Synthesis* **2007**, 2660–2666.
15. J. M. Frechet, H. H. Baer, *Can. J. Chem.* **1975**, *53*, 670–679.
16. H. M. Zuurmond, P. A. M. Van der Klein, G. H. Veeneman, J. H. Van Boom, *Recl. Trav. Chim. Pays Bas* **1990**, *109*, 437–441.
17. W. Peng, A. B. Jayasuriya, A. Imamura, T. L. Lowary, *Org. Lett.* **2011**, *13*, 5290–5293.
18. P. Norris, D. Horton, Dialkyl dithioacetals of sugars, in *Preparative Carbohydrate Chemistry*, S. Hanessian, Ed. Marcel Dekker: New York, **1997**; pp. 35–52.

19 Improved Large-Scale Synthesis of β-D-Arabinofuranose 1,2,5-Orthobenzoate

Nikita M. Podvalnyy, Alexander I. Zinin,
Boddu Venkateswara Rao,[†] and
Leonid O. Kononov[*]

CONTENTS

Arabinofuranose 1,2,5-orthobenzoate[1–6] (**5**, see Scheme 19.1) is a convenient precursor for a number of mono- and oligosaccharide building blocks,[7–9] which are useful for the synthesis of oligosaccharides related to the fragments of mycobacterial lipoarabinomannan and arabinogalactan.[10]

Recently, we have undertaken a comparison[6] of the existing synthetic routes to **5** and have published an *overall* procedure for the reliable preparation of the tricyclic orthoester **5** starting from the readily available[11–13] methyl 2,3,5-tri-*O*-benzoyl-α-D-arabinofuranoside (**1**). Methyl glycoside **1** is converted into the bromide **2**,[11,12] which is treated with methanol in presence of an appropriate base. The benzoyl groups in the bicyclic 1,2-orthobenzoate **3** thus obtained are removed under basic conditions, and diol **4** is then converted into tricyclic orthoester **5** in the presence of an acid.

---

[*] Corresponding author; e-mail: kononov@ioc.ac.ru.
[†] Checker, under the supervision of Srinivas Hotha; e-mail: s.hotha@iiserpune.ac.in.

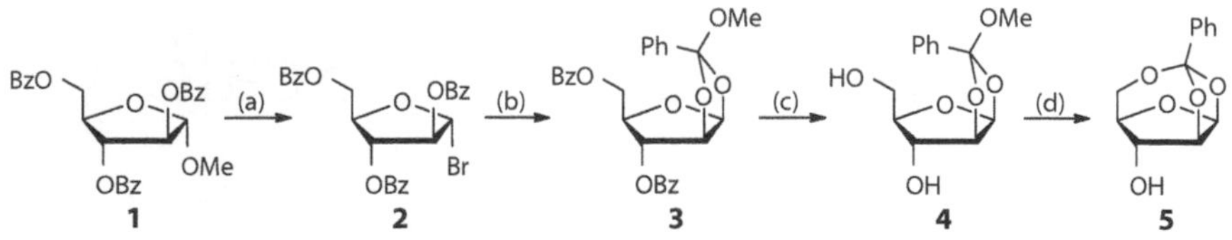

SCHEME 19.1 *Reagents and conditions*: (a) AcBr, MeOH, CH$_2$Cl$_2$, 0°C; (b) 2,4,6-collidine, MeOH, ~20°C; (c) MeONa, MeOH, ~20°C; and (d) TsOH, C$_2$H$_4$Cl$_2$, reflux with 4 Å molecular sieves in a Soxhlet-type adapter (88% from **1**).

Herein, we report a detailed, improved procedure of synthesis of orthobenzoate **5**. For an effective synthesis of bromide **2**, we used a procedure that relied on the use of HBr in relatively nonpolar CH$_2$Cl$_2$ (generated *in situ* from AcBr and MeOH)[14–17,*] rather than HBr in AcOH described in the original procedure[11,12] for the preparation of glycosyl bromide **2**.[†] The use of CH$_2$Cl$_2$ as the solvent significantly simplifies isolation of glycosyl bromide **2** (no need to neutralize excess AcOH) without compromising overall yield of **2**, which can be used in subsequent steps without crystallization.[‡] Another essential point we wish to emphasize is the importance of using 2,4,6-collidine instead of 2,6-lutidine[6] as the base to effect conversion of bromide **2** to orthoester **3**. When lutidine was used,[6] the organic layer turned black during the aqueous workup (washing with NaHCO$_3$ solution). Liquid–solid extraction of the orthoester **3** from the insoluble black by-product and subsequent filtration of the extract through a layer of silica gel were required. This significantly complicated the workup and slightly lowered the yield. When collidine is used, no colored products are formed and, after aqueous washings, the organic layer can be concentrated and used in the next step without further purification. Contrary to our recently published procedure,[6] here, we totally exclude benzene[§] and use toluene instead of benzene for chromatographic purification of diol **4** as well as for crystallization of the target orthobenzoate **5**. The yield of **5** obtained by this method is 88% over 4 steps.

---

* Similar procedures have been described for generation of other labile glycosyl bromides under mild conditions.[15,17]

† There might be a mistake in the original procedure.[11,12] According to our experience, crystalline methyl 2,3,5-tri-*O*-benzoyl-α-D-arabinofuranoside (**1**) shows limited solubility in AcOH and does not dissolve at *room temperature* (20°C) at the concentration indicated by the authors.

‡ The earlier described procedure for the preparation of **2**[12] relies on the use of large volumes of aqueous solutions (e.g., 1 L of water on a 10 g scale), which makes scale-up difficult, and leads to a low yield of pure bromide **2** (36% of α-bromide **2α** and 2% of β-bromide **2β** after crystallization along with other crystalline fractions containing unspecified admixtures). Using the data provided,[12] the total yield of all crystalline fractions of the bromide **2** obtained by this procedure can be roughly estimated as ~75%. The low yield of pure bromide **2α** obtained seems to be related not only to the losses during crystallization. In our hands, crystallization of the crude bromide **2** (1.0 g), prepared as described here in almost quantitative yield but crystallized as reported earlier,[12] gave crystalline α-bromide **2α** (829 mg, 83%) and β-bromide **2β** (38 mg, 4%) containing minor admixture of α-isomer **2α** (2%–3% according to ¹H NMR).

§ Benzene is a proven carcinogen and its use should be avoided.

EXPERIMENTAL METHODS

GENERAL METHODS

The reactions were performed using purified solvents and commercial reagents (Aldrich and Fluka). Methyl 2,3,5-tri-O-benzoyl-α-D-arabinofuranoside (**1**) was synthesized according to the known[13] procedure. Column chromatography was performed on silica gel 60 (40–63 μm, Merck). Thin-layer chromatography was carried out on silica gel 60 F_{254}–coated aluminum foil (Merck). Spots were visualized by heating the plates after immersion in a 1:10 (v/v) mixture of 85% aqueous H_3PO_4 and 95% EtOH. The ^{1}H and ^{13}C NMR spectra were recorded for solutions in CDCl$_3$ with a Bruker AM-300 instrument (300.13, and 75.48 MHz, respectively). The ^{1}H chemical shifts are referenced to the signal of the residual CHCl$_3$ (δ_H 7.27) and the ^{13}C chemical shifts are referenced to the central signal of CDCl$_3$ (δ_C 77.0). High-resolution mass spectrum (electrospray ionization, ESI-MS) was measured in a positive mode on a Bruker micrOTOF II mass spectrometer for 2×10^{-5} M solution in MeCN. Optical rotation was measured at 25°C using a PU-07 automatic digital polarimeter (Russia).

2,3,5-TRI-O-BENZOYL-α-D-ARABINOFURANOSYL BROMIDE (2)

Methyl 2,3,5-tri-O-benzoyl-α-D-arabinofuranoside[13] (**1**) (20.0 g, 42.0 mmol) was dissolved in anhydrous CH$_2$Cl$_2$ (120 mL) and cooled to 0°C (ice–water bath), then acetyl bromide (17.25 mL, 233 mmol) was added in one portion. A solution of MeOH (7.65 mL, 189 mmol) in CH$_2$Cl$_2$ (20 mL) was added dropwise from a pressure-equalising dropping funnel during 15 min. The reaction flask was equipped with a Bunsen valve,* and the reaction mixture was stirred at 0°C (ice–water bath) for 2 h.

The reaction mixture was diluted with CH$_2$Cl$_2$ (~350 mL) and transferred, through a wide plain conical funnel (diameter, 100 mm) filled with large pieces of ice, into a 2 L separatory funnel containing ice-cold water (~400 mL) and large amount of crushed ice (total volume of the funnel content is ~600 mL).[†,‡] The reaction flask and the plain conical funnel were washed with small portions of CH$_2$Cl$_2$ (3×20 mL), and the washings were added into the separatory funnel. The mixture in the separatory funnel was immediately shaken vigorously (~100 quick shakings), and the organic (lower) layer was drained immediately after the separation of layers into a beaker containing ice-cold water (500 mL) with large pieces of ice (total volume ~600 mL). The ice remaining in the plain conical funnel was also transferred into the beaker. Aqueous layer that remained in the separatory funnel was extracted with CH$_2$Cl$_2$ (50 mL), and the organic layer from this second extraction[§] was combined with the extract in the beaker. The content of the beaker was transferred into the

---

* To allow excess HBr to escape. This operation must be carried out in a well-ventilated hood.

† Workup of the reaction mixture requires a 2 L separatory funnel, a 2 L thick-walled porcelain beaker with a handle (handling more than 1 L of liquids with ice using a glass beaker, prone to breakage, would be less convenient), two 100 mm wide conical plain funnels, and a 2 L round-bottom flask.

‡ Ice should always be present in the separatory funnel and the beaker (see below) during all steps of aqueous extraction!

§ The aqueous layer after the second extraction was discarded.

same separatory funnel and additional amount of crushed ice (~100 mL) was added, the mixture in the separatory funnel was immediately shaken vigorously (~100 quick shakings), and, immediately after the separation of layers, the organic layer was poured into the same beaker containing ice-cold saturated aqueous $NaHCO_3$ solution (~500 mL) with large pieces of ice (total volume ~700 mL). The ice remaining in the plain conical funnel was also transferred into the beaker. The aqueous layer remaining in the separation funnel was extracted with CH_2Cl_2 (50 mL), and the organic layer from this second extraction was added to the content of the beaker. The content of the beaker was transferred into the same (empty) separatory funnel followed by an additional amount of crushed ice (~100 mL), the beaker and the plain conical funnel with the remaining pieces of ice were washed with CH_2Cl_2 (3 × 20 mL), and the washings were added into the separatory funnel. The mixture in the separatory funnel was well shaken*,† (~100 quick shakings), and, after the separation of the layers, the organic layer was filtered through a cotton plug‡ into a 2 L round-bottom flask. The basic (pH ~8) aqueous layer§ was washed with CH_2Cl_2 (50 mL), the organic layer from this extraction was also filtered through the same cotton plug into the 2 L round-bottom flask, and the cotton plug was washed with CH_2Cl_2 (3 × 20 mL).¶ The combined organic extracts (~700 mL) in the 2 L flask were concentrated under reduced pressure (bath temperature 30°C) using rotary evaporator and the residue was dried *in vacuo* at ~20°C to give 2,3,5-tri-*O*-benzoyl-α-D-arabinofuranosyl bromide (**2**) as a white foam** (22.58 g), which was used in the next step without further purification.

3,5-Dɪ-*O*-ʙᴇɴᴢᴏʏʟ-β-ᴅ-ᴀʀᴀʙɪɴᴏꜰᴜʀᴀɴᴏꜱᴇ 1,2-(ᴍᴇᴛʜʏʟ)ᴏʀᴛʜᴏʙᴇɴᴢᴏᴀᴛᴇ (3)

To a 2 L round-bottom flask containing the foamy crude glycosyl bromide **2** obtained as described earlier (22.58 g, 42.0 mmol), anhydrous 2,4,6-collidine (20 mL, 180 mmol) was added. Immediately after dissolution of most of **2**, anhydrous MeOH (100 mL) was added in one portion; the flask was flushed with argon and stoppered. The mixture was manually stirred until homogeneous†† and kept at ~20°C for 1 h. The volatiles were evaporated under reduced pressure (35–40°C (bath)/~10 mbar), the residue was dissolved in CH_2Cl_2 (400 mL) and washed with water (400 mL), 1 M aqueous $NaHSO_4$ (200 mL), and saturated aqueous $NaHCO_3$ (400 mL). Additional extraction of the aqueous layers with CH_2Cl_2 (50 mL each) was performed after

---

* All aqueous extractions before neutralization must be performed within 15 min because the bromide **2** is unstable in aqueous acid.
† Since the pressure in the separatory funnel would increase due to liberation of CO_2, precautions should be undertaken and protective wear should be used! The stopcock of the separatory funnel must be periodically opened (when in upward position) to equalize the pressure.
‡ Placed in another 100 mm plain conical funnel.
§ After neutralization of the acid, it is vitally important to ensure at this point that the aqueous layer is no longer acidic. Otherwise, the second washing with the $NaHCO_3$ solution is required.
¶ The overall amount of CH_2Cl_2 used for all extractions and washings was 550–600 mL.
** The resulting arabinofuranosyl bromide **2** forms voluminous white foam, which tends to fill all the flask volume during evacuation.
†† A slight exothermic effect of the reaction was observed.

every aqueous extraction.* The combined organic extracts were filtered through a cotton plug, concentrated under reduced pressure, and dried *in vacuo*, to give dibenzoate **3** (21.64 g, R_f=0.63, toluene–EtOAc, 10:1) as a slightly yellow syrup, which was used in the next step without additional purification.

β-D-ARABINOFURANOSE 1,2-(METHYL)ORTHOBENZOATE (4)

To a 1 L round-bottom flask containing the foregoing dibenzoate **3** (21.64 g, 42.0 mmol), anhydrous MeOH (100 mL) was added, followed by methanolic sodium methoxide (1 M, 4.5 mL, 4.5 mmol), and the mixture was manually stirred until homogeneous.† The mixture was kept at ~20°C for 16 h, Dowex 50W×8 resin ($HNEt_3^+$ form, 4.5 g) was added, and mixture was stirred manually for a few minutes.‡ The mixture was filtered and the solids were thoroughly washed with MeOH (several 10–20 mL portions, the total amount of MeOH used for the washing was 170 mL).§ The combined filtrate was concentrated under reduced pressure (bath temperature 35°C–40°C) and dried *in vacuo*. A solution of the syrupy residue in toluene that contained 0.2% NEt_3 (50 mL) was purified by chromatography¶ (stepwise gradient elution: 10% acetone in toluene + 0.2% NEt_3 → 40% acetone in toluene + 0.2% NEt_3 with stepwise increment of 10% of acetone per each 200 mL of eluent). The fractions containing orthoesters **4** and **5** were combined,** concentrated under reduced pressure (bath temperature 35°C), and dried *in vacuo* (~0.15 mbar) for 1 h†† to give a mixture (12.26 g) of β-D-arabinofuranose 1,2-(methyl)orthobenzoate (**4**) (R_f=0.32, toluene–acetone, 3:2) and β-D-arabinofuranose 1,2,5-orthobenzoate (**5**) (R_f=0.51, toluene–acetone, 3:2) as a colorless syrup.

β-D-ARABINOFURANOSE 1,2,5-ORTHOBENZOATE (5)

Solution of TsOH in 1,2-Dichloroethane

A 50 mL flask containing mixture of TsOH·H_2O‡‡ (7.2 mg, 0.038 mmol) and anhydrous 1,2-dichloroethane (12 mL) was equipped by a small Soxhlet-type adapter

---

* To minimize the time of contact between the acid labile product and the $NaHSO_4$ solution, the *organic layer* was poured, immediately after the corresponding extraction, into a beaker containing saturated aqueous $NaHCO_3$ solution.

† A magnetic stirring bar is ineffective in keeping the syrupy starting material in motion.

‡ Magnetic stirring should not be applied, lest the resin might disintegrate into fine particles, which are difficult to remove.

§ The washing should continue until the characteristic methyl benzoate odor can no longer be detected.

¶ The column with silica gel (v=200 mL) was washed with 0.2% NEt_3 in toluene (200 mL) prior to chromatography.

** Orthoester **4** undergoes partial spontaneous cyclization into **5** during silica gel chromatography despite the presence of NEt_3.

†† Prolonged drying of this mixture in high vacuum is not recommended because the orthoester **4** starts to undergo cyclization to tricyclic orthoester **5** with concomitant release of methanol under these conditions.

‡‡ Equimolar amount of camphorsulfonic acid might be used instead of anhydrous TsOH.

with the thimble filled with 4 Å molecular sieves* and a reflux condenser, and the solution was refluxed[†] under Ar for 15 min[‡] and then cooled to ~20°C to give a solution of anhydrous TsOH in 1,2-dichloroethane.

Cyclization

A mixture of the foregoing β-D-arabinofuranose 1,2-(methyl)orthobenzoate (**4**) and β-D-arabinofuranose 1,2,5-orthobenzoate (**5**) (12.26 g) was co-concentrated with anhydrous toluene (3×20 mL), to remove traces of NEt$_3$, dried *in vacuo* for 30 min, dissolved in anhydrous 1,2-dichloroethane (100 mL), and refluxed for 30 min under Ar with the Soxhlet-type adapter, which was used for preparation of TsOH solution,[§] equipped with a reflux condenser. Heating was discontinued, and the Soxhlet-type adapter was removed as soon as the boiling stopped, the previously prepared solution of TsOH was added in one portion, the Soxhlet-type adapter was immediately replaced, and the mixture was refluxed under Ar for an additional 20 min and cooled to ~20°C. Pyridine (0.2 mL, 2.4 mmol) was added and the mixture was concentrated to dryness under reduced pressure (bath temperature ~35°C). The residue was dissolved in hot ethyl acetate (30 mL) and cooled to 20°C and toluene (50 mL) was added, followed by slow addition of light petroleum (100 mL). The crystallized product was filtered off and dried in air and then *in vacuo* to give β-D-arabinofuranose 1,2,5-orthobenzoate (**5**) as white crystals. Yield 8.79 g (88% from methyl glycoside **1**), mp 147°C–150°C (recrystallized from EtOAc–light petroleum), [α]$_D$ −28.2 (*c* 1.0 CHCl$_3$) [lit.[1] 148°C–149°C, for L-enantiomer; (lit.[4] −28 (*c* 0.94, CHCl$_3$)]. ^{1}H NMR (CDCl$_3$) δ: 7.61–7.68 (m, 2 H, Ph), 7.37–7.42 (m, 3 H, Ph), 6.13 (d, 1 H, $J_{1,2}$ 3.8 Hz, H-1), 4.71 (dd, 1 H, $J_{1,2}$ 3.8 Hz, $J_{2,3}$ 1.5 Hz, H-2), 4.50 (d, 1 H, $J_{3,OH}$ 7.2 Hz, H-3), 4.39–4.43 (m, 1 H, H-4), 4.14 (dd, 1 H, $J_{5a,5b}$ 13.0 Hz, $J_{4,5b}$ 1.1 Hz, H-5b), 4.06 (dd, 1 H, $J_{5a,5b}$ 13.0 Hz, $J_{4,5a}$ 3.0 Hz, H-5a), 2.10 (d, 1 H, $J_{3,OH}$ 7.2 Hz, 3-OH). ^{13}C NMR (CDCl$_3$) δ: 135.9, 129.5, 128.1, 125.9 (Ph), 121.1 (PhC), 103.7 (C-1), 83.8, 83.7 (C-2, C-4), 77.2 (C-3), 68.3 (C-5). ESI-MS [M+K]$^+$ calcd for C$_{12}$H$_{12}$O$_5$: 275.0316. Found: 275.0324. Anal. calcd for C$_{12}$H$_{12}$O$_5$: C, 61.01; H, 5.12. Found: C, 61.07; H, 5.27.

ACKNOWLEDGMENTS

This work was supported by the Russian Foundation for Basic Research (Project Nos. 13-03-00666, 14-03-31479) and by the president program of supporting of leading scientific groups and young scientists (Project No. MK-6405.2014.3).

---

* A 30 mL pressure-equalising dropping funnel filled with freshly activated 4Å molecular sieves (15 g, 8–12 mesh, Aldrich) placed over a small silanized glass wool (Serva) plug was used as a substitute for Soxhlet-type adapter.

[†] It is very convenient to perform refluxing by careful heating of the flask with a stream of hot air from a powerful heat gun while stirring with a magnetic stirring bar.

[‡] The condensate from the reflux condenser should flow through the layer of molecular sieves.

[§] There is no need to replace molecular sieves in the Soxhlet-type adapter. The same portion of 4Å molecular sieves (15 g) used for the preparation of TsOH solution can be used for this step.

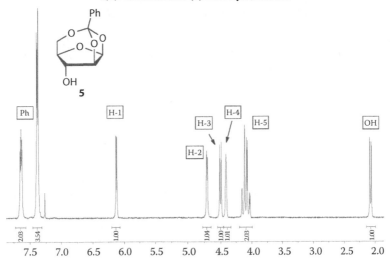

^{1}H NMR spectrum (obtained at 300 MHz) of β-D-arabinofuranose 1,2,5-orthobenzoate (**5**) after crystallization.

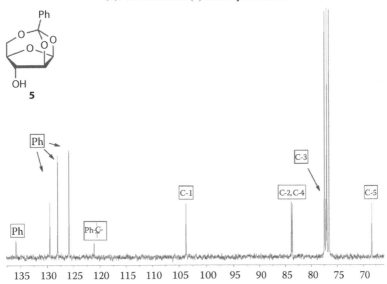

^{13}C NMR spectrum (obtained at 75 MHz) of β-D-arabinofuranose 1,2,5-orthobenzoate (**5**) after crystallization.

REFERENCES

1. Kochetkov, N. K.; Khorlin, A. Y.; Bochkov, A. F.; Yazlovetskii, I. G. *Izv. Akad. Nauk. SSSR, Ser. Khim.* **1966**, 2030–2032 (in Russian) [*Bull. Acad. Sci. USSR, Div. Chem. Sci.* (Engl. Transl.), **1966**, *15*, 1966–1968].

2. Bochkov, A. F.; Voznyi, Y. V.; Chernetskii, V. N.; Dashunin, V. M.; Rodionov, A. V. *Izv. Akad. Nauk. SSSR, Ser. Khim.* **1975**, 420–423 (in Russian) [*Bull. Acad. Sci. USSR, Div. Chem. Sci.* (Engl. Transl.), **1975**, *24*, 348–351].

3. Kochetkov, N. K.; Bochkov, A. F.; Yazlovetskii, I. G. *Carbohydr. Res.* **1969**, *9*, 49–60.

4. Sanchez, S.; Bamhaoud, T.; Prandi, J. *Eur. J. Org. Chem.* **2002**, 3864–3873.

5. Podvalnyy, N. M.; Zinin, A. I.; Abronina, P. I.; Torgov, V. I.; Kononov, L. O. *Russ. Chem. Bull.* **2009**, *58*, 482–483.

6. Podvalnyy, N. M.; Sedinkin, S. L.; Abronina, P. I.; Zinin, A. I.; Fedina, K. G.; Torgov, V. I.; Kononov, L. O. *Carbohydr. Res.* **2011**, *346*, 7–15.

7. Marotte, K.; Sanchez, S.; Bamhaoud, T.; Prandi, J. *Eur. J. Org. Chem.* **2003**, 3587–3598.

8. Liu, C.; Lowary, T. L.; Richards, M. R. *Org. Biomol. Chem.* **2011**, *9*, 165–176.

9. Abronina, P. I.; Podvalnyy, N. M.; Sedinkin, S. L.; Fedina, K. G.; Zinin, A. I.; Chizhov, A. O.; Torgov, V. I.; Kononov, L.O. *Synthesis* **2012**, *44*, 1219–1225.

10. Tam, P.-H.; Lowary, T. L. Mycobacterial lipoarabinomannan fragments as haptens for potential anti-tuberculosis vaccines. In *Carbohydrate Chemistry*, Rauter, A. P., Lindhorst, T. K., Eds.; Royal Society of Chemistry, London, U.K., **2010**, Vol. 36, pp. 38–63.

11. Ness, R. K.; Fletcher, H. G. *J. Am. Chem. Soc.* **1958**, *80*, 2007–2010.

12. Fletcher, H.G. The anomeric tri-*O*-benzoyl-D-arabinofuranosyl bromides. In *Methods in Carbohydrate Chemistry*, Whistler, R. L., Wolfrom, M. L, Eds.; Academic Press Inc., San Diego, CA, **1963**, Vol. 2, 228–230.

13. Callam, C. S.; Lowary, T. L. *J. Chem. Educ.* **2001**, *78*, 73–74.

14. Betaneli, V. I.; Ovchinnikov, M. V.; Backinowsky, L. V.; Kochetkov, N. K. *Carbohydr. Res.* **1980**, *84*, 211–224.

15. Ovchinnikov, M. V.; Byramova, N. E.; Backinowsky, L. V.; Kochetkov, N. K. *Bioorg. Khim.* **1983**, *9*, 391–400 (in Russian).

16. Kochetkov, N. K., Byramova, N. E.; Tsvetkov, Y. E.; Backinowsky, L. V. *Tetrahedron* **1985**, *41*, 3363–3375.

17. Byramova, N. E.; Tuzikov, A. B.; Bovin, N. V. *Carbohydr. Res.* **1992**, *237*, 161–175.

20 Synthesis of Benzohydroxamic Acid Glucosides and Galactosides

*Mikaël Thomas, Isabelle Opalinski, Brigitte Renoux,
Thibaut Chalopin,† and Sébastien Papot**

CONTENTS

Recently, *O*-glycosyl hydroxamates have emerged as a novel class of compounds with potential biological interest. For instance, Trichostatin D, the α-glucoside of Trichostatin A, is an inducer of phenotypic reversion in oncogene-transformed cells that could be used as a new anticancer agent.[1] The glucuronides of Trocade and Vorinostat are well-known metabolites of the corresponding drugs.[2,3] Furthermore, the suberoylanilide hydroxamic acid β-galactoside was reported as a promising pro-drug for selective cancer chemotherapy (Scheme 20.1).[4]

However, the synthesis of *O*-glycosyl hydroxamates *via* direct *O*-glycosylation of hydroxamic acids has remained an unsolved problem. For many years, the only known methodology for preparing such carbohydrate derivatives relied on the *N*-acylation of *O*-glycosyl hydroxylamines by the corresponding carboxylic acids.[2,4,5]

---

* Corresponding author; e-mail: sebastien.papot@univ-poitiers.fr.
† Checker; under the supervision of Sébastien Gouin; e-mail: sebastien.gouin@univ-nantes.fr.

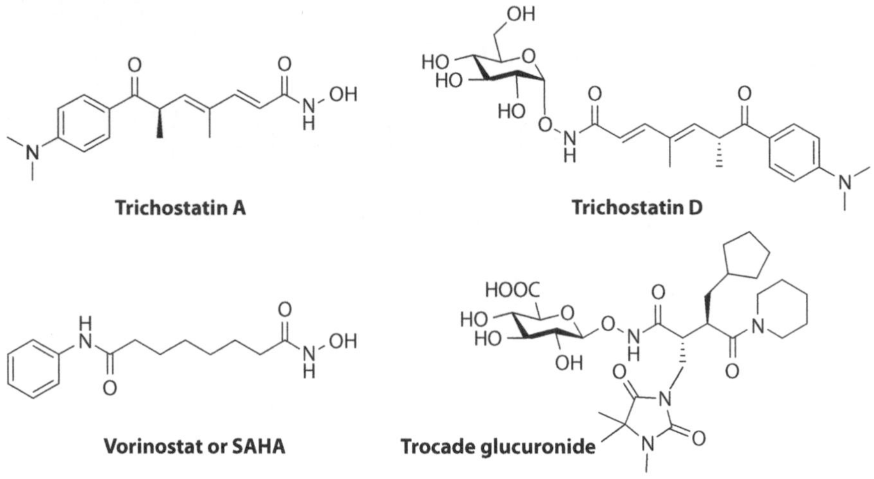

Trichostatin A

Trichostatin D

Vorinostat or SAHA

Trocade glucuronide

SCHEME 20.1   Some Hydroxamates of biological interest.

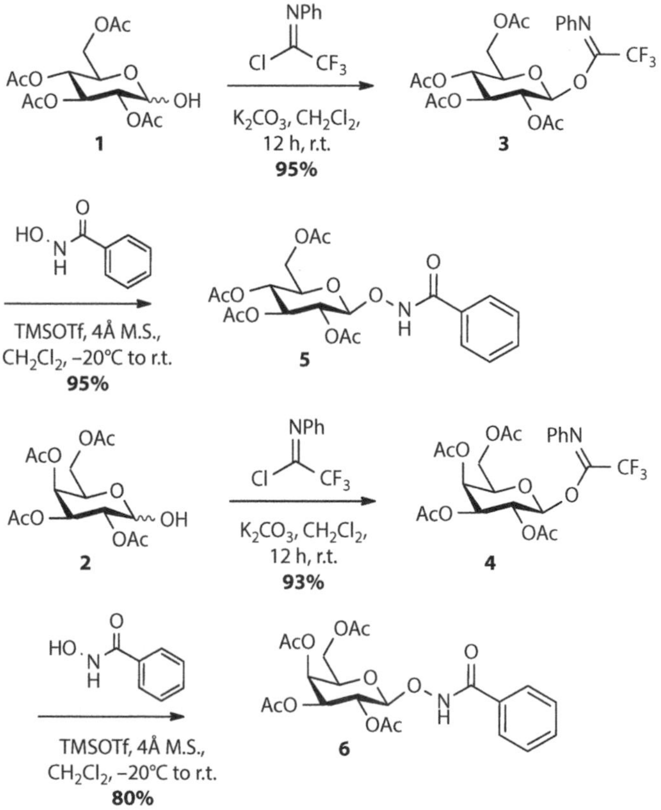

SCHEME 20.2   Synthesis of glycosyl hydroxamates

In 2007, we have developed the first direct O-glycosylation of hydroxamic acids using glycosyl N-phenyl trifluoroacetimidates as carbohydrate donors.[6] When treated with 1.5 equivalents of hydroxamic acid and TMSOTf as promoter, these donors allowed efficient and stereoselective access to a wide range of glycosyl hydroxamates. Because glycosyl N-phenyl trifluoroacetimidates are not quite stable, the best results were obtained when these glycosyl donors were used directly after their purifications by flash column chromatography. Here, we illustrate the procedure through the glycosylation of benzohydroxamic acid with either glucosyl or galactosyl N-trifluorophenyl acetimidate.[7] Donors 3 and 4 were prepared from anomerically unprotected glycosides 1 and 2 using N-phenyl trifluoroacetimidoyl chloride[8] (2 equiv.) and K_2CO_3 (2 equiv.) in dichloromethane. Compounds 3 and 4 were converted to glycosyl hydroxamates 5 and 6 with good to excellent yields (95% and 80%, respectively) (Scheme 20.2).

EXPERIMENTAL METHODS

GENERAL METHODS

Compounds 1 and 2 were prepared according to known procedure.[9] 2,2,2-Trifluoro-N-phenylacetimidoyl chloride was prepared following the procedure by Tamura et al.[8] and was found stable up to 6 months at −18°C. Benzohydroxamic acid and molecular sieves (4Å) were purchased from Aldrich. All reactions were performed under N_2 atmosphere. These reactions were monitored by TLC on precoated silica gel plates (Macherey-Nagel ALUGRAM® SIL G/UV$_{254}$, 0.2 mm silica gel 60 Å). Spots were visualized under 254 nm UV light and/or charring with a solution of 3 g of phosphomolybdic acid in 100 mL of ethanol. Flash column chromatography was performed using Macherey-Nagel silica gel 60 (15–40 μm). ^1H and ^{13}C NMR were performed on an Avance 300 Bruker spectrometer. Chemical shifts are expressed in part per million (ppm) relative to TMS ($\delta = 0$ ppm) and the coupling constants J in hertz (Hz). NMR multiplicities are reported as b = broad, s = singlet, d = doublet, t = triplet, q = quadruplet, and m = multiplet. Proton and carbon signals have been assigned through DEPT 135, ^1H–^1H correlation, and ^1H–^{13}C correlation experiments. Stereochemistry at the anomeric position for compounds 3 and 4 were determined by NOESY experiment and proved to be β-glycosides. Optical rotation was measured on a Schmidt + Haensch polartronic HH8 polarimeter.

2,3,4,6-TETRA-O-ACETYL-β-D-GLUCOPYRANOSYL N-PHENYL-2,2,2-TRIFLUOROACETIMIDATE (3)

N-Phenyl trifluoroacetimidoyl chloride (2.64 g, 12.6 mmol, 2 equiv.) and K_2CO_3 (1.74 g, 12.6 mmol, 2 equiv.) were added to a solution of 1 (2.2 g, 6.3 mmol, 1 equiv.) in anhydrous dichloromethane (30 mL). The solution was stirred overnight* at room

---

* The reaction time could be sometimes longer (up to 5 days) in order to reach completion. This parameter could not be controlled carefully but the outcome of the reaction was identical in terms of yields and purity.

temperature and concentrated, and chromatography (2:3 EtOAc–PE) afforded the desired compound as amorphous white solid (3.11 g, 95%). $[\alpha]^{25}_D$ +41 (c 0.1, CHCl$_3$). ^{1}H NMR (CDCl$_3$, 300 MHz): 7.32 (t, 2H, J = 7.7 Hz, H-aromatic meta), 7.14 (t, 1H, J = 7.5 Hz, H-aromatic para), 6.84 (d, 2H, J = 7.7 Hz, H-aromatic ortho), 5.78 (br s, 1H, H-1), 5.29–5.13 (m, 3H, H-2, H-3, H-4), 4.26 (dd, 1H, J = 12.5, 4.4 Hz, H-6), 4.12 (d, 1H, J = 11.4 Hz, H-6), 3.76 (br s, 1H, H-5), 2.07 (s, 6H, CH$_3$), 2.02 (s, 6H, CH$_3$). ^{13}C NMR (CDCl$_3$, 75 MHz): 171.0 (C=O), 170.6 (C=O), 169.7 (C=O), 169.4 (C=O), 143.3 (C-aromatic), 129.2 (C-aromatic meta), 125.1 (C-aromatic para), 119.6 (C-aromatic ortho), 94.9 (C-1), 73.1–72.9 (C-3, C-5), 70.6 (C-2), 68.1 (C-4), 61.8 (C-6), 21.1 (CH$_3$), 21.0 (CH$_3$), 20.9 (CH$_3$), 20.8 (CH$_3$).

2,3,4,6-Tetra-O-acetyl-β-D-galactopyranosyl N-Phenyl-2,2,2-trifluoroacetimidate (4)

N-Phenyl trifluoroacetimidoyl chloride (156 mg, 0.75 mmol, 2 equiv.) and K$_2$CO$_3$ (103 mg, 0.75 mmol, 2 equiv.) were added to a solution of **2** (130 mg, 0.37 mmol, 1 equiv.) in anhydrous dichloromethane (3 mL). The solution was stirred overnight* at room temperature and concentrated, and chromatography (4:6 EtOAc–PE) afforded amorphous white solid (179 mg, 93%). $[\alpha]^{25}_D$ +59.1 (c 0.22, CHCl$_3$). ^{1}H NMR (CDCl$_3$, 300 MHz): 7.32 (t, 2H, J = 7.8 Hz, H-aromatic meta), 7.14 (t, 1H, J = 7.7 Hz, H-aromatic para), 6.83 (d, 2H, J = 7.7 Hz, H-aromatic ortho), 5.73 (br s, 1H, H-1), 5.51–5.31 (m, 2H, H-2, H-4), 5.07 (d, 1H, J = 9.0 Hz, H-3), 4.15 (d, 2H, J = 6.4 Hz, H-6), 3.96 (br s, 1H, H-5), 2.17 (s, 3H, CH$_3$), 2.07 (s, 3H, CH$_3$), 2.00 (s, 3H, CH$_3$), 1.99 (s, 3H, CH$_3$). ^{13}C NMR (CDCl$_3$, 75 MHz): 170.6 (C=O), 170.5 (C=O), 170.3 (C=O), 169.4 (C=O), 143.3 (C-aromatic), 129.2 (C-aromatic meta), 125.0 (C-aromatic para), 119.5 (C-aromatic ortho), 95.4 (C-1), 72.2 (C-5), 71.0 (C-3), 68.2 (C-2), 67.1 (C-5), 61.4 (C-6), 21.0 (CH$_3$), 20.9 (CH$_3$), 20.9 (CH$_3$), 20.9 (CH$_3$).

2,3,4,6-Tetra-O-acetyl-1-O-benzohydroxamoyl-β-D-glucopyranose (5)

A mixture of freshly† prepared glycosyl trifluoroacetimidate **3** (260 mg, 0.5 mmol, 1 equiv.), benzohydroxamic acid (103 mg, 0.75 mmol, 1.5 equiv.), and 4 Å molecular sieves‡ (2 g) in anhydrous dichloromethane (5 mL) was stirred at room temperature for 1 h. After cooling to −25°C, TMSOTf§ (91 μL, 0.5 mmol, 1 equiv.) was added and, after 15 min, the mixture was allowed to warm to room temperature and stirred for an additional 2 h. Triethylamine (1 mL) was added, the mixture was filtered through

---

* The reaction time could be sometimes longer (up to 5 days) in order to reach completion. This parameter could not be controlled carefully but the outcome of the reaction was identical in terms of yields and purity.

† Compounds **3** and **4** show limited stability and were used the same day.

‡ Activation was performed under vacuum at 300°C and molecular sieves were stored at 100°C for up to 3 days. Same comment for compound **6**.

§ The quality of TMSOTf greatly influences outcome of the reaction and reagent from a newly opened container is required for optimum results. Same comment for compound **6**.

a pad of Celite®, and the filtrate was concentrated. Chromatography* of material in the filtrate (1% MeOH in DCM) provided **5** (227 mg, 0.484 mmol, 95%, β-anomer as judged by ^{1}H NMR) as a white solid.† Crystals growth: a 3 mL vial containing a solution of 21 mg of **5** in 0.4 mL of CHCl$_3$ was placed on the bottom of a 250 mL screw-caped glass jar containing 25 mL of PE. The jar was closed with the screw cap and kept at room temperature. Slow vapor diffusion overnight produced colorless crystals suitable for x-ray crystallography; mp 83.0°C ± 0.5°C; $[\alpha]^{25}_D$ −24.5 (c 0.2, CHCl$_3$). ^{1}H NMR (CDCl$_3$, 300 MHz): 9.61 (br s, 1H, NH), 7.68 (d, 2H, J = 7.8 Hz, H-aromatic ortho), 7.45 (t, 1H, J = 7.3 Hz, H-aromatic para), 7.35 (t, 2H, J = 7.7 Hz, H-aromatic meta), 5.23 (t, 1H, J = 9.4 Hz, H-3), 5.10 (t, 1H, J = 8.15 Hz, H-2), 5.04 (t, 1H, J = 9.4 Hz, H-4), 4.91 (d, 1H, J = 8.0 Hz, H-1), 4.22 (dd, 1H, J = 12.5, 4.4 Hz, H-6), 4.10 (dd, 1H, J = 12.4, 2.3 Hz, H-6), 3.73 (ddd, 1H, J = 9.9, 4.4, 2.3 Hz, H-5), 2.08 (s, 3H, CH$_3$), 1.99 (s, 3H, CH$_3$), 1.95 (s, 6H, CH$_3$). ^{13}C NMR (CDCl$_3$, 75 MHz): 171.1 (C=O), 170.7 (C=O), 170.4 (C=O), 169.8 (C=O), 166.7 (C=O), 132.6 (C-aromatic), 131.5 (C-aromatic para), 129.1 (C-aromatic meta), 127.6 (C-aromatic ortho), 103.9 (C-1), 72.7 (C-5), 72.6 (C-3) 69.7 (C-2), 68.4 (C-4), 62.1 (C-6), 21.1 (CH$_3$), 21.0 (CH$_3$), 20.9 (s, 2C, CH$_3$). HMRS(ESI) calcd for C$_{21}$H$_{25}$NO$_{11}$Na: 490.1325. Found: 490.1337.

2,3,4,6-TETRA-O-ACETYL-1-O-BENZOHYDROXAMOYL-β-D-GALACTOPYRANOSE (6)

A mixture of freshly prepared glycosyl trifluoroacetimidate **4** (260 mg, 0.5 mmol, 1 equiv.), benzohydroxamic acid (103 mg, 0.75 mmol, 1.5 equiv.), and 4Å molecular sieves (2 g) in anhydrous dichloromethane (5 mL) was stirred at room temperature for 1 h. After cooling to −25°C, TMSOTf (91 μL, 0.5 mmol, 1 equiv.) was added and, after 15 min, the mixture was allowed to warm to room temperature. After an additional 3 h, triethylamine (1 mL) was added, the mixture was filtered through Celite, and chromatography of material in the filtrate (4:6→5:5 EtOAc–PE,) gave **6** (188 mg, 0.40 mmol, 80%) as amorphous white solid. $[\alpha]^{25}_D$ −25 (c 0.2, CHCl$_3$). ^{1}H NMR (CDCl$_3$, 300 MHz): 9.24 (s, 1H, NH), 7.76 (d, 2H, J = 7.2 Hz, H-aromatic ortho), 7.55 (t, 1H, J = 7.2 Hz, H-aromatic para), 7.46 (t, 2H, J = 7.3 Hz, H-aromatic meta), 5.45 (d, 1H, J = 3.5 Hz, H-4), 5.36 (t, 1H, J = 8.8 Hz, H-2), 5.12 (dd, 1H, J = 10.5, 2.8 Hz, H-3), 4.93 (d, 1H, J = 8.2 Hz, H-1), 4.24–4.14 (m, 2H, H-6), 4.00 (t, 1H, J = 6.4 Hz, H-5), 2.22 (s, 3H, CH$_3$), 2.17 (s, 3H, CH$_3$), 2.05 (s, 3H, CH$_3$), 2.01 (s, 3H, CH$_3$). ^{13}C NMR (CDCl$_3$, 75 MHz): 171.1 (C=O), 170.8 (C=O), 170.5 (C=O), 170.4 (C=O), 167.0 (C=O), 132.9 (C-aromatic), 131.8 (C-aromatic para), 129.2 (C-aromatic meta), 127.6 (C-aromatic ortho), 104.6 (C-1), 71.7 (C-5), 70.9 (C-3), 67.3–67.1 (C-2, C-4), 61.5 (C-6), 21.3 (CH$_3$), 21.1 (CH$_3$), 21.0 (s, 2C, CH$_3$). HMRS(ESI) calcd for C$_{21}$H$_{25}$NO$_{11}$Na: 490.1325. Found: 490.1325.

---

* Adsorption on silica should be avoided to allow optimal purity and yields of the reaction since partial degradation of the compound could be observed under these conditions. Same comment for compound **6**.

† Partial degradation was observed by ^{1}H NMR analysis after 6 months and storage of the solid at +4°C. Same comment for compound **6**. Because of the limited stability of these compounds, correct combustion analysis figures could not be obtained.

X-Ray Analysis of 2,3,4,6-tetra-O-acetyl-1-O-
benzohydroxamoyl-β-d-glucopyranose (5)

Crystal of **5** was obtained as a dimeric structure where one molecule of CHCl₃ was trapped in between two carbohydrates.

Compound reference	mt-205a-frx
Bond precision	C–C=0.0095 Å
a	13.000(3)
b	14.260(3)
c	27.218(5)
alpha	90°
beta	90°
gamma	90°
Temperature	100 K
Volume	5045.7(18) Å³
Space group	P 21 21 21
Hall group	P 2ac 2ab
Moiety formula	2(C21 H25 N O11), CHCl3
Sum formula	C43 H51 Cl3 N2 O22 C43 H51 Cl3 N2 O22
Mr	1054.21
Dx	1.388 g cm⁻³
Z	4
Mu	2.351 mm⁻¹
F000	2200.0
F000'	2211.86
h,k,lmax	15,17,32
Nref	9295
Tmin,Tmax	0.864,1.000
Correction method	MULTI-SCAN
Data completeness	1.77/0.98
Theta(max)	69.359
R(reflections)	0.0783(9000)
wR2(reflections)	0.2232(9295)
S	1.030
Npar	668

Crystallographic data have been deposited with the Cambridge Crystallographic Data Centre as supplementary publication number CCDC 965971.

ACKNOWLEDGMENTS

The authors thank CNRS, La Ligue Contre le Cancer (comités Deux Sèvres, Charente, Charente-Maritime, Vienne), and the region Poitou-Charentes for providing financial support of this study.

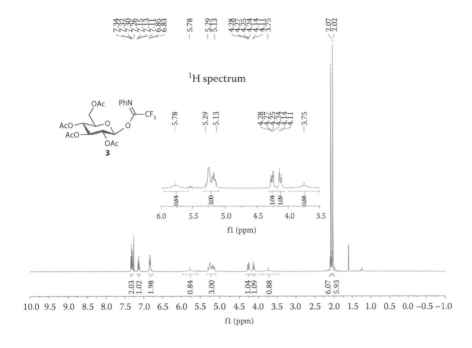

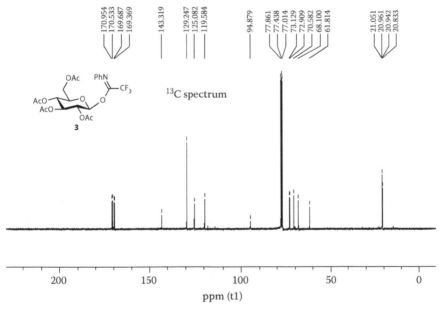

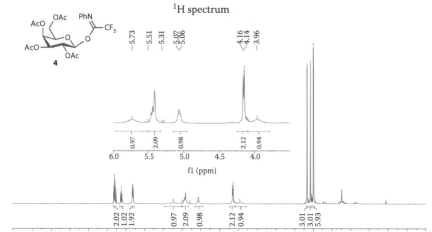

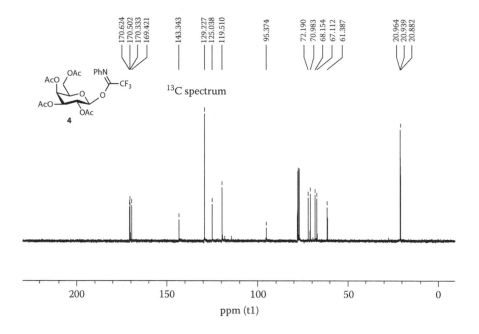

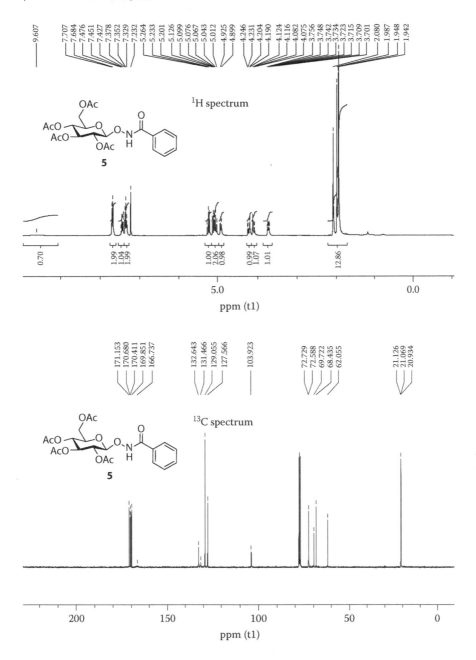

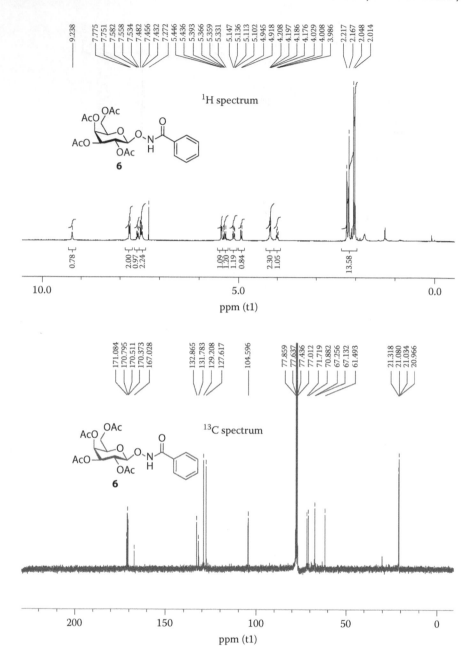

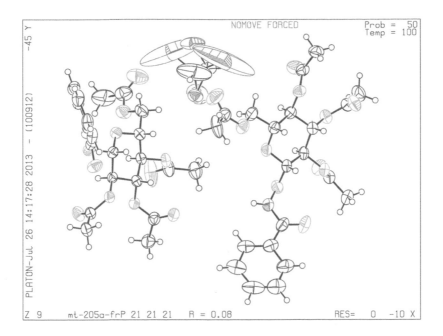

REFERENCES

1. Hayakawa, Y.; Nakai, M.; Furihata, K.; Shin-ya, K.; Seto, H. *J. Antibiot.* **2000**, *53*, 179–183.
2. Mitchell, M.B.; Whitcombe, I.A. *Tetrahedron Lett.* **2000**, *41*, 8829–8834.
3. Parise, R.A; Holleran, J.L.; Beumer, J.H.; Ramalingam, S.; Egorin, M.J. *J. Chromatogr. B* **2006**, *840*, 108–115.
4. Thomas, M.; Rivault, F.; Tranoy-Opalinski, I.; Roche, J.; Gesson, J.P.; Papot, S. *Bioorg. Med. Chem. Lett.* **2007**, *17*, 983–986.
5. Hosokawa, S.; Ogura, T.; Togashi, H.; Tatsuta, K. *Tetrahedron Lett.* **2005**, *46*, 333–337.
6. Thomas, M.; Gesson, J.P.; Papot, S. *J. Org. Chem.* **2007**, *72*, 4262–4264.
7. Yu. B.; Tao, H. *Tetrahedron Lett.* **2001**, *42*, 2405–2407.
8. Tamura, K.; Mizukami, H.; Maeda, K.; Watanabe, H.; Uneyama, K. *J. Org. Chem.* **1993**, *58*, 32–35.
9. Sim, M.M.; Kondo, H.; Wong, C.-H. *J. Am. Chem. Soc.* **1993**, *115*, 2260–2267.

21 Hexa-O-benzoyl-4′,6′-O-benzylidene- and p-Methoxybenzylidene-β-lactose

*Xiaowei Lu, Deepak Sail, Ján Hirsch,[†]
and Pavol Kováč**

CONTENTS

INTRODUCTION

Ylidene derivatives of sugars are important intermediates in synthetic carbohydrate chemistry. They are most often formed from sugars and carbonyl compounds or their dialkyl acetals under anhydrous conditions in presence of acid catalyst. Deprotection is done frequently by acid hydrolysis.[1,2] Benzylidene protecting groups are of special importance because they can be removed under nonhydrolytic conditions, by catalytic hydrogenolysis. O-p-methoxybenzylidene group offers possibility for more sophisticated orthogonal protection, as this group can be selectively removed in the presence of benzylidene acetal under both nonhydrolytic and nonreductive conditions by treatment with 2,3-dichloro-5,6-dicyano-1,4-benzoquinone (DDQ),[3] ceric ammonium nitrate (CAN),[4] or $SnCl_4$ and thiophenol.[5]

Unlike 4,6-O-ylidene derivatives of alkyl lactosides, alkylidene derivatives of lactose have not been duly utilized because efficient preparation of pure compounds of this class has not been described (10% yield was reported[6] for 4,6-O-benzylidenelactose, isolated as β-per-O-acetate; yields of pure compounds in similar preparations of

---

* Corresponding author; e-mail: kpn@helix.nih.gov.
[†] Checker; e-mail: chemhirs@savba.sk.

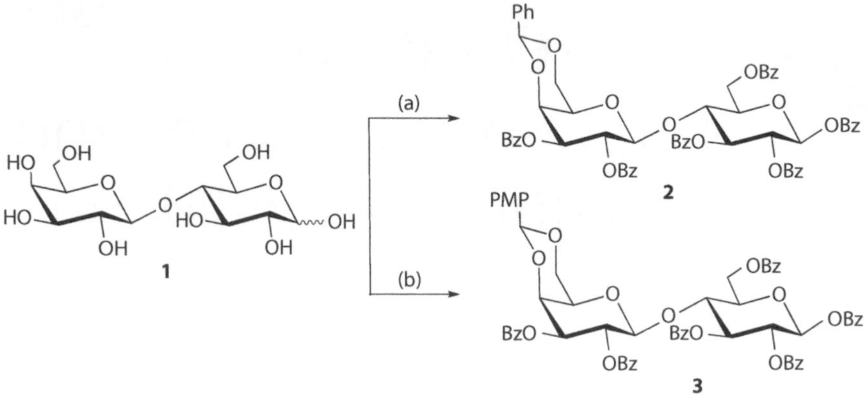

SCHEME 21.1 (a) 1. PhCH(OMe)$_2$, CSA, DMF; 2. Bz$_2$O, TEA. (b) 1. pMeOPhCH(OMe)$_2$, CSA, DMF; 2. Bz$_2$O, TEA.

hexa-O-acetyl-4′,6′-O-benzylidene- and 4′,6′-O-p-methoxybenzylidene-α,β-lactose[7] were not reported). We describe here a simple, one-pot preparation of pure, crystalline, benzoylated β-lactose derivatives **2** and **3** (Scheme 21.1). Preparation of benzyl penta-O-benzyl-4′,6′-O-benzylidene-β-cellobioside has been described previously in this series.[8]

EXPERIMENTAL METHODS

GENERAL METHODS

Optical rotations were measured at ambient temperature in CHCl$_3$ with a Jasco automatic polarimeter, Model P-2000. All reactions were monitored by thin-layer chromatography (TLC) on silica gel 60–coated glass slides (Analtech, Inc.). Column chromatography was performed by elution from columns of silica gel with Biotage Isolera chromatograph. Solvent mixtures less polar than those used for TLC were used at the onset of separations. Nuclear magnetic resonance (NMR) spectra were measured at 600 MHz (^{1}H) and 150 MHz (^{13}C) with a Bruker Avance 600 spectrometer in CDCl$_3$ as solvent. ^{1}H and ^{13}C chemical shifts are referenced to signals of TMS (0 ppm) and CDCl$_3$ (77.0 ppm). Assignments of NMR signals were made by homonuclear and heteronuclear 2D correlation spectroscopy. Lactose monohydrate, camphorsulfonic acid (CSA), Dry dimethylformamide (DMF), benzaldehyde dimethyl acetal (PhCH(OMe)$_2$), p-methoxybenzaldehyde dimethyl acetal (pMeOPhCH(OMe)$_2$), triethylamine (TEA), and benzoic anhydride (Bz$_2$O) were purchased from Sigma Chemical Company. Solutions in CH$_2$Cl$_2$ were dried with Na$_2$SO$_4$ and concentrated at 2 kPa/40°C.

1,2,3,6,2′,3′-HEXA-O-BENZOYL-4′,6′-O-BENZYLIDENE-β-LACTOSE (2)

A suspension of lactose monohydrate that had been dried in a vacuum oven at 40°C overnight (1.8 g, 5 mmol), CSA (0.29 g, 1.25 mmol), and PhCH(OMe)$_2$

(1.9 mL, 12.5 mmol) in DMF (25 mL) was stirred at room temperature for 2 days, when a clear solution was formed. TLC (4:1 CH_2Cl_2–MeOH) showed that virtually no starting material was present. The mixture was treated with TEA (16.7 mL, 120 mmol) and benzoic anhydride (18.1 g, 80 mmol) and stirred at 55°C for 2 days, when examination of the yellow mixture by TLC (9:1 toluene–acetone) showed presence of a major product ($R_f \sim 0.4$). Methanol was added to quench the reaction, and after 0.5 h, the solvent was removed. The crude mixture was partitioned between CH_2Cl_2 and brine, the organic phase was concentrated, and chromatography $CH_2Cl_2 \rightarrow 5\%$ MeOH in CH_2Cl_2 yielded pure product **2** (3 g, 57% in two steps), mp 266°C–267°C (needles, CH_2Cl_2–EtOH) $[\alpha]_D$ +103.5 (c 1.0, $CHCl_3$). ^1H NMR (600 MHz, $CDCl_3$) δ: 8.05–7.15 (m, 35 H, aromatic protons), 6.14 (d, 1 H, $J_{1,2}=8.0$ Hz, H-1), 5.97 (t, 1 H, $J=8.9$ Hz, H-3), 5.82 (dd, 1 H, $J_{1',2'}=8.0$ Hz, $J_{2',3'}=10.4$ Hz, H-2′), 5.67 (t, 1 H, $J=8.5$ Hz, H-2), 5.31 (s, 1 H, C*H*Ph), 5.18 (dd, 1 H, $J_{3',4'}=3.5$ Hz, $J_{2',3'}=10.5$ Hz, H-3′), 4.88 (d, 1 H, $J_{1',2'}=8.0$ Hz, H-1′), 4.64 (dd, 1 H, $J_{5,6a}=1.7$ Hz, $J_{6a,6b}=12.1$ Hz, H-6a), 4.42 (dd, 1 H, $J_{5,6b}=3.5$ Hz, $J_{6a,6b}=11.6$ Hz, H-6b), 4.42 (t, 1 H, $J=9.3$ Hz, H-4), 4.33 (bd, 1 H, $J=3.4$ Hz, H-4′), 4.12 (ddd, 1 H, $J_{5,6a}=2.1$ Hz, $J_{5,6b}=3.2$ Hz, $J_{4,5}=9.9$ Hz, H-5), ~3.81 (bd, 1 H, $J=12.2$ Hz, H-6′a), 3.61 (dd, 1 H, $J_{5',6'b}=1.4$ Hz, $J_{6'a,6'b}=12.2$ Hz, H-6′b), 3.02 (m. 1H, H-5′). ^{13}C NMR (150 MHz, $CDCl_3$) δ: 166.1, 165.5, 165.3, 165.0, 164.8, 164.4 (6×COPh), 137.3, 133.6, 133.3, 133.2, 133.0, 130.1, 130.0, 129.7, 129.6, 129.5, 129.4, 128.8, 128.7, 128.5, 128.4, 128.3, 128.2, 128.0, 126.3 (aromatic carbon), 101.5 (C-1′), 100.6 (C*H*Ph), 92.2 (C-1), 76.1 (C-4), 73.8 (C-3), 73.4 (C-5), 73.0 (C-4′), 72.5 (C-3′), 71.1 (C-2), 69.2 (C-2′), 67.9 (C-6′), 66.5 (C-5′), 62.0 (C-6); ESI-MS [M + Na]⁺ calcd for $C_{61}H_{50}O_{17}Na$, 1077.2946; found: 1077.2892; Anal. calcd for $C_{61}H_{50}O_{17}$: C, 69.44; H, 4.78. Found: C, 69.49; H, 4.77.

1,2,3,6,2′,3′-HEXA-*O*-BENZOYL-4′,6′-*O*-P-METHOXYBENZYLIDENE-β-LACTOSE (3)

A suspension of lactose monohydrate that had been dried in a vacuum oven at 40°C overnight (1.8 g, 5 mmol), pMeOPhCH(OMe)₂ (2.13 mL, 12.5 mmol), and CSA (0.29 g, 1.25 mmol) in DMF (25 mL) was stirred at room temperature for 5 h. A clear solution was obtained, and TLC (4:1 CH_2Cl_2–MeOH) showed that amount of remaining starting material was negligible. TEA (16.7 mL, 120 mmol) and Bz_2O (18.1 g, 80 mmol) were added, and the mixture was stirred at 55°C for 2 days. TLC (9:1 toluene–acetone) showed that one major product ($R_f \sim 0.4$) was formed and that materials with slower chromatographic mobility—which were present at 16 h of the reaction time—had been consumed. Methanol was added to quench the reaction, and after 0.5 h, the mixture was concentrated. The residue was dissolved in CH_2Cl_2 (200 mL), and the solution was partitioned between brine and CH_2Cl_2. After concentration of the organic phase, chromatography of the residue ($CH_2Cl_2 \rightarrow 5\%$ MeOH in CH_2Cl_2) yielded pure product **3** (3.4 g, 63%, in 2 steps), mp 267°C–269°C (needles, CH_2Cl_2–MeOH), $[\alpha]_D$ +101.7 (c 1.0, $CHCl_3$). ^1H NMR (600 MHz, $CDCl_3$) δ: 8.05–6.82 (m, 34 H, aromatic protons), 6.15 (d, 1 H, $J_{1,2}=7.9$ Hz, H-1), 5.98 (t, 1 H, $J=8.8$ Hz, H-3), 5.82 (dd, 1 H, $J_{1',2'}=7.9$ Hz, $J_{2',3'}=10.4$ Hz, H-2′), 5.68 (dd, 1 H, $J_{1,2}=7.9$ Hz, $J_{2,3}=8.8$ Hz, H-2), 5.27 (s, 1 H, C*H*Ph), 5.18 (dd, 1 H, $J_{3',4'}=3.6$ Hz,

$J_{2',3'}$ = 10.4 Hz, H-3'), 4.88 (d, 1 H, $J_{1',2'}$ = 7.9 Hz, H-1'), 4.64 (dd, 1 H, $J_{5,6a}$ = 2.0 Hz, $J_{6a,6b}$ = 12.1 Hz, H-6a), 4.43 (dd, 1 H, $J_{5,6b}$ = 3.7 Hz, $J_{6a,6b}$ = 12.2 Hz, H-6b), 4.42 (dd, 1 H, $J_{3,4}$ = 8.8 Hz, $J_{4,5}$ = 9.8 Hz, H-4), 4.31 (dd, 1 H, $J_{4',5'}$ = 0.8 Hz, $J_{3',4'}$ = 3.6 Hz, H-4'), 4.12 (ddd, 1 H, $J_{5,6a}$ = 2.0 Hz, $J_{5,6b}$ = 3.7 Hz, $J_{4,5}$ = 9.8 Hz, H-5), ~3.79 (dd, p.o., 1 H, $J_{5',6'a}$ = 1.3 Hz, H-6'a), 3.79 (s, 3 H, OCH$_3$), 3.66 (dd, 1 H, $J_{5',6'b}$ = 1.7 Hz, $J_{6'a,6'b}$ = 12.2 Hz, H-6'b), 3.02 (m. 1 H, H-5'). ^{13}C NMR (150 MHz, CDCl$_3$) δ: 166.1, 165.5, 165.3, 165.0, 164.8, 164.4 (6×COPh), 159.9 (aromatic carbon), 133.6, 133.3, 133.2, 133.1, 132.9, 130.0, 129.8, 129.7, 129.6, 129.5, 129. 4, 113.2 (aromatic carbons), 101.4 (C-1'), 100.6 (CHPh), 92.2 (C-1), 76.0 (C-4), 73.8 (C-3), 73.4 (C-5), 73.0 (C-4'), 72.5 (C-3'), 71.1 (C-2), 69.3 (C-2'), 67.9 (C-6'), 66.5 (C-5'), 62.0 (C-6), 55.2 (OCH$_3$); ESI-MS [M + Na]$^+$ calcd for C$_{62}$H$_{52}$O$_{18}$Na, 1107.3051; found: 1107.3135; Anal. calcd for C$_{62}$H$_{52}$O$_{18}$: C, 68.63; H, 4.83. Found: C, 68.36; H, 4.77.

ACKNOWLEDGMENT

The checking process was partially supported by the VEGA 2/0101/11 and SAS-NSC JRP 2012/8 grants.

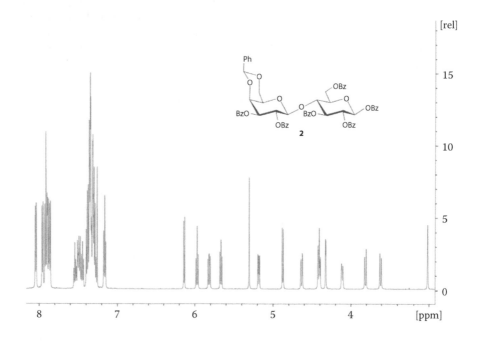

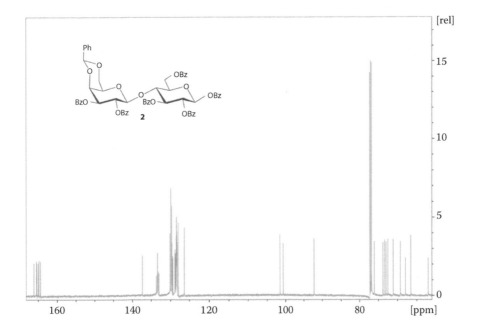

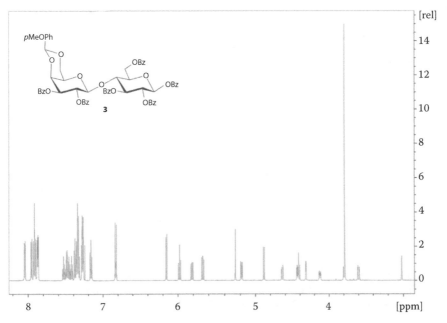

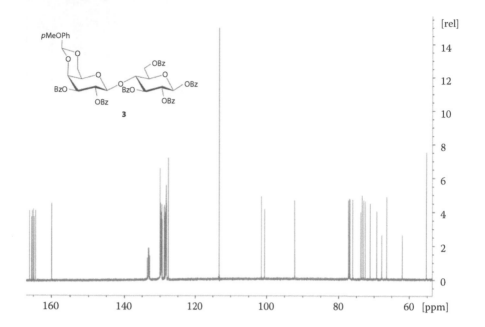

REFERENCES

1. Fletcher, H. G. Benzylidene derivatives, in *Methods in Carbohydrate Chemistry*; Whistler, R. L. and Wolfrom, M. L., Eds.; Academic Press: New York, **1963**; Vol. 2, pp. 307–308.
2. a. Calinaud, P.; Gelas, J. Synthesis of isopropylidene, benzylidene and related acetals, in *Preparative Carbohydrate Chemistry*; Hanessian, S., Ed.; Marcel Dekker: New York, **1997**, pp. 3–33; b. David, S. Selective O-substitution and oxidation using stannylene acetals and stannyl ethers, in *Preparative Carbohydrate Chemistry*; Hanessian, S., Ed.; Marcel Dekker, Inc.: New York, **1996**, pp. 69–83.
3. Oikawa, Y.; Yoshioka, T.; Yonemitsu, O. *Tetrahedron Lett.* **1982**, *23*, 885–888.
4. Classon, B.; Garegg, P. J.; Samuelsson, B. *Acta Chem. Scand.* **1984**, *38B*, 419–422.
5. Yu, W.; Su, M.; Gao, X.; Yang, Z.; Jin, Z. *Tetrahedron Lett.* **2000**, *41*, 4015–4017.
6. Ly, H. D.; Lougheed, B.; Wakarchuk, W. W.; Withers, S. G. *Biochemistry* **2002**, *41*, 5075–5085.
7. Tamerlani, G.; Lombardi, I.; Bartalucci, D.; Danesi, A.; Salsini, L.; Manoni, M.; Cipoletti, G., 6'-Sialyllactose salts and process for their synthesis and for the synthesis of other α-sialyl oligosaccharides, International Patent WO 2010/116317 A1, October 14, 2010.
8. Sail, D.; da Silva, C. P.; Kováč, P. 2,3,6,2',3',6'-Hexa-O-benzyl-β-cellobioside, in *Carbohydrate Chemistry: Proven Synthetic Methods*; Kováč, P., Ed.; CRC/Taylor & Francis: Boca Raton, FL, **2011**; Vol. 1, pp. 221–230.

22 Conversion of Allyl 2-acetamido-2-deoxy-β-D-glucopyranoside to Allyl 2-acetamido-2-deoxy-4,6-di-O-pivaloyl-β-D-galactopyranoside

Nicolò Marnoni, Monica Varese,
*Kottari Naresh,† and Luigi Panza**

CONTENTS

The biological role of oligosaccharides in cell signaling is well established, and interest in such compounds as therapeutics is growing.[1] Therefore, it has become increasingly important to gain access to pure, well-defined, complex carbohydrates. Chemical synthesis is one among the tools to obtain this class of compounds. *N*-acetyl-D-galactopyranose glycosylated at O-3 is commonly found in natural,

* Corresponding author; e-mail: luigi.panza@pharm.unipmn.it.
† Checker, under supervision of Prof. René Roy; e-mail: roy.rene@uqam.ca.

biologically active mammal glycoconjugates and mucin polysaccharides, both in physiological and in pathological circumstances.

The commercially available N-acetyl-D-galactosamine is quite expensive and requires chemical manipulations before it can be used in oligosaccharide synthesis.

Herein we present a procedure that allows ready access to allyl 2-acetamido-2-deoxy-4,6-di-O-pivaloyl-β-D-galactopyranoside **3** from easily available[2] allyl 2-acetamido-2-deoxy-β-D-glucopyranoside **1**. Allyl 2-acetamido-2-deoxy-β-D-glucopyranoside **1** can be easily converted to the corresponding crystalline 3,6-di-O-pivaloyl derivative **2** by treatment with pivaloyl chloride in pyridine/dichloromethane. After triflation of the free 4-OH group in glucoside **2**, warming a solution of the product in wet dichloromethane promotes the intramolecular displacement of the triflate by the carbonyl oxygen of the pivaloyl group at C-3 to give a dioxolenium ion intermediate. The intermediate can be intercepted by a water molecule to give a hemiorthoester that collapses to the axial ester providing compound **3**. The thus formed free 3-OH in galactoside **3** becomes amenable for further transformations.

The presence of selectively removable allyl group at the anomeric position allows access to new glycoconjugates,[2] for example, after conversion of **3** to the corresponding oxazoline, which has been demonstrated to be a quite efficient glycosyl donor.[3]

The protocol described here can be applied also to other *gluco-* derivatives, either β- or α-,[4-7] although conflicting results have been obtained with the same compound, namely, allyl 2-acetamido-2-deoxy-3,6-di-O-pivaloyl-α-D-glucopyranoside.[6,8]

EXPERIMENTAL

GENERAL METHODS

Reagents and dry solvents were added through septa *via* oven-dried syringe. Thin-layer chromatography (TLC): *Merck* silica gel *60 F_{254}* plates; detection by spraying with a 1:1 mixture of 20% H_2SO_4 soln. and a soln. of I_2 (10 g) and KI (100 g) in H_2O (500 mL) followed by heating. Flash column chromatography (FC): *Merck* silica gel *60* (230–400 mesh); m.p.: *Buchi* apparatus; uncorrected. Specific rotations ($[α]_D$): *PerkinElmer 241* polarimeter; at 20°C. ^1H and ^{13}C NMR spectra: *JEOL ECP300* instrument. Proton and carbon signals have been assigned through COSY and HETCOR experiments, respectively.

ALLYL 2-ACETAMIDO-2-DEOXY-3,6-DI-O-PIVALOYL-β-D-GLUCOPYRANOSIDE (2)

A mixture of **1**[2] (2.3 g, 8.8 mmol) and pivaloyl chloride (3.25 mL, 26.8 mmol) in dry CH_2Cl_2/pyridine (30 mL, 1:2) was stirred for 2 h at 0°C, and the reaction was monitored by TLC (solvent: toluene/acetone 7:3, R_f=0.45 for the desired intermediate **2**). MeOH (1 mL) was added, and the mixture diluted with CH_2Cl_2 (100 mL). The organic phase was washed with 5% HCl soln. (2×30 mL), 5% $NaHCO_3$ soln. (2×30 mL), and H_2O (2×30 mL), dried (Na_2SO_4), and concentrated, and the residue directly crystallized from AcOEt/hexane. Yield 2.87 g (76%), white solid, m.p.

135°C–137°C. $[\alpha]_D = -42.0$ (c 1, CHCl$_3$). ^{1}H-NMR (300 MHz, CDCl$_3$) $\delta = 5.84$ (dddd, 1H, $J_{2',3'a}$ 17.5 Hz, $J_{2',3'b}$ 10.3 Hz, $J_{2',1'a}$ 6.4 Hz, $J_{2',1'b}$ 4.6 Hz, CH=CH$_2$), 5.69 (d, 1H, $J_{\text{NH},2}$ 10.4 Hz, NH), 5.24 (dddd, 1H, $J_{3'a,2'}$ 17.2 Hz, $J_{3'a,1'a}$=$J_{3'a,1'b}$=$J_{3'a,3'Z}$ 1.5 Hz, CH=CH_2-cis), 5.17 (dddd, 1H, $J_{3'b,2'}$ 10.3 Hz, $J_{3'b,1'a}$=$J_{3'b,1'b}$=$J_{3'b,3'a}$ 1.5 Hz, CH=CH_2-trans), 5.08 (dd, 1H, $J_{3,2}$ 10.7 Hz, $J_{3,4}$ 8.3 Hz, H-3), 4.53 (d, 1H, $J_{1,2}$ 8.3 Hz, H-1), 4.40–4.35 (m, 2H, 2 H-6), 4.30 (dddd, $J_{1'a,1'b}$ 13.2 Hz, $J_{1'a,2'}$ 4.9 Hz, $J_{1'a,3'E}$=$J_{1'a,3'Z}$ 1.5 Hz, one of OCH_2CH =), 4.07 (dddd, 1H, $J_{1'b,1'a}$ 13.2 Hz, $J_{1'b,2'}$ 6.1 Hz, $J_{1'b,3'E}$=$J_{1'b,3'Z}$ 1.5 Hz, one of OCH_2CH =), 3.97 (dt, J_{NH}=$J_{2,3}$ 10.6 Hz, $J_{2,1}$ 8.4 Hz, H-2), 3.57–3.45 (m, 2H H-4, H-5), 3.09 (br. s, OH), 1.92 (s, 3H, Ac), 1.22;1.19 (2s, 9H each, 2 t-BuCO). ^{13}C-NMR (75 MHz, CDCl$_3$) $\delta = 179.9$ (t-BuCO), 179.0 (t-BuCO), 170.4 (CH$_3$$C$O), 134.0 (C-2'), 117.4 (C-3'), 99.9 (C-1), 75.2 (C-3), 74.1 (C-5), 69.7 (C-1'), 69.6 (C-4), 63.7 (C-6), 53.9 (C-2), 39.0 (C(CH$_3$)$_3$), 38.9 (C(CH$_3$)$_3$), 27.2 (C(CH$_3$)$_3$), 27.1 (C(CH$_3$)$_3$), 23.2 (CH$_3$CO). Anal. calcd for C$_{21}$H$_{35}$NO$_8$ (429.51): C 58.73, H 8.21, N 3.26. Found: C 58.55, H 8.44, N 3.19.

ALLYL 2-ACETAMIDO-2-DEOXY-4,6-DI-O-PIVALOYL-β-D-GALACTOPYRANOSIDE (3)

Under N$_2$, Tf$_2$O (0.45 mL, 2.8 mmol) was added dropwise to a soln. of glucoside **2** (1 g, 2.3 mmol) in CH$_2$Cl$_2$ (20 mL) and pyridine (1 mL) at −35°C. The mixture was allowed to warm to 0°C and stirred for 2 h. H$_2$O (1 mL) was added, and the mixture refluxed (50°C, bath temperature) for 4 h and then diluted with CH$_2$Cl$_2$ (80 mL). The reaction was monitored by TLC (solvent: ethyl acetate, R_f=0.41 for the desired compound **3**). The org. phase was washed with 5% HCl soln. (2× 20 mL), 5% NaHCO$_3$ soln. (2×20 mL), and H$_2$O (2×20 mL), dried (Na$_2$SO$_4$), and concentrated. Flash chromatography (SiO$_2$, 2:3→3:7 hexane–AcOEt) afforded 860 mg (86%) of galactoside **3** as a syrup that solidified on standing. M.p. 48°C–50°C. Attempts to crystallize the compound from a number of solvents (Et$_2$O/hexane, EtOAc/hexane, cyclohexane, methanol, and methanol/water) failed. $[\alpha]_D = -32.5$ (c 1.05, CHCl$_3$). ^{1}H-NMR (300 MHz, CDCl$_3$) $\delta = 5.91$ (dddd, $J_{2',3'E}$ 17.2 Hz, $J_{2',3'Z}$ 10.4 Hz, $J_{2',1'a}$ 6.7 Hz, $J_{2',1'a}$ 5.1 Hz, 1H, CH=CH$_2$), 5.73 (d, $J_{\text{NH},2}$ 5.2 Hz, 1H, NH), 5.30 (dd, 1H $J_{4,3}$ 3.4 Hz, $J_{4,5}$ 1.0 Hz, H-4), 5.4–5.2 (m, 2H, CH=CH_2), 4.58 (d, 1H, $J_{1,2}$ 8.3 Hz, H-1), 4.37 (ddt, $J_{1'a,1'b}$ 12.9, $J_{1'a,2b}$ 5.2, $J_{1'a,3'E}$=$J_{1'a,3'b}$ 1.5 Hz, 1H, one of OCH_2CH =), 4.16 (dd, J = 11.2, 7.3 Hz, 1H, H-6a); 4.12–3.95 (m, 4 H, H-6b, H-3, one of OCH_2CH =, OH), 3.88 (ddd, 1H, $J_{5,6a}$ 7.3, $J_{5,6b}$ 6.1, $J_{5,4}$ 1.4 Hz, H-5), 3.67 (ddd, 1H, $J_{2,3}$ 10.5 Hz, $J_{2,1}$ 8.3 Hz, $J_{2,\text{NH}}$ 5.2 Hz, H-2), 2.04 (s, 3H, Ac), 1.26;1.18 (2s, 9H each, 2 t-BuCO). ^{13}C-NMR (75 MHz, CDCl$_3$) $\delta = 177.9$ (2 C, 2 t-BuCO), 172.8 (CH$_3$$C$O), 133.4 (C-2'), 118.0 (C-3'), 99.3 (C-1), 71.5 (C-3), 71.2 (C-5), 69.7 (C-1'), 68.4 (C-4), 61.9 (C-6), 55.0 C-2), 39.1 (C(CH$_3$)$_3$), 38.6 (C(CH$_3$)$_3$), 27.1 (C(CH$_3$)$_3$), 27.0 (C(CH$_3$)$_3$), 23.2 (CH$_3$CO). Anal. calcd for C$_{21}$H$_{35}$NO$_8$ (429.51): C 58.73, H 8.21, N 3.26. Found: C 58.50, H 8.05, N 3.13.

ACKNOWLEDGMENT

We thank Università del Piemonte Orientale for the financial support.

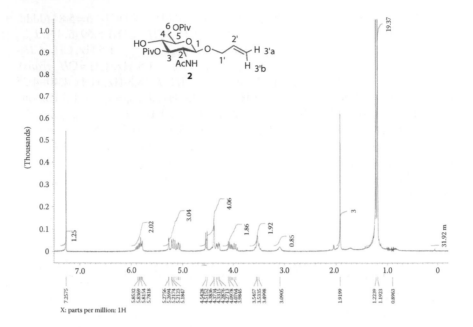

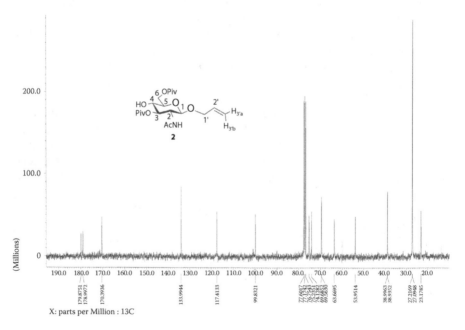

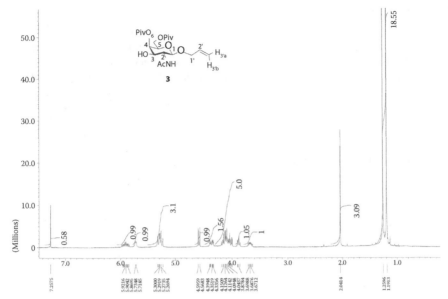

X: parts per Million : 1H

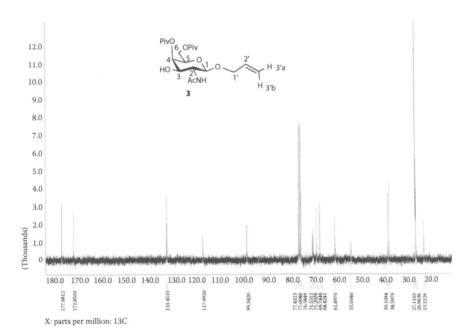

X: parts per million: 13C

REFERENCES

1. Doores, K. J.; Gamblin, D. P.; Davis, B. G. *Chem. Eur. J.* **2006**, *12*, 656–665.
2. Lin, Y. A.; Chalker, J. M.; Davis, B. G. *J. Am. Chem. Soc.* **2010**, *132*, 16805–16811.
3. Colombo, D.; Panza, L.; Ronchetti, F. *Carbohydr. Res.* **1995**, *276*, 437–441.
4. Prosperi, D.; Panza, L.; Haltrich, D.; Nonini, M.; Riva, S. *J. Carbohydr. Chem.* **2003**, *22*, 267–274.
5. Cai, Y.; Ling, C.-C.; Bundle, D. R. *J. Org. Chem.* **2008**, *74*, 580–589.
6. Arosio, D.; Vrasidas, J.; Valentini, P.; Liskamp, R. M. J.; Pieters R. J.; Bernardi, A. *Org. Biomol. Chem.* **2004**, *2*, 2113–2124.
7. Bouvet, V. R.; Ben, R. N. *J. Org. Chem.* **2006**, *71*, 3617–3622.
8. Amer, H.; Hofinger, A.; Kosma P. *Carbohydr. Res.* **2003**, *338*, 35–45.

23 Synthesis of 2,3:4,5-di-O-isopropylidene-D-Arabinose from D-Gluconolactone

Michał Kowalski, Katarzyna Łęczycka, Arkadiusz Listkowski,[†] and Sławomir Jarosz*

CONTENTS

2,3:4,5-Di-O-isopropylidene-D-arabinose[1] (4) is a convenient building block for the synthesis of complex optically pure derivatives, such as KDO[2] or tunicamycin[3] or higher carbon sugars.[4] It is usually synthesized from D-arabinose (1) as shown in Scheme 23.1.[5] The synthesis requires conversion of the parent pentose into the dithioacetal 2, conversion of the latter to diacetonide 3, and deprotection at the C1 to give 2,3:4,5-di-O-isopropylidene-D-arabinose (4).

Although the overall yield of 4 prepared through the aforementioned method is satisfactory, and the route is suitable for large-scale preparation, this procedure requires the use of environmentally unfriendly ethanethiol and toxic heavy metal salts (Scheme 23.1).

Being involved in the synthesis of higher carbon sugars with more than 10 carbon atoms in the chain by coupling of two properly activated monosaccharide building blocks,[6] we were interested in the efficient preparation of carbohydrate aldehydes. We have found that multigram quantities of 2,3:4,5-di-O-isopropylidene-D-arabinose (4) can be conveniently prepared following the odorless, three-step synthesis described in the following, which can be carried out from D-gluconolactone without any chromatographic purification (Scheme 23.2).[7]

---

* Corresponding author; e-mail: slawomir.jarosz@icho.edu.pl.
† Checker, e-mail: areklist@ichf.edu.pl.

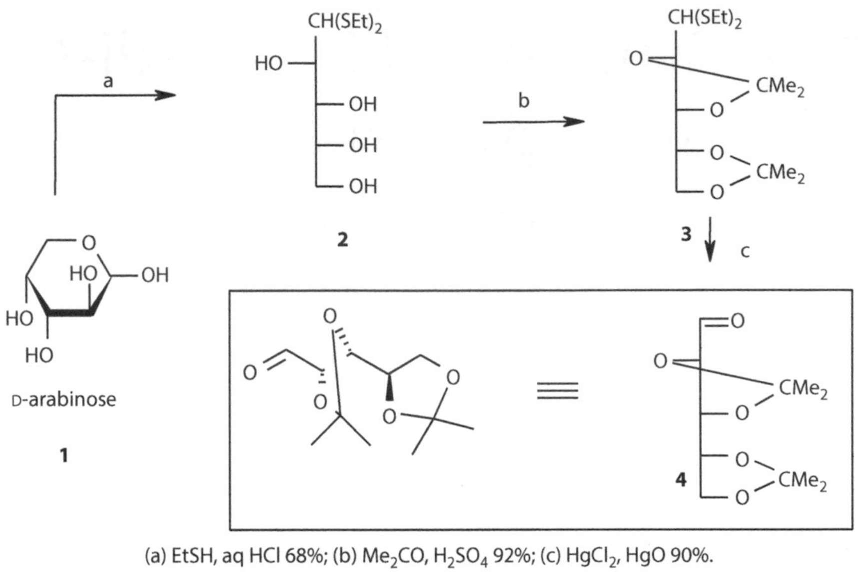

(a) EtSH, aq HCl 68%; (b) Me$_2$CO, H$_2$SO$_4$ 92%; (c) HgCl$_2$, HgO 90%.

SCHEME 23.1   Synthesis of 2,3:4,5-di-O-isopropylidene-D-arabinose through the "thioacetal" route.

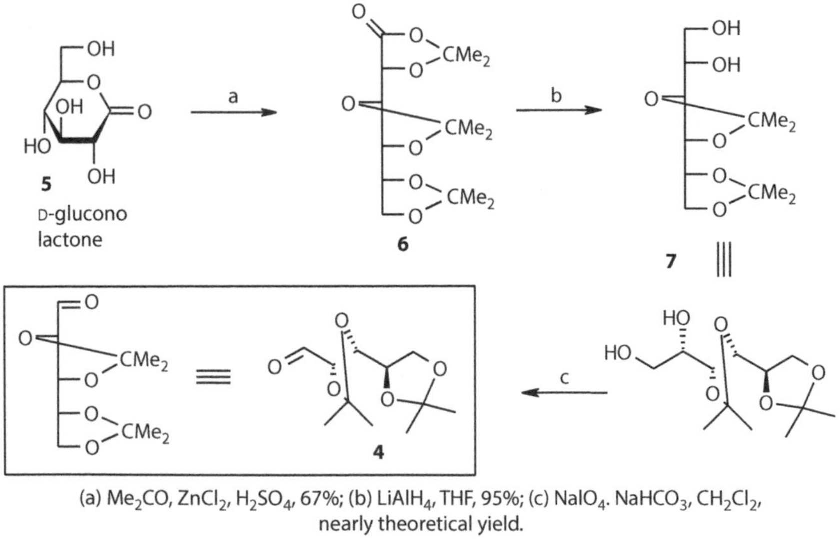

(a) Me$_2$CO, ZnCl$_2$, H$_2$SO$_4$, 67%; (b) LiAlH$_4$, THF, 95%; (c) NaIO$_4$. NaHCO$_3$, CH$_2$Cl$_2$,
nearly theoretical yield.

SCHEME 23.2   Synthesis of 2,3:4,5-di-O-isopropylidene-D-arabinose from D-gluconolactone.

Accordingly, ZnCl$_2$-catalyzed reaction of D-gluconolactone (**5**) with acetone affords the known[8] triacetonide **6**, which can be isolated and purified by crystallization.[7] Its reduction with LiAlH$_4$ provides 3,4:5,6-di-*O*-isopropylidene-D-sorbitol (**7**) whose cleavage with periodate affords the desired 2,3:4,5-di-*O*-isopropylidene-D-arabinose (**4**). Although the aldehyde is relatively stable, it is recommended to prepare it directly from the diol **7** before further synthetic steps.

EXPERIMENTAL METHODS

GENERAL METHODS

The NMR spectra were recorded at room temperature for solutions in CDCl$_3$ (internal Me$_4$Si) with a Varian AM-600 spectrometer (600 MHz ^{1}H, 150 MHz ^{13}C). Chemical shifts (δ) are reported relative to Me$_4$Si (δ 0.00) for ^{1}H and residual CDCl$_3$ (δ 77.00) for ^{13}C. The resonances were assigned by COSY (^{1}H–^{1}H), HSQC (^{1}H–^{13}C), and HMBC (^{1}H–^{13}C) experiments. Reagents were purchased from Sigma-Aldrich and Alfa Aesar and used without further purification. Commercial THF was kept over KOH (for 2 days) and distilled over benzophenone/sodium. Dichloromethane was distilled over CaH$_2$. Solutions in organic solvents were dried over MgSO$_4$. Optical rotations were measured with a Jasco P 1020 polarimeter (c = 1; CH$_2$Cl$_2$). Melting point was measured with a SRS EZ-Melt apparatus and is uncorrected.

1,2:3,4:5,6-TRI-*O*-ISOPROPYLIDENE-D-GLUCONATE[7] (**6**)

Powdered zinc chloride (11.5 g, 84.4 mmol) was dissolved in acetone (60 mL) and cooled in an ice-water bath. Concentrated sulfuric acid was added dropwise (0.1 mL). D-Glucono-1,5-lactone (**5**) was added in one portion (5.25 g, 29.5 mmol), and the mixture was stirred overnight at room temperature. The brown-orange mixture was diluted with toluene (125 mL) and brine (25 mL), the phases were separated, and the organic layer (the color changed from dark red to yellow) was washed with brine (4 × 30 mL). Combined solutions in toluene were dried and concentrated, and crystallization from MeOH (12 mL) at ~4°C gave the first crop of **6**. Two more crops were obtained by crystallization from the concentrated mother liquor (3–5 mL of MeOH). The title product was obtained as a white crystalline solid (total yield, 6.3 g, 67%, 19.3 mmol), m.p. 107°C–109°C (MeOH); [α]$_D$ + 29.1. (lit.[8] 115°C, [α]$_D$ + 34.8 (c 2.3, CHCl$_3$). ^{1}H NMR δ: 4.62 (d, 1 H, $J_{2,3}$ = 1.4 Hz, H-2), 4.27 (dd, 1 H, $J_{3,4}$ = 8.3 Hz, H-3), 4.14 (dd, 1 H, $J_{6,6'}$ = 8.5, $J_{5,6}$ = 6.0 Hz, H-6), 4.10 (ddd, 1 H, $J_{4,5}$ = 8.9, $J_{5,6'}$ = 4.0 Hz, H-5), 3.99 (dd, 1 H, H-6'), 3.94 (t, 1 H, H-4), 1.64, 1.57, 1.42, 1.40, 1.39, 1.33 (6s, 18 H, 3 × C*Me*$_2$). ^{13}C NMR δ: 170.6 (C-1), 111.5, 110.3, 109.7 (3s, 3 × C*Me*$_2$), 78.6 (C-3), 77.16 (C-5), 76.3 (C-4), 73.7 (C-2), 67.8 (C-6), 27.1, 26.97, 26.89, 26.7, 26.5, 25.2 (6s, 6 × C*Me*$_2$).

3,4:5,6-DI-*O*-ISOPROPYLIDENE-D-GLUCITOL (**7**)

A solution of 1,2:3,4:5,6-tri-*O*-isopropylidene-D-gluconate (**6**, 3.0 g, 9.54 mmol) in anhydrous THF (20 mL) was added dropwise at room temperature to the suspension of lithium aluminum hydride (1.09 g, 28.7 mmol) in THF (20 mL). The reaction mixture was stirred for 45 min, and the excess of lithium aluminum hydride was carefully decomposed with 10% solution of NaOH in water. The mixture was filtered through Celite pad; the filtrate was washed with brine, dried, and concentrated to afford the title compound as a colorless oil (2.31 g, 8.80 mmol, 92%), $[\alpha]_D + 15.8$ (lit.[9] $[\alpha]_D + 7.2$; c 1.75; CHCl$_3$). ^{1}H NMR δ: 4.16 (dd, 1 H, $J_{6,6'}$ = 8.7, $J_{5,6}$ = 6.1 Hz, H-6), 4.06 (ddd, 1 H, $J_{4,5}$ = 8.6, $J_{5,6'}$ = 6.0 Hz, H-5), 4.02 (dd, 1 H, $J_{3,4}$ = 7.6, $J_{2,3}$ = 3.1 Hz, H-3), 3.97 (dd, 1 H, H-4), 3.97 (dd, 1 H, H-6'), 3.82 (ddd, 1 H, $J_{1,2}$ = 4.9, $J_{1',2}$ = 4.3 Hz, H-2), 3.78 (dd, 1 H, $J_{1,1'}$ = 11.5 Hz, H-1), 3.75 (dd, 1 H, H-1'), 2.24–2.64 (bs, 2 H, OH), 1.43, 1.42, 1.39, 1.35 (4s, 12 H, 4×CMe_2). ^{13}C NMR δ: 109.9, 109.8 (2×CMe_2), 81.8 (C-3), 77.4 (C-4), 77.3 (C-5), 70.4 (C-2), 68.0 (C-6), 64.9 (C-1), 27.1, 26.8, 26.5, 25.2 (4×CMe_2).

2,3:4,5-DI-O-ISOPROPYLIDENE-D-ARABINOSE (8)

Sodium periodate (7.20 g, 34.0 mmol) was added to a solution of 3,4;5,6-di-O-iso-propylidene-D-glucitol (7, 3.15 g, 12.0 mmol) in a mixture of diethyl ether (50 mL) and water (20 mL), and the mixture was stirred at room temperature for 1 h. Aqueous layer was separated and extracted with ether (20 mL). Combined organic phases were dried and concentrated to afford the title product as a colorless oil (2.1 g, 9.12 mmol, 76%), $[\alpha]_D$ −15.2 (lit.[10] $[\alpha]_D$ −17.9; c 1, CHCl$_3$). ^{1}H NMR δ: 9.76 (d, 1 H, $J_{1,2}$ = 1.0 Hz, H-1), 4.41 (dd, 1 H, $J_{2,3}$ = 6.2 Hz, H-2), 4.17 (ddd, 1 H, $J_{3,4}$ = 7.1, $J_{4,5}$ = 6.2, $J_{4,5'}$ = 4.0 Hz, H-4), 4.14 (dd, 1 H, $J_{5,5'}$ = 8.4 Hz, H-5), 4.07 (dd, 1 H, H-3), 3.98 (dd, 1 H, H-5'), 1.48, 1.43, 1.38, 1.35 (4s, 12 H, 2×CMe_2). ^{13}C NMR δ: 199.8 (C-1), 111.8. 110.0 (2×CMe_2), 83.2 (C-2), 77.7 (C-3), 76.4 (C-4), 66.9 (C-5), 26.9, 26.6, 26.2, 25.0 (4×CMe_2).

The compound thus obtained is sufficiently pure for further transformations.

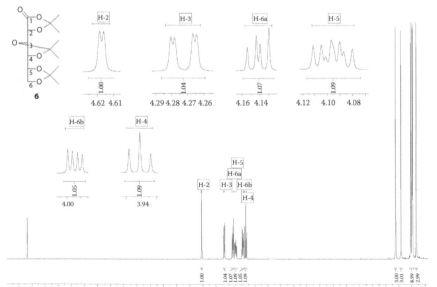

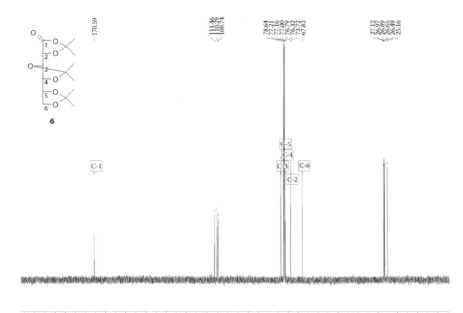

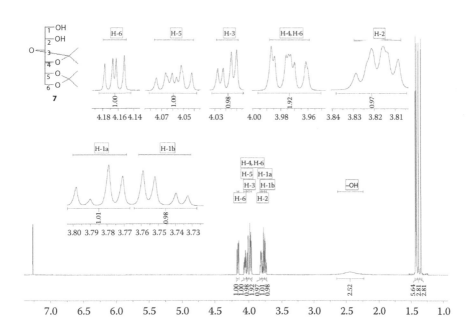

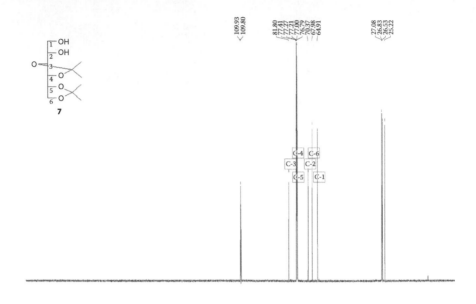

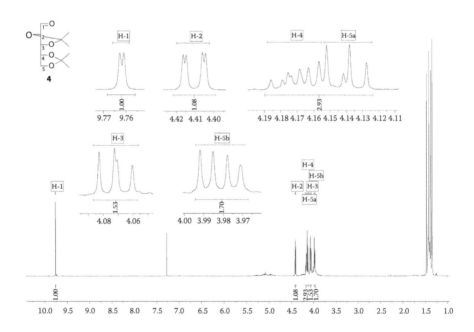

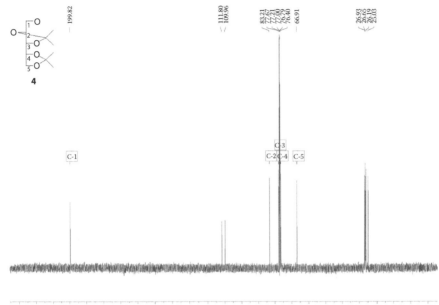

REFERENCES

1. (a) Zinner, H.; Wittenburg, E.; Rembarz, G. *Chem. Ber.*, **1959**, *92*, 1614–1617; *review:* (b) Zhdanov, Yu. A.; Alexeev, E.; Alexeeva, V. G. *Adv. Carbohydr. Chem. Biochem.*, **1972**, *27*, 227–299; *see also:* (c) Regeglink, H.; de Rouville, E.; Chittenden, G. J. F. *Recl. Trav. Chim. Pays-Bas*, **1987**, *106*, 461–464.
2. (a) Esswein A.; Betz, R.; Schmidt, R. R. *Helv. Chim. Acta*, **1989**, *72*, 213–223. (b) Gao, J.; Hlirter, R.; Gordon, D. M.; Whitesides, G. M. *J. Org. Chem.*, **1994**, *59*, 3714–3715.
3. (a) Fukuda, Y.; Sasai, H.; Suami, T., *Bull. Chem. Soc. Jpn.*, **1981**, *54*, 1830–1833. (b) Fukuda, Y.; Kitasato, H.; Sasai, H. Suami, T. *Bull. Chem. Soc. Jpn.*, **1982**, *55*, 880–886.
4. *Recent examples:* (a) Jarosz, S.; Skóra, S.; Kościołowska, I. *Carbohydr. Res.*, **2003**, *338*, 407–413. (b) Cieplak, M.; Jarosz, S. *Tetrahedron: Asymmetry*, **2011**, *22*, 1757–1762. (c) Cieplak, M.; Ceborska, M.; Cmoch, P.; Jarosz, S. *Tetrahedron: Asymmetry*, **2012**, *23*, 1213–1217.
5. (a) White, J. D.; Jensen, M. S. *J. Am. Chem. Soc.*, **1995**, *117*, 6224–6233. (b) Tang, Ch.-J.; Wu, Y., *Tetrahedron*, **2007**, *63*, 4887–4906.
6. Jarosz, S. *Curr. Org. Chem.*, **2008**, *12*, 985–994.
7. Synthesis of tri-acetonide **6**: Jarosz, S.; Zamojski, A. *J. Carbohydr. Chem.*, **1993**, *12*, 1223–1228.
8. Yadav, J. S.; Barma, D. K. *Tetrahedron*, **1996**, *52*, 4457–4466.
9. Yadav, J. S.; Madhava Rao, B.; Sanjeeva Rao, K. *Tetrahedron: Asymmetry*, **2009**, *20*, 1725–1730.
10. Allevi, P.; Ciuffreda, P.; Tarocco, G.; Anastasia, M. *Tetrahedron: Asymmetry*, **1995**, *6*, 2357–2364.

24 Phenyl 2-O-acetyl-3-O-allyl-4-O-benzyl-1-thio-β-D-glucopyranoside, a Versatile, Orthogonally Protected Building Block

László Kalmár, Zoltán Szurmai,
János Kerékgyártó, András Guttman,
Marie Bøjstrup,[†] and Károly Ágoston*

CONTENTS

D-Glucose is a cheap raw material available in large quantities and carries multiple chiral centers. The multifunctionality around the ring provides a platform for creation of molecular diversity,[1] but the secondary hydroxyl groups show only minor differences in reactivity. The traditional approach to differentiate among the hydroxyl functions is *via* selective protecting group manipulations.

It has always been of high interest to prepare orthogonally protected monosaccharides. To date, there are only a few examples in the literature of D-glucose protected with four different functionalities.[2] We have recently developed a practical eight-step synthesis (Scheme 24.1) of such substance from commercially available diacetone-D-glucose, which does not require chromatographic purification.

---

* Corresponding author; e-mail: agoston.karoly@ttk.mta.hu.
[†] Checker; e-mail: Marie.Bojstrup@carlsberg.dk.

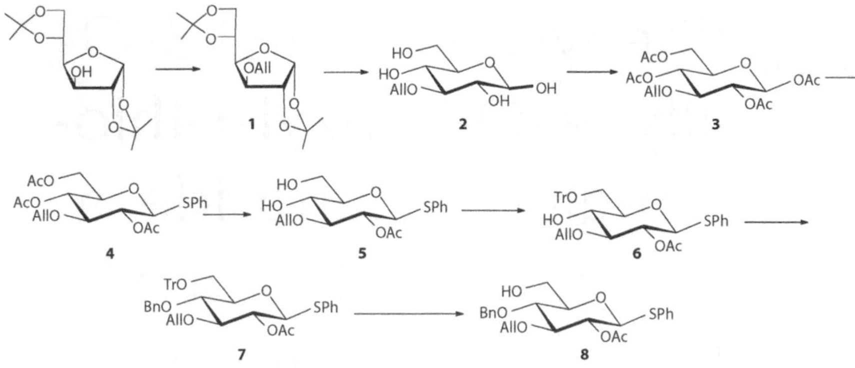

SCHEME 24.1   Preparation of phenyl 2-*O*-acetyl-3-*O*-allyl-4-*O*-benzyl-1-thio-β-D-gluco-pyranoside.

Most of the intermediates are crystalline; crude products from each conversion were sufficiently pure for the next step. The overall yield of the eight-step procedure is 27% (~85% per step). Because the crystalline, fully characterized derivative **4** is known,[3] here we describe only the last four steps of the synthetic path.

A carefully controlled Zemplén deacetylation[4] of derivative **4** leaves the 2-*O*-acetyl group intact, giving diol **5**. The progress of the transformation must be monitored frequently, and the reaction should be terminated when the conversion of the starting material into the desired diol is optimal (no starting material left, and less than 5% of the undesired triol is formed, as indicated by TLC). Prolonged reaction time results in excessive formation of the triol. Selective protection of the primary hydroxyl function in **5** as a trityl ether afforded **6**. The crude oily product was used for benzylation in the presence of NaH (→**7**). The ester functionality at O-2 remained intact during the reaction due to steric hindrance. To avoid deacetylation during work-up, the decomposition of the excess of NaH was done in a biphasic system. The trityl protecting group was removed from compound **7** by acid hydrolysis yielding the final derivative **8** in crystalline form. The only purification step required during the whole process was the crystallization of the final product from *iso*propyl alcohol. Starting from **4**, the target compound **8** was obtained in 58% overall yield in four steps (average yield 87% per step).

EXPERIMENTAL METHODS

GENERAL METHOD

Commercially available (Carbosynth, UK), starting 1,2:5,6-di-*O*-isopropylidene-α-D-glucofuranose was used without further purification. Solvents were dried according to standard methods. Melting points (uncorrected) were determined by a Griffin apparatus. Optical rotations were measured with a Jasco optical activity AA-10R polarimeter. NMR spectra were recorded on a Varian Gemini 2000 (200 MHz for ^{1}H and 50 MHz for ^{13}C) and on a Varian Unity-Inova (300 MHz for

[1]H and 75 MHz for ^{13}C) spectrometer in CDCl₃ as solvent. All chemical shifts are quoted in ppm downfield from the characteristic signals ([1]H: 0.00 ppm [TMS], [13]C: 77.00 ppm [CDCl₃]). DC-Alufolien Kieselgel 60 F_{256} plates were used for TLC. MS spectra were recorded on an Applied Biosystems 3200 QTRAP spectrometer.

PHENYL 2-O-ACETYL-3-O-ALLYL-1-THIO-β-D-GLUCOPYRANOSIDE (5)

Phenyl 2,4,6-tri-O-acetyl-3-O-allyl-1-thio-β-D-glucopyranoside[3] (4) (37.70 g, 86.0 mmol) was dissolved in a mixture of MeOH (50 mL) and THF (50 mL) and cooled to +4°C. NaOMe (300 µL, 25% solution) was added at +4°C, and the mixture was stirred for 3 h keeping the temperature at +4°C. The progress of the reaction was monitored by TLC (95:5 CH₂Cl₂–MeOH). The reaction was stopped by the addition of Amberlite IR 120 (H⁺) resin (1.9 g), the mixture was stirred for 5 min, then the resin was filtered off, and the filtrate was concentrated and then coconcentrated with toluene (2×50 mL). The residue that contains approximately 5% triol impurity crystallized upon standing. The crude crystalline material was used for the next step without further purification. One hundred milligrams of the crude material was purified by column chromatography for analytical purposes. Pure **5** had m.p. 108°C–110°C (toluene–MeOH), $[\alpha]_D$ −35.5 (c 0.62, CHCl₃). ^{1}H NMR δ: 7.50–7.20 (m, 5 H, aromatic), 5.94–5.78 (m, 1 H, –CH₂CHCH₂), 5.28–5.12 (m, 2 H, –CH₂CHCH₂), 4.93 (dd, 1 H, $J_{1,2}$ 10.1 Hz, $J_{2,3}$ 9.8 Hz, H-2), 4.68 (d, 1 H, H-1), 4.25–4.12 (m, 2 H, –CH₂CHCH₂), 3.90 and 3.79 (each m, 2 H, H-6), 3.65 (ddd, 1 H, $J_{3,4}$ 9.4 Hz, $J_{4,5}$ 9.4 Hz, $J_{4,OH}$ 3.8 Hz, H-4), 3.46 (dd, 1 H, H-3), 3.40 (m, 1 H, H-5), 3.34 (d, 1 H, 4-OH), 2.56 (dd, 1 H, $J_{6,OH}$ 6.5 Hz 6-OH), 2.14 (s, 3 H, OAc). ^{13}C NMR (CDCl₃) δ: 169.5 (C=O), 134.5 (–CH₂CHCH₂), 133.0, 131.8, 129.0 and 127.8 (aromatic), 117.3 (–CH₂CHCH₂), 86.2 (C-1), 83.5 (C-3), 79.4 (C-5), 73.5 (–CH₂CHCH₂), 71.6 (C-2), 69.8 (C-4), 62.2 (C-6), 21.0 (OAc); MS: calcd for: C₁₇H₂₂O₆S; 354, found: 355 [M+H]⁺, 372 [M+NH₄]⁺, 377 [M+Na]⁺, 393 [M+K]⁺.

PHENYL 2-O-ACETYL-3-O-ALLYL-6-O-TRITYL-1-THIO-β-D-GLUCOPYRANOSIDE (6)

Et₃N (31.5 ml, 230 mmol) was added to a solution of the foregoing crude phenyl 2-O-acetyl-3-O-allyl-1-thio-β-D-glucopyranoside (5) in CH₂Cl₂ (100 mL), followed by trityl chloride (31.2 g, 112 mmol), and the reaction mixture was stirred for 20 min. DMAP (500 mg, 4.1 mmol) was added, and stirring was continued for 12 h at room temperature. The progress of the reaction was monitored by TLC (7:3 hexane–EtOAc). The reaction mixture was diluted with CH₂Cl₂ (300 mL), washed with water (2×50 mL), dried over MgSO₄, filtered, and concentrated. The crude yellow syrup (73.4 g) was used for the next reaction without further purification. One hundred milligrams of the crude material was purified by column chromatography for analytical purposes. $[\alpha]_D$ −23.2 (c 0.7, CHCl₃); ^{1}H NMR δ: 7.60–7.10 (m, 20 H, aromatic), 5.90–5.76 (m, 1 H, –CH₂CHCH₂), 5.24–5.10 (m, 2 H, –CH₂CHCH₂), 4.95 (dd, 1 H, $J_{1,2}$ 10.1 Hz, $J_{2,3}$ 9.7 Hz, H-2), 4.64 (d, 1 H, H-1), 4.25–4.10 (m, 2 H, –CH₂CHCH₂), 3.65 (ddd, 1 H, $J_{3,4}$ 9.4 Hz, $J_{4,5}$ 9.4 Hz, $J_{4,OH}$ 2.3 Hz, H-4), 3.40 (m, 4 H, H-3, H-5 and H-6), 2.45 (d, 1 H, 4-OH), 2.14 (s, 3 H, OAc). ^{13}C NMR (CDCl₃) δ: 169.4 (C=O), 134.5 (–CH₂CHCH₂), 143.6, 132.9, 132.2, 128.6, 127.8, 127.7 and 127.0 (aromatic), 117.2

(–CH$_2$CHCH$_2$), 86.8 (C-1), 85.9 (-CPh$_3$), 83.7 (C-3), 78.5 (C-5), 73.5 (–CH$_2$CHCH$_2$), 71.4 (C-2), 70.9 (C-4), 63.4 (C-6), 21.1 (OAc); MS: calcd. for: C$_{36}$H$_{36}$O$_6$S; 596, found: 597 [M+H]$^+$, 614 [M+NH$_4$]$^+$.

PHENYL 2-O-ACETYL-3-O-ALLYL-4-O-BENZYL-6-O-TRITYL-1-THIO-β-D-GLUCOPYRANOSIDE (7)

Crude **6** (73.4 g) was dissolved in DMF (200 mL), and the solution was cooled in an ice bath. NaH (60% in oil) (4.34 g, 113 mmol) was added to the mixture, and the resulting suspension was stirred for 1 h in an ice bath. A solution of benzyl bromide (13.4 mL, 113 mmol) in DMF (40 mL) was added to the suspension over a period of 1 h while keeping the reaction mixture in ice bath. The reaction mixture was stirred further for 3 h, and TLC (7:3 hexane–EtOAc) was used to monitor the reaction. Upon completion of the reaction, the mixture was diluted with toluene (800 mL), and the excess of NaH was decomposed by careful addition of water (10 mL). The mixture was washed with water (2 × 100 mL), and the organic phase was concentrated and used for the next step without further purification. One hundred milligrams of the crude material was purified by column chromatography for analytical purposes. [α]$_D$ −13.8 (c 0.6, CHCl$_3$); ^{1}H NMR δ: 7.65–6.88 (m, 25 H, aromatic), 5.95–5.78 (m, 1 H, –CH$_2$CHCH$_2$), 5.25–5.10 (m, 2 H, –CH$_2$CHCH$_2$), 5.05 (dd, 1 H, J$_{1,2}$ 10.1 Hz, J$_{2,3}$ 9.8 Hz, H-2), 4.70 (d, 1 H, H-1), 4.62 and 4.28 (each d, 2 H, J$_{gem}$ 10.3 Hz, –CH$_2$Bn), 4.24 and 4.11 (each m, 2 H, –CH$_2$CHCH$_2$), 3.73 (dd, 1 H, J$_{3,4}$ 9.4 Hz, J$_{4,5}$ 9.4 Hz, H-4), 3.58 and 3.25 (each m, 2 H, J$_{gem}$ 10.1 Hz, H-6), 3.50 (dd, 1 H, H-3), 3.46 (m, 1 H, H-5), 2.14 (s, 3 H, OAc); ^{13}C NMR (CDCl$_3$) δ 169.5 (C=O), 146.8, 143.8 and 137.5 (aromatic), 134.5 (–CH$_2$CHCH$_2$), 133.0, 132.3, 128.9, 128.7, 128.2, 128.2, 127.9, 127.8, 127.7, 127.7, 127.2, and 126.9 (aromatic), 117.0 (–CH$_2$CHCH$_2$), 86.4 (C-1), 86.0 (-CPh$_3$), 84.1 (C-3), 79.0 (C-5), 77.4 (C-4), 75.0 (–CH$_2$CHCH$_2$), 74.2 (–CH$_2$Bn), 71.9 (C-2), 62.3 (C-6), 21.1 (OAc); MS: calcd. for: C$_{43}$H$_{42}$O$_6$S; 686, found: 704 [M+NH$_4$]$^+$.

PHENYL 2-O-ACETYL-3-O-ALLYL-4-O-BENZYL-1-THIO-β-D-GLUCOPYRANOSIDE (8)

Crude **7** was dissolved in CH$_2$Cl$_2$ (150 mL), then MeOH (50 mL) and HCl (30 mL, aqueous, 20%) were added to the solution, and the biphasic reaction mixture was stirred for 12 h at room temperature. When the reaction was complete (TLC, 7:3 hexane–EtOAc), the mixture was diluted with CH$_2$Cl$_2$ (300 mL), washed with water (3 × 100 mL), dried (MgSO$_4$), and filtered, and the filtrate was concentrated. At this point, the crude residue slowly started to crystallize. Hexane (250 mL) was added, and the mixture was heated to 50°C with stirring. The resulting white suspension was allowed to cool at room temperature and kept at 4°C for 10 h. The solid was filtered off yielding 26.32 g of white crystals showing ~90% purity. The pure final product was obtained by recrystallization from iPrOH (40 mL) to afford 22.17 g of **8** (58% over four steps). M.p.: 112°C–114°C (iPrOH), [α]$_D$ −8.2 (c 0.6, CHCl$_3$). ^{1}H NMR δ: 7.50–7.20 (m, 10 H, aromatic), 5.94–5.80 (m, 1 H, –CH$_2$CHCH$_2$), 5.29–5.12 (m, 2 H, –CH$_2$CHCH$_2$), 4.95 (m, 1 H, J$_{1,2}$ 10.1 Hz, J$_{2,3}$

9.8 Hz, H-2), 4.84 and 4.63 (each d, 2 H, J_{gem} 10.8 Hz, –*CH*$_2$Bn), 4.67 (d, 1 H, H-1), 4.28 and 4.16 (each m, 2 H, –*CH*$_2$CHCH$_2$), 3.86 and 3.68 (each m, 2 H, J_{gem} 12.0 Hz, H-6), 3.54 (m, 2 H, H-3 and H-4), 3.40 (m, 1 H, H-5), 2.14 (s, 3 H, OAc), 1.95 (dd, 1 H, $J_{6,OH}$ 7.5 and 6.2 Hz, OH); ^{13}C NMR (CDCl$_3$) δ 169.4 (C=O), 137.7 (aromatic), 134.5 (–CH$_2$*CH*CH$_2$), 132.8, 132.0, 129.0, 128.5, 128.1, 128.0, and 127.9 (aromatic), 117.0 (–CH$_2$CH*CH*$_2$), 85.9 (C-1), 83.5 (C-3), 79.2 (C-5), 77.1 (C-4), 75.1 (–*CH*$_2$CHCH$_2$), 74.0 (–*CH*$_2$Bn), 71.9 (C-2), 61.9 (C-6), 21.0 (OAc); MS: calcd for: C$_{24}$H$_{28}$O$_6$S; 444, found: 462 [M+NH$_4$]$^+$ Anal. calcd for C$_{24}$H$_{28}$O$_6$S: C, 64.84 H, 6.35. Found C, 64.76 H, 6.40.

ACKNOWLEDGMENTS

This research was realized in the frames of TÁMOP 4.2.4. A/2-11-1-2012-0001 "National Excellence Program - Elaborating and operating an inland student and researcher personal support system convergence program." The project was subsidized by the European Union and co-financed by the European Social Fund.

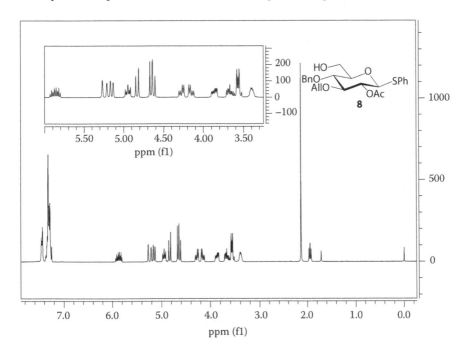

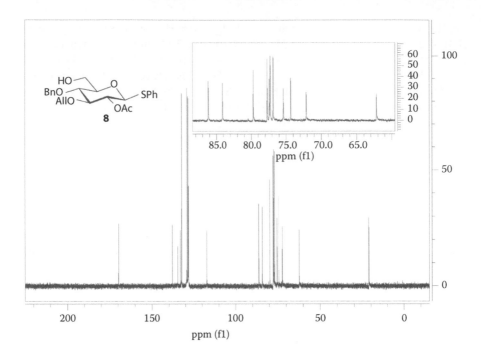

REFERENCES

1. (a) Thanh, G. L.; Abbenante, G.; Becker, B.; Gratwohl, M.; Halliday, J.; Tometzki, G.; Zuegg, J.; Meutermans, W., *Drug Discov. Today*, **2003**, *8*, 701–709. (b) Filice, M.; Palomo, J. M., *RSC Adv.*, **2012**, *2*, 1729–1742.
2. Recent examples: (a) Wang, C. C.; Lee, J. C.; Yuo, S. Y.; Hung, S. C.; Lee, C. C.; Chang, K. L.; Hung, S. C., *Nature*, **2007**, *446*, 896–899. (b) Français, A.; Urban, D.; Beau, J. M., *Angew. Chem. Int. Ed.*, **2007**, *46*, 8662–8665. (c) Demizu, Y.; Kubo, Y.; Miyoshi, H.; Maki, T.; Matsumura, Y.; Moriyama, N.; Onomura, O., *Org. Lett.*, **2008**, *10*, 5075–5077.
3. Peri, F.; Nicotra, F.; Leslie, C. P.; Micheli, F.; Seneci, P.; Marchioro, C. *J. Carbohydr. Chem.*, **2003**, *22*, 57–71.
4. (a) Lipták, A.; Szurmai, Z.; Nánási, P.; Neszmélyi, A. *Carbohydr. Res.*, **1982**, *99*, 13–21. (b) Eby, R.; Schuerch, C. *Carbohydr. Res.*, **1981**, *92*, 149–153. (c) Dmitriev, B. A.; Nikolaev, A. V.; Shashkov, A. S.; Kochetkov, N. K. *Carbohydr. Res.*; **1982**, *100*, 195–206. (d) Takeo, K.; Nakaji, T.; Shinmitsu, K. *Carbohydr. Res.*, **1984**, *133*, 275–287. (e) Szurmai, Z.; Kerékgyártó, J.; Harangi, J.; Lipták, A. *Carbohydr. Res.*, **1987**, *164*, 313–325. (f) Lipták, A.; Szabó, L. *J. Carbohydr. Chem.*, **1989**, *8*, 629–644.

25 Synthesis of Methyl 4-azido-2,3-di-O-benzoyl-4-deoxy-α-L-arabino-pyranoside

Bernhard Müller, Sameh E. Soliman,[†]
*Markus Blaukopf, Ralph Hollaus, and Paul Kosma**

CONTENTS

(i) Bu$_2$SnO, toluene, reflux, 5.5 h; (ii) TsCl, DMAP, 1,4-dioxane, r.t., 96 h;
(iii) BzCl, pyridine, DMAP r.t., 5 h; (iv) NaN$_3$, DMSO, 110°C, 2.5 h.

4-Amino-4-deoxy-L-arabinose (Ara4N) is a component of numerous bacterial lipopolysaccharides and is associated with bacterial resistance mechanisms against antimicrobial agents.[1] For the preparation of Ara4N, methyl β-D-xylopyranoside **1** serves as commercially available starting material that needs to be regioselectively protected in order to allow introduction of the azido group at position 4 with inversion of configuration. The previously reported formation of a 2,3-O-isopropylidene intermediate from **1** is frequently hampered by the formation of by-products, which renders scale-up and purification difficult.[2] A regioselective tosylation of **1** using catalytic amounts of dimethyltin dichloride, Hünig's

---

* Corresponding author; e-mail: paul.kosma@boku.ac.at.
† Checker, under supervision of Pavol Kováč; e-mail: kpn@helix.nih.gov.

base, and THF has been reported to provide compound **2** in 81% after chromatographic purification.[3] In an alternative approach, the conversion of an intermediate 3,4-*O*-stannylene acetal into tosylate **2** is near quantitative.[4,5] Benzoylation of **2** produces crystalline **3** in 95% yield. Subsequent azide introduction can be accomplished at 90°C and 15 h reaction time or by increasing the reaction temperature to 110°C (safety shield required!) within 2.5 h, which provides the target compound **4** in high (91%) yield.[6] Compound **4** is a versatile precursor for Ara4N glycosides *via* transglycosylation of the methyl glycoside and allows easy transformation into variably protected Ara4N glycosyl donors.[7]

EXPERIMENTAL METHODS

GENERAL METHODS

Methyl β-D-xylopyranoside **1** was purchased from Sigma. Melting points were measured on a Kofler hot stage. Solvents (DMF, pyridine, 1,4-dioxane, DMSO) were dried over activated 4 Å molecular sieves. Solutions in organic solvents were dried with Na_2SO_4 and concentrated under reduced pressure at <40°C. Optical rotations were measured with a PerkinElmer 243 B polarimeter. Thin-layer chromatography (TLC) was performed on Merck precoated glass plates: 5 × 10 cm, layer thickness 0.25 mm, silica gel $60F_{254}$, or HPTLC plates with 2.5 cm concentration zone. Spots were visualized by dipping reagent (anisaldehyde–H_2SO_4) and heating (~250°C) or UV light (254 nm). For column chromatography, silica gel (0.040–0.063 mm) was used. NMR spectra were recorded in $CDCl_3$ with Bruker Avance III 600 instrument (600.2 MHz for ^{1}H and 150.9 MHz for ^{13}C) using standard Bruker NMR software. ^{1}H NMR spectra were referenced to TMS (δ 0), and ^{13}C NMR spectra were referenced to $CDCl_3$ (δ 77.00). Tosyl chloride was freshly crystallized from petroleum ether, to give colorless material.

METHYL 4-*O*-*p*-TOLUENESULFONYL-β-D-XYLOPYRANOSIDE (2)

Methyl β-D-xylopyranoside **1** (15.48 g, 94.1 mmol) and dibutyltin oxide (23.66 g, 95 mmol) were suspended in dry toluene (300 mL) and heated to reflux. Full conversion was accomplished when ~1.7 mL of water had been collected *via* a Dean–Stark trap (~5.5 h). The mixture was cooled to 40°C and concentrated. The remaining, slightly yellow solid was dried under reduced pressure overnight. A solution of tosyl chloride (21.74 g, 114.1 mmol) in dry 1,4-dioxane (180 mL) was added dropwise during 40 min at 13°C–15°C (inside temperature, controlled with ice-water bath) to a suspension of the foregoing stannylidene intermediate and a catalytic amount of DMAP (10 mg, 0.08 mmol) in dry 1,4-dioxane (140 mL). The suspension was stirred for 96 h at r.t., when TLC (EtOAc) indicated that no starting material was present. The colorless solution was concentrated, and the crude product was dissolved in EtOAc (100 mL). Silica gel (250 g) was added, the solvent was removed, and the remaining off-white solid was directly applied on top of a flash chromatography silica gel (500 g) column. The column was first eluted with toluene (2 L), to remove most of the tin impurities, followed by elution with EtOAc. Fractions containing the carbohydrate were combined and concentrated. The residue was suspended in *n*-hexane and sonicated for 30 min at r.t. The solid was filtered off, the extraction with hexane under sonication was repeated two more times, and the last solid was washed thoroughly with *n*-hexane (to remove residual tin salts) and dried to give a colorless solid (28.8 g, 95%). The material can be directly used for the next step. Crystallization from EtOH gave **2** as colorless needles, m.p. 134.5°C–135°C, [α]$_D^{20}$ −58.3 (*c* 1.05, CHCl$_3$); Lit.[4] 135°C–136°C; [α]$_D$ not reported. R_f=0.55 (EtOAc). ^{1}H NMR δ: 7.83 (d, 2 H, *J* 12.0 Hz, arom. H), 7.36 (d, 2 H, *J* 12.0 Hz, arom. H), 4.36 (ddd, 1 H, $J_{4,5eq}$ 4.9, $J_{3,4}$ 8.0 Hz, H-4), 4.23 (d, 1 H, $J_{1,2}$ 6.5 Hz, H-1), 4.00 (dd, 1 H, $J_{5ax,5eq}$ 12.2 Hz, H-5*eq*), 3.68 (t, 1 H, $J_{3,2}$ 8.0 Hz, H-3), 3.49 (s, 3 H, OMe), 3.41 (dd, 1 H, $J_{5ax,4}$ 8.6 Hz, H-5*ax*), 3.38 (dd, 1 H, H-2), 3.08 and 2.85 (br. s, each 1 H, 2-OH, 3-OH), 2.45 (s, 3 H, *Me*Ph). ^{13}C NMR δ: 145.4 (C-1 Ar), 132.8 (C-4 Ar), 130.0 (C-3, C-5 Ar), 128.0 (C-2, C-6 Ar), 103.2 (C-1), 77.6 (C-4), 72.4 (C-2, C-3), 62.0 (C-5), 56.9 (OMe), 21.7 (*Me*Ph).

METHYL 2,3-DI-*O*-BENZOYL-4-*O*-*P*-TOLUENESULFONYL-β-D-XYLOPYRANOSIDE (3)

A solution of benzoyl chloride (26.1 mL, 226 mmol) was added dropwise during 5 min at 0°C to a solution of **2** (28.8 g, 90.5 mmol) in dry pyridine (80 mL). DMAP (10 mg, 0.08 mmol) was added, the mixture was allowed to warm up to room temperature, stirred for 5 h, and diluted with CH$_2$Cl$_2$ (150 mL). The organic phase was successively washed with aq. 2 M HCl (2×250 mL), water (150 mL), satd. aq. NaHCO$_3$ (150 mL), and brine (150 mL) and concentrated. A solution of the residue in toluene (200 mL) was concentrated, and the residue was crystallized from EtOH, to afford **3** as colorless needles (45.4 g 95%); m.p. 139°C–141°C; [α]$_D^{20}$ +40 (*c* 1, CHCl$_3$), Lit.[8] 141°C–142°C; [α]$_D$ value not reported; R_f=0.58 (2:1 *n*-hexane–acetone). ^{1}H NMR δ: 7.90–7.89 (m, 2 H, arom. H), 7.70–7.69 (d, 2 H, arom. H), 7.62 (d, 2 H, arom. H), 7.52–7.48 (m, 2 H, arom. H), 7.36–7.31 (m, 4 H, arom. H), 6.96 (d, 2 H, arom. H), 5.56 (t, 1 H, $J_{2,3}=J_{3,4}$ 8.6 Hz, H-3), 5.23 (dd, 1 H, $J_{1,2}$ 6.7 Hz, H-2), 4.64 (ddd, 1 H, H-4), 4.57 (d, 1 H, *J* 6.7 Hz, H-1), 4.38 (dd, 1 H, $J_{4,5eq}$ 5.1, $J_{5ax,5eq}$ 12.2 Hz, H-5*eq*), 3.69 (dd, 1 H, $J_{4,5ax}$ 8.8 Hz, H-5*ax*), 3.49 (s, 3 H, OMe), 2.19 (s, 3 H, *Me*Ph). ^{13}C NMR δ: 165.0, 164.8 (C=O), 144.9, 133.3, 133.2, 132.6, 129.8, 129.8, 129.75, 129.0, 128.7, 128.3, 128.2, 127.65 (C Ar), 101.65 (C-1), 75.0 (C-4), 70.8 (C-2, C-3), 62.9 (C-5), 56.9 (OMe), 21.55 (*Me*Ph).

METHYL 4-AZIDO-2,3-DI-*O*-BENZOYL-4-DEOXY-α-L-ARABINOPYRANOSIDE (4)

A suspension of **3** (45.4 g, 86.3 mmol) and sodium azide (14.0 g, 215 mmol) in dry DMSO (85 mL) was stirred at 110°C for 2.5 h,* when TLC (2:1 *n*-hexane–acetone) showed that the conversion was complete. The mixture was cooled to r.t., diluted with CH$_2$Cl$_2$ (300 mL), and washed with water (2×250 mL). The organic layer was washed with satd. aq. NaHCO$_3$ (200 mL), dried, and concentrated to give **4** as a slightly off-white solid, which was crystallized from EtOH. Yield: 31.2 g (91%),

---

* This operation should be performed behind a safety shield. Alternatively, the reaction can be performed at 90°C for 15 h.

colorless needles. A portion was recrystallized from EtOH, to give analytical sample, m.p. 106°C–107°C, [α]$_D^{20}$ +18.3 (c 1.0, CHCl$_3$); Lit.[6]: 104°C–105°C; [α]$_D$ +18.6 (c 0.3, CHCl$_3$); R_f=0.64 (2:1 n-hexane/acetone). ^{1}H NMR δ: 8.05–8.03 (m, 2 H, arom. H), 8.01–8.00 (m, 2 H, arom. H), 7.56–7.53 (m, 2 H, arom. H), 7.43–7.40 (m, 4 H, arom. H), 5.57 (dd, 1 H, $J_{1,2}$ 5.6, $J_{2,3}$ 8.0 Hz, H-2), 5.51 (dd, 1 H, $J_{3,4}$ 3.4 Hz, H-3), 4.60 (d, 1 H, J 5.6 Hz, H-1), 4.20–4.16 (m, 2 H, H-4, H-5eq), 3.81–3.77 (m, 1 H, $J_{4, 5ax}$ 4.3, $J_{5eq, 5ax}$ 13.8 Hz, H-5ax), 3.50 (s, 3 H, OMe). ^{13}C NMR δ: 165.64, 165.04 (C=O), 133.59, 133.34, 130.02, 129.81, 129.28, 128.75, 128.50, 128.40 (C Ar), 101.06 (C-1), 71.48 (C-3), 69.17 (C-2), 61.68 (C-5), 57.57 (C-4), 56.49 (OMe). Anal. calcd for C$_{20}$H$_{19}$N$_3$O$_6$: C, 60.45; H, 4.82; N, 10.57: Found: C, 60.58; H, 4.78; N, 10.36.

ACKNOWLEDGMENTS

The authors thank the Austrian Science Fund FWF for the financial support (project P 19295), Dr. Andreas Hofinger for measuring the NMR spectra, and Christian Schuster for the technical assistance.

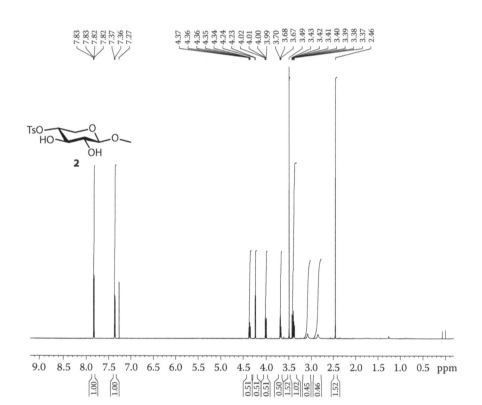

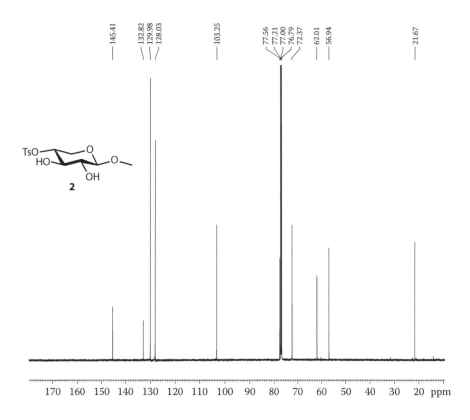

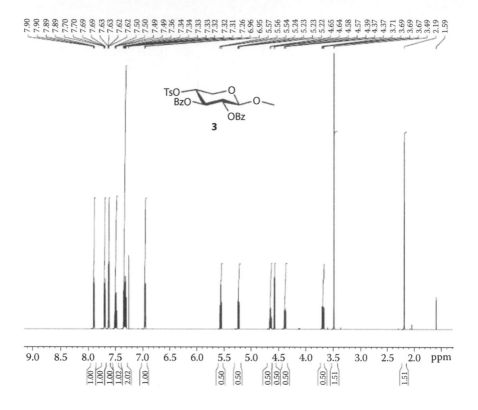

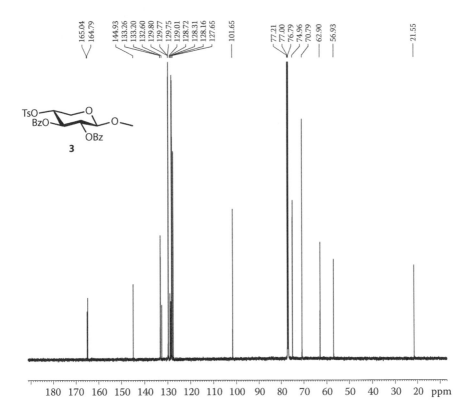

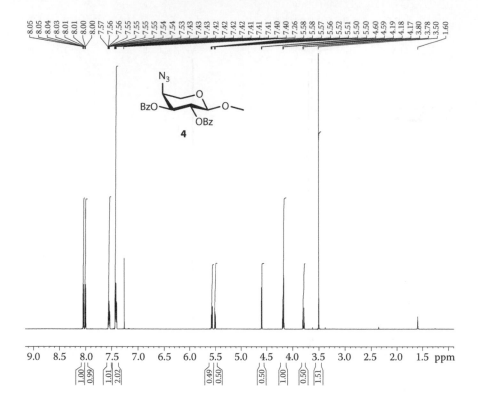

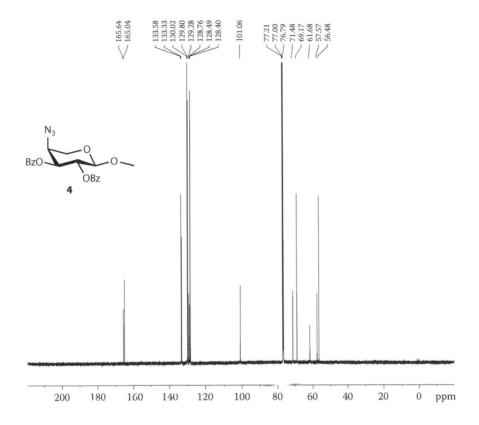

REFERENCES

1. (a) Trent, M. S.; Ribeiro, A. A.; Lin, S.; Cotter, R. S.; Raetz, C. H. *J. Biol. Chem.* **2001**, *276*, 43122–43131. (b) Hamad, M. A.; Di Lorenzo, F.; Molinaro A.; Valvano, M. A. *Mol. Microbiol.* **2012**, *85*, 962–974.

2. (a) Naleway, J. J.; Raetz, C. H. R.; Anderson, L. *Carbohydr. Res.* **1988**, *179*, 199–209. (b) Kline, T.; Trent, M. S.; Stead, C. M.; Lee, M. S.; Sousa, M. C.; Felise, H. B.; Nguyen, H. V.; Miller, S. I., *Bioorg. Chem. Lett.* **2008**, *18*, 1507–1510.

3. Demizu, Y.; Kubo, Y.; Miyoshi, H.; Maki, T.; Matsumara, Y.; Moriyama, N.; Onomura, O. *Org. Lett.* **2008**, *10*, 5075–5077.

4. Tsuda, Y.; Nishimura, M.; Kobayashi, T.; Sato, Y.; Kanemitsu, K. *Chem. Pharm. Bull.* **1991**, *39*, 2883–2887.

5. (a) Martinelli, M. J.; Vaidyanathan, R.; Khau, V. V. *Tetrahedron Lett.* **2000**, *41*, 3773–3776. (b) Martinelli, M. J.; Vaidyanathan, R.; Pawlak, J. M.; Nayyar, N. K.; Dhokte, U. P.; Doecke, C. W.; Zollars, L. M. H.; Moher, E. D.; Khau, V. V.; Košmrlj, B. *J. Am. Chem. Soc.* **2002**, *124*, 3578–3585.

6. Müller, B.; Blaukopf, M.; Hofinger, A.; Zamyatina, A.; Brade, H.; Kosma, P. *Synthesis*, **2010**, 3143–3151.

7. (a) Blaukopf, M.; Müller, B.; Hofinger, A.; Kosma, P. *Eur. J. Org. Chem.* **2012**, 119–131. (b) Yang, Y.; Oishi, S.; Martin, C. E.; Seeberger, P. H. *J. Am. Chem. Soc.* **2013**, *135*, 6262–6271.

8. Kondo, Y. *Carbohydr. Res.* **1982**, *110*, 339–344.

26 One-Pot Synthesis and Benzylation of Galactose, Glucose, and Mannose 1,2-Orthoesters

Xiao G. Jia, Abhijeet K. Kayastha, Scott J. Hasty,
Juan A. Ventura,[†] and Alexei V. Demchenko*

CONTENTS

1,2-Orthoesters[1–4] as well as structurally similar thioorthoesters[5] and cyanoethylidene derivatives[6] have found broad application in carbohydrate chemistry. For instance, orthoesters were discovered as one of the first glycosyl donors beyond the classical halides, hemiacetals, and acetates.[7] However, due to the 1,2-cis bicyclic structure of orthoesters, their use as glycosyl donors is limited to the synthesis of 1,2-trans-glycosides only. The common feature of orthoesters is that they can be selectively activated in the presence of stable leaving groups, such as thioglycosides and pentenyl glycosides. This offers the utility of orthoesters in multistep sequential oligosaccharide synthesis via selective activation protocol.[8–11] Orthoesters are also useful intermediates for the synthesis of various building blocks with differential protecting group pattern. For instance, orthoesters offer the direct entry into glycosyl donors of the superarmed series equipped with 2-*O*-acyl-3,4,6-tri-*O*-benzyl

---

* Corresponding author; e-mail: demchenkoa@umsl.edu.
[†] Checker, under supervision of Prof. J. Cristobal López; e-mail: jc.lopez@csic.es.

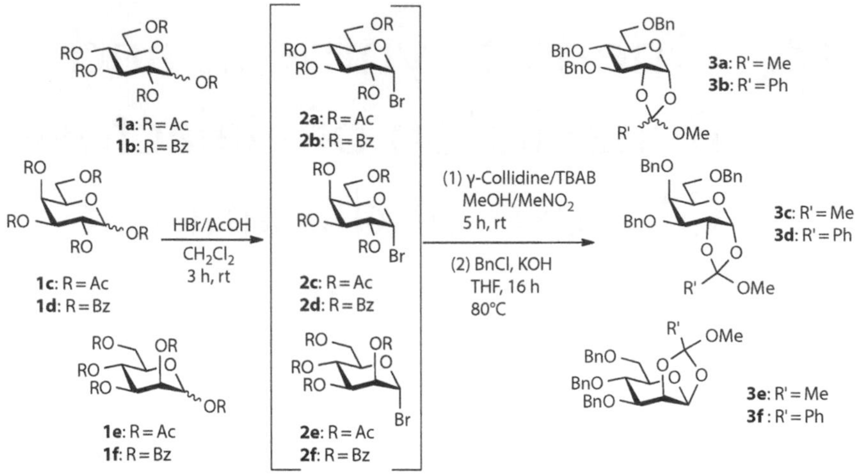

SCHEME 26.1   One-Pot Synthesis and Benzylation of 1,2-Orthoesters.

protecting group pattern.[12–14] Described herein is the synthesis of orthoesters of common hexoses from their pentaacetates or pentabenzoates. Sequential bromination of the latter, orthoesterification, deacylation, and benzylation can be all performed without purification of intermediates (Scheme 26.1). Conventions established previously for the synthesis of orthoesters[5] and reprotection thereof[15] were very useful in designing this convenient protocol.

EXPERIMENTAL METHODS

GENERAL METHODS

The reactions were performed using commercial reagents (Aldrich and Acros), and solvents were purified according to standard procedures. Peracetates were purchased from commercial sources and used as supplied; perbenzoates have been prepared from the respective unprotected sugars with benzoyl chloride in the presence of pyridine as previously described.[16] Molecular sieves (3 Å), used for reactions, were crushed and activated first *in vacuo* at 390°C during 8 h and then for 2–3 h at 390°C directly prior to application. Potassium hydroxide was crushed using mortar and pestle immediately before use. Reactions were monitored by TLC on Kieselgel 60 F_{254} (EM Science). The compounds were detected by UV light and by charring with 10% sulfuric acid in methanol. Solvents were dried over Na_2SO_4 and removed under reduced pressure at <40°C. Column chromatography was performed on silica gel 60 (70–230 mesh). Melting points were measured with Thomas–Hoover Uni-Melt capillary melting point apparatus. Optical rotations were measured with a Jasco P-1020 polarimeter. ^{1}H and ^{13}C NMR spectra were recorded at 300 and 75 MHz, respectively (Bruker Avance). The chemical shifts are referenced to the signal of the residual CHCl$_3$ ($\delta_H = 7.27$ ppm, $\delta_C = 77.23$ ppm) for solutions in CDCl$_3$. HRMS were measured with a JEOL MStation (JMS-700) mass spectrometer.

3,4,6-Tri-*O*-benzyl-1,2-*O*-(1-methoxyethylidene)-α-d-glucopyranose* (3a)[17]

A 33% solution of HBr in acetic acid (2.8 mL, 15.37 mmol) was added dropwise to a solution of 1,2,3,4,6-penta-*O*-acetyl-d-glucopyranose (**1a**, 1.00 g, 2.56 mmol) in dry CH_2Cl_2 (5.0 mL), and the resulting mixture was stirred under argon for 3 h at r.t. The mixture was diluted with CH_2Cl_2 (~125 mL) and washed with cold water (20 mL), cold sat. aq. $NaHCO_3$ (2×20 mL), and cold water (3×20 mL). The organic phase was separated and concentrated, and the crude residue containing 2,3,4,6-tetra-*O*-acetyl-α-d-glucopyranosyl bromide[18] (**2a**, ~2.56 mmol) was dissolved in nitromethane (8.0 mL). Molecular sieves (3 Å, 1.0 g) were added, and the resulting mixture was stirred under argon for 1 h at r.t. The flask was then covered with aluminum foil to block direct light; γ-collidine (0.61 mL, 4.60 mmol), dry methanol (130 μL, 3.06 mmol), and tetrabutylammonium bromide (660 mg, 2.04 mmol) were added sequentially; and the resulting mixture was stirred under argon for 6 h at r.t. Triethylamine (0.5 mL) was added, the solid was filtered off, and the filtrate was concentrated. A solution of the residue in CH_2Cl_2 (~150 mL) was washed with water (2×20 mL), and the organic phase was separated and concentrated. The residue was dissolved in THF (7.0 mL), benzyl chloride (5.14 mL, 44.68 mmol) and KOH (1.86 g, 33.19 mmol) were added, and the mixture was stirred for 16 h at reflux (~80°C). The mixture was allowed to cool to room temperature (r.t.), and the volatiles were removed. A solution of the residue in CH_2Cl_2 (~200 mL) was washed with water (3×50 mL); the organic phase was separated, dried, and concentrated. Chromatography (isocratic, 15:85 EtOAc–hexane) afforded the title compound (0.798–0.940 g), 66%–74% over three steps (typical endo/exo ratio 1:15) as a light-yellow syrup; R_f=0.6 (3:7 EtOAc–hexanes). ^{1}H NMR δ: 1.58 (s, 3H, COCH$_3$), 3.21 (s, 3H, CH$_3$), 3.57–3.59 (m, 2H, H-6a, 6b), 3.63 (dd, 1H, $J_{4,5}$=9.5 Hz, H-4), 3.73 (m, 1H, H-5), 3.81 (dd, 1H, $J_{3,4}$=4.3 Hz, H-3), 4.41 (d, 2H, ^2J=11.4 Hz, C*H$_2$*Ph), 4.35 (dd, $J_{2,3}$=3.6 Hz, H-2), 4.46 (dd, 2H, ^2J=12.2 Hz, C*H$_2$*Ph), 4.58 (dd, 2H, ^2J=11.9 Hz, C*H$_2$*Ph), 5.75 (d, 1H, $J_{1,2}$=5.3 Hz, H-1), 7.30–7.16 (m, 15H, aromatic) ppm. ^{13}C NMR δ: 21.3, 50.7, 65.3, 69.2, 70.5, 71.9, 73.0, 73.5, 74.9, 75.9, 78.7, 97.9, 121.4, 127.0, 127.6, 127.7, 127.9 (×3), 128.0, 128.1 (×4), 128.4 (×4), 128.5 (×2), 128.6, 137.7, 137.9, 138.1, 141.0 ppm; HR-FAB MS [M+Na]$^+$ calcd for $C_{30}H_{34}O_7Na$, 529.2202; found: 529.2208.

3,4,6-Tri-*O*-benzyl-1,2-*O*-(1-methoxyethylidene)-
α-d-galactopyranose (3c)[19]

This compound was prepared from **1c** (1.00 g, 2.56 mmol) as described for the synthesis of **3a**. The title compound was isolated in 82%–89% yield (0.593–0.634 g, typical endo/exo ratio 1:5) over three steps as a light-yellow syrup; R_f=0.5 (3:7 EtOAc–hexanes). ^{1}H NMR δ: 1.53 (s, 3H, COCH$_3$), 3.22 (s, 3H, CH$_3$), 3.23–3.60 (m, 3H, H-3, 6a, 6b), 3.95–4.00 (m, 2H, H-4, 5), 4.41 (dd, 2H, ^2J=12.2 Hz, C*H$_2$*Ph), 4.43 (dd, 1H, $J_{2,3}$=5.0 Hz, H-2), 4.69 (dd, 2H, ^2J=11.8 Hz, C*H$_2$*Ph), 4.71 (dd, 2H, ^2J=11.5 Hz, C*H$_2$*Ph), 5.69 (d, 1H, $J_{1,2}$=4.2 Hz, H-1), 7.19–7.35 (m, 15H, aromatic) ppm. ^{13}C NMR δ: 24.6, 49.7, 68.1, 71.4, 73.0, 73.2, 73.6, 74.6, 79.9, 80.3, 97.8, 122.1, 127.6 (×2),

---

* Alternative name accepted by IUPAC is 1,2-*O*-(methyl orthoacetate).

127.7 (×2), 127.8 (×2), 127.9 (×2), 128.0 (×2), 128.3 (×2), 128.5 (×8), 137.9, 138.1, 138.4 ppm; HR-FAB MS [M+Na]⁺ calcd for $C_{30}H_{34}O_7Na$, 529.2202; found: 529.2208.

3,4,6-Tri-O-benzyl-1,2-O-(1-methoxyethylidene)-β-D-mannopyranose (3e)[20,21]

This compound was prepared from 1e (1.00 g, 2.56 mmol) as described for the synthesis of 3a.* The title compound was isolated as a light-yellow syrup (0.89 g, 68% over three steps). Crystallization (CH₂Cl₂–hexanes) gave white crystals (0.73g, 56%), m.p. 75°C–76°C, $[\alpha]_D^{23}$ +30.3 (c = 1, CHCl₃); Lit.[20] m.p. 73°C–76°C (Et₂O–hexanes); Lit.[21] $[\alpha]_D^{25}$ +29.8 (c = 1.04, CHCl₃); R_f = 0.7 (3:7 ethyl acetate–hexanes); ¹H NMR: δ, 1.66 (s, 3H, COCH₃), 3.21 (s, 3H, CH₃), 3.34 (m, 1H, H-5), 3.60–3.69 (m, 3H, H-3, 6a, 6b), 3.85 (dd, 1H, $J_{3,4} = J_{4,5} = 9.3$ Hz, H-4), 4.32 (dd, 1H, $J_{2,3} = 3.8$ Hz, H-2), 4.49 (dd, 2H, $^2J = 10.6$ Hz, CH₂Ph), 4.46, (dd, 2H, $^2J = 10.8$ Hz, CH₂Ph), 4.71 (s, 2H, CH₂Ph), 5.27 (d, 1H, $J_{1,2} = 2.5$ Hz, H-1), 7.15–7.31 (m. 15H, aromatic) ppm. ¹³C NMR δ: 24.4, 49.7, 68.9, 72.4, 73.3, 74.1, 74.2, 75.2, 77.1, 79.0, 97.5, 124.0, 127.5, 127.8, 128.0 (×2), 128.3, 128.4, 128.5, 137.8, 138.2 ppm; HR-FAB MS [M+Na]⁺ calcd for $C_{30}H_{34}O_7Na$, 529.2202; found: 529.2208.

3,4,6-Tri-O-benzyl-1,2-O-(1-methoxybenzylidene)-α-D-glucopyranose† (3b)[22]

A 33% solution of HBr in acetic acid (1.6 mL, 8.56 mmol) was added dropwise to a solution of 1,2,3,4,6-penta-O-benzoyl-D-glucopyranose (1b, 1.00 g, 1.43 mmol) in dry CH₂Cl₂ (5.0 mL), and the resulting mixture was stirred under argon for 3 h at r.t. The mixture was diluted with CH₂Cl₂ (~125 mL) and washed with cold water (20 mL), cold sat. aq. NaHCO₃ (2 × 20 mL), and cold water (3 × 20 mL). The organic phase was separated, dried, and concentrated, and molecular sieves (3 Å, 1.0 g) were added to a solution of the residue containing 2,3,4,6-tetra-O-benzoyl-α-D-glucopyranosyl bromide[23] (2b, ~1.43 mmol) in nitromethane (8.0 mL). The mixture was stirred under argon for 1 h at r.t.; the flask was covered with aluminum foil; γ-collidine (0.34 mL, 2.57 mmol), dry methanol (70 µL, 1.71 mmol), and tetra-butylammonium bromide (368 mg, 1.14 mmol) were added sequentially; and the mixture was stirred under argon for 6 h at r.t. After the addition of triethylamine (0.5 mL), the solid was filtered off, and the filtrate was concentrated. The residue was diluted with CH₂Cl₂ (~150 mL) and washed with water (2 × 20 mL); the organic phase was separated, dried, and concentrated. Benzyl chloride (2.87 mL, 25.0 mmol) and potassium hydroxide (1.04 g, 18.5 mmol) were added to a solution of the residue in tetrahydrofuran (7.0 mL), and the mixture was stirred for 16 h at reflux (~80°C). The mixture was allowed to cool to r.t., and the volatiles were removed. CH₂Cl₂ (~200 mL) was added to the residue, and the solution was washed with water (3 × 50 mL). The organic phase was separated, dried, and concentrated. Chromatography (gradient hexanes → ethyl acetate) afforded the title compound

---

* This synthesis was performed in the absence of TBAB.
† Alternative name accepted by IUPAC is 1,2-O-(methylorthobenzoate).

(0.521–0.536 g, 64%–66% over three steps) as yellow syrup; $R_f = 0.6$ (3:7 ethyl acetate–hexanes); $[\alpha]_D$ +25.6 (c 1.00, CHCl$_3$); Lit.[24] $[\alpha]_D$ +35 (c 2.0, CHCl$_3$). ^{1}H NMR δ: 3.13 (s, 3H, CH$_3$), 3.57 (d, 2H, $J = 3.6$ Hz, H-6a, 6b), 3.69–3.74 (m, 2H, H-4, 5), 3.88 (dd, 1H, $J_{3,4} = 3.7$ Hz, H-3), 4.41 (dd, 2H, $^2J = 11.4$ Hz, CH_2Ph), 4.45 (dd, 2H, $^2J = 12.1$ Hz, CH_2Ph), 4.53 (dd, 1H, $J_{2,3} = 3.7$ Hz, H-2), 4.63 (dd, 2H, $^2J = 11.9$ Hz, CH_2Ph), 5.92 (d, 1H, $J_{1,2} = 5.3$ Hz, H-1), 7.31–7.10 (m, 18H, aromatic), 7.63–7.60 (m, 2H, aromatic) ppm. ^{13}C NMR δ: 51.3, 69.6, 70.4, 72.1, 72.8, 73.2, 75.2, 75.6, 78.0, 98.3, 120.7, 126.5 (×2), 127.7, 127.8, 127.9 (×5), 128.0, 128.1 (×2), 128.3 (×5), 128.6 (×2), 129.4, 136.0, 137.8, 138.1, 138.2 ppm; HR-FAB MS [M+Na]$^+$ calcd for C$_{35}$H$_{36}$O$_7$Na, 591.2359; found: 591.2344.

3,4,6-TRI-O-BENZYL-1,2-O-(1-METHOXYBENZYLIDENE)-
α-D-GALACTOPYRANOSE (3D)

This compound was prepared from **1d** (1.00 g, 1.43 mmol) as described for the synthesis of **3b** (0.431–0.457 g, 53%–56% over three steps, typical endo/exo ratio 1:6.5) as a yellow syrup, $R_f = 0.63$ (3:7 ethyl acetate/hexanes). ^{1}H NMR δ: 3.25 (s, 3H, CH$_3$), 3.68 (d, 2H, $J = 2.7$ Hz, H-6a, 6b), 3.68–3.83 (m, 2H, H-4, 5), 4.00 (dd, 1H, $J_{2,3} = 3.6$ Hz, H-3), 4.52 (dd, 2H, $^2J = 11.3$ Hz, CH_2Ph), 4.55 (dd, 2H, $^2J = 12.1$ Hz, CH_2Ph), 4.69 (dd, 1H, $J_{2,3} = 4.0$ Hz, H-2), 4.74 (dd, 2H, $^2J = 11.9$ Hz, CH_2Ph), 6.01 (d, 1H, $J_{1,2} = 5.3$ Hz, H-1), 7.21–7.49 (m, 18H, aromatic), 7.71–7.74 (m, 2H, aromatic) ppm. ^{13}C NMR δ: 51.3, 69.1, 70.4, 72.0, 72.1, 73.2, 75.2, 75.6, 76.8, 78.0, 80.0, 98.3, 120.7, 126.5 (×2), 127.7, 127.8, 127.9 (×3), 128.0, 128.1 (×2), 128.4 (×3), 128.6 (×3), 129.4, 129.8, 136.0, 137.8, 138.0, 138.2 ppm; HR-FAB MS [M+Na]$^+$ calcd for C$_{35}$H$_{36}$O$_7$Na, 591.2359; found: 591.2336.

3,4,6-TRI-O-BENZYL-1,2-O-(1-METHOXYBENZYLIDENE)-
β-D-MANNOPYRANOSE* (3F)[24]

This compound was prepared from **1f** (1.00 g, 1.43 mmol) as described for the synthesis of **3b** (0.357–0.504 g, 44%–62% over three steps) as a yellow syrup $R_f = 0.50$ (3:7 EtOAc–hexanes); $[\alpha]_D$ –23.3 (c 1.00, CHCl$_3$); Lit[24] $[\alpha]_D$ –26.5 (c 0.56, CHCl$_3$). ^{1}H NMR δ: 3.32 (s, 3H, CH$_3$), 3.54 (m, 1H, H-5), 3.61 (dd, 1H, $J_{5,6a} = 2.2$ Hz, $J_{6a,6b} = 10.9$ Hz, H-6a), 3.71 (dd, 1H, $J_{5,6b} = 4.8$ Hz, H-6b), 3.89 (dd, 1H, $J_{3,4} = 9.2$ Hz, H-3), 3.99 (dd, 1H, $J_{4,5} = 9.0$ Hz, H-4), 7.62–7.33 (m, 18H, aromatic), 7.88–7.85 (m, 2H, aromatic) ppm. ^{13}C NMR δ: 51.6, 65.3, 69.1, 72.1, 73.3, 74.4, 75.0, 75.1, 76.4, 78.5, 97.9, 122.6, 126.8 (×2), 127.0 (×2), 127.5 (×2), 127.7, 127.8, 128.0, 128.1 (×2),

---

* We have observed that the reaction time and the amount of HBr–AcOH required for the reaction to go to completion depend on the anomeric ratio of penta-O-benzoyl mannopyranoside (**1f**). When the pure β-anomer (^{1}H NMR CDCl$_3$: 6.44 ppm (H-1)) is used as the starting material, the bromination step takes place with 6 equiv of HBr–AcOH in 3 h, as reported. However, if an α/β-mixture of anomers is used instead, higher amounts of HBr–AcOH and longer reaction times were required, for example, for a sample containing an anomeric mixture (α/β, 0.4:1; α-anomer ^{1}H NMR CDCl$_3$: 6.63 ppm (H-1), 9 equiv of HBr–AcOH and 4 h of reaction time were required. Nevertheless, we have observed that the excess of HBr–AcOH added to the reaction, in the case of anomeric mixtures, did not have any deleterious effect on the reported yields.

128.3 (×2), 128.4, 128.6 (×3), 129.1, 136.6, 137.9, 138.3, 138.4, 140.1 ppm; HR-FAB MS [M+Na]$^+$ calcd for $C_{35}H_{36}O_7Na$, 591.2359; found: 591.2349.

ACKNOWLEDGMENTS

This work was supported by an award from the National Science Foundation (CHE-1058112).

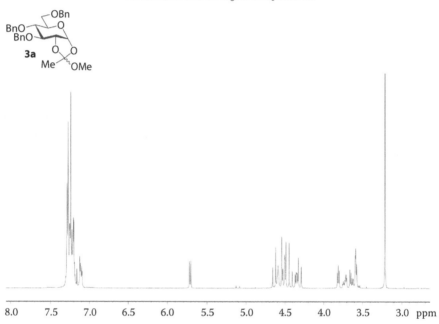

^{1}H NMR (300 Mz, CDCl$_3$) of compound **3a**.

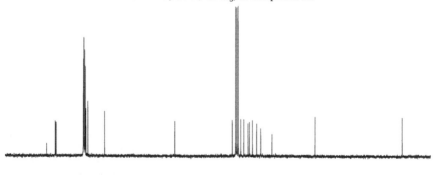

^{13}C NMR (75 Mz, CDCl$_3$) of compound **3a**.

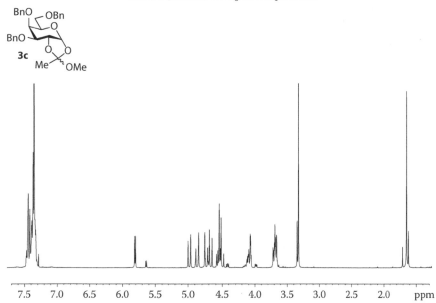

^{1}H NMR (300 Mz, CDCl$_3$) of compound **3c**.

^{13}C NMR (75 Mz, CDCl$_3$) of compound **3c**.

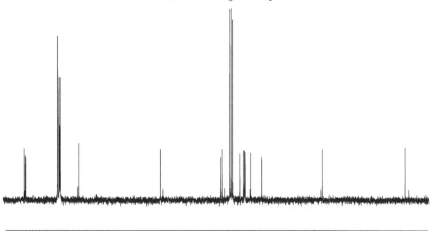

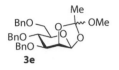

3e

^{1}H NMR (300 Mz, CDCl$_3$) of compound **3e**.

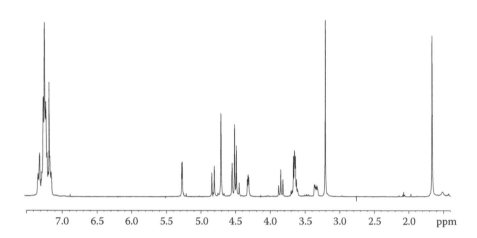

^{13}C NMR (75 Mz, CDCl$_3$) of compound **3e**.

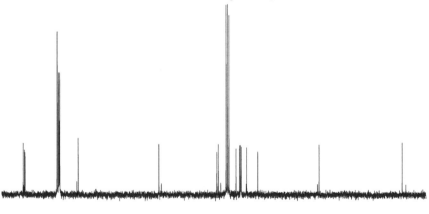

3b

^{1}H NMR (300 Mz, CDCl$_3$) of compound **3b**.

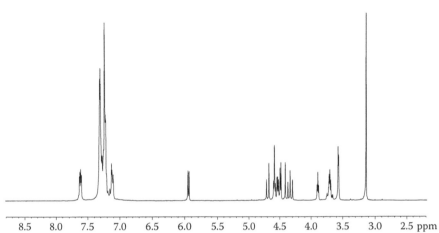

^{13}C NMR (75 Mz, CDCl$_3$) of compound **3b**.

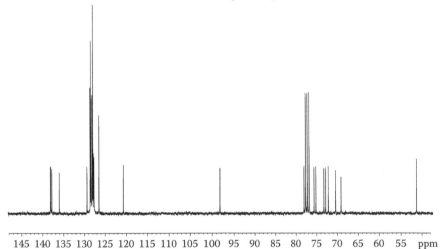

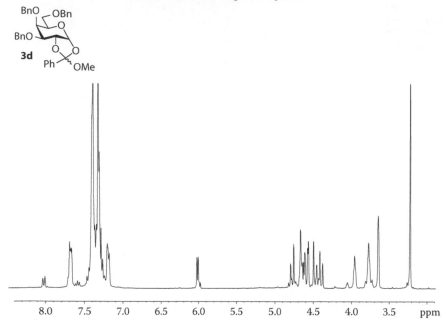

^{1}H NMR (300 Mz, CDCl$_3$) of compound **3d**.

^{13}C NMR (75 Mz, CDCl$_3$) of compound **3d**.

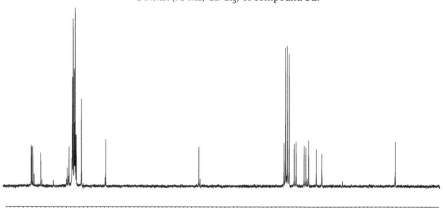

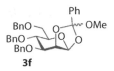

3f

^{1}H NMR (300 Mz, CDCl$_3$) of compound **3f**.

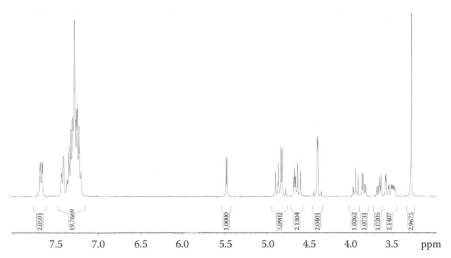

^{13}C NMR (75 Mz, CDCl$_3$) of compound **3f**.

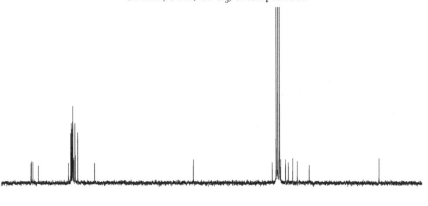

REFERENCES

1. Bochkov, A. F.; Voznyi, Y. V.; Kalinevich, V. M.; Shashkov, A. S.; Kochetkov, N. K. *Izv. Akad. Nauk, Ser. Khim.* **1975**, 415–420.
2. Lemieux, G. A.; Morgan, A. R. *Can. J. Chem.* **1965**, *43*, 2199–2204.
3. Isbell, H. S.; Frush, H. L. *J. Res. Natl. Bur. Stand.* **1949**, *43*, 161–171.
4. Pigman, W.; Isbell, H. *J. Res. Natl. Bur. Stand.* **1937**, *19*, 189–213.
5. Kochetkov, N. K.; Backinowsky, L. V.; Tsvetkov, Y. E. *Tetrahedron Lett.* **1977**, *41*, 3681–3684.
6. Kochetkov, N. K. *Tetrahedron* **1987**, *43*, 2389–2436.
7. Fraser-Reid, B.; Lopez, J. C. In *Handbook of Chemical Glycosylation*; Demchenko, A. V., Ed.; Wiley-VCH: Weinheim, Germany, **2008**, p. 381–415.
8. Uriel, C.; Ventura, J.; Gomez, A. M.; Lopez, J. C.; Fraser-Reid, B. *Eur. J. Org. Chem.* **2012**, 3122–3131.
9. Vidadala, S. R.; Thadke, S. A.; Hotha, S. *J. Org. Chem.* **2009**, *74*, 9233–9236.
10. Lopez, J. C.; Agocs, A.; Uriel, C.; Gomez, A. M.; Fraser-Reid, B. *Chem. Commun.* **2005**, 5088–5090.
11. Mach, M.; Schlueter, U.; Mathew, F.; Fraser-Reid, B.; Hazen, K. C. *Tetrahedron* **2002**, *58*, 7345–7354.
12. Mydock, L. K.; Demchenko, A. V. *Org. Lett.* **2008**, *10*, 2107–2110.
13. Mydock, L. K.; Demchenko, A. V. *Org. Lett.* **2008**, *10*, 2103–2106.
14. Premathilake, H. D.; Mydock, L. K.; Demchenko, A. V. *J. Org. Chem.* **2010**, *75*, 1095–1100.
15. Franks, N. E.; Montgomery, R. *Carbohydr. Res.* **1968**, *6*, 286–298.
16. Bhowmik, S.; Maitra, U. *Chem. Commun.* **2012**, *48*, 4624–4626.
17. Draghetti, V.; Poletti, L.; Prosperi, D.; Lay, L. *J. Carbohydr. Chem.* **2001**, *20*, 813–819.
18. Lemieux, R. U. In *Methods in Carbohydrate Chemistry*; Whistler, R. L., Wolform, M. L., Eds.; Academic Press Inc.: New York, **1963**; Vol. 2, p. 221–222.
19. Asai, N.; Fusetani, N.; Matsunaga, S. *J. Natl. Prod.* **2001**, *64*, 1210–1215.
20. Ammann, H.; Dupuis, G. *Can. J. Chem.* **1988**, *66*, 1651–1655.
21. Patel, M. K.; Vijayakrishnan, B.; Koeppe, J. R.; Chalker, J. M.; Doores, K. J.; Davis, B. G. *Chem. Commun.* **2010**, *46*, 9119–9121.
22. Ekborg, G.; Glaudemans, C. P. J. *Carbohydr. Res.* **1984**, *129*, 287–292.
23. Lemieux, R. U. In *Methods in Carbohydrate Chemistry*; Whistler, R. L., Wolform, M. L., Eds.; Academic Press Inc.: New York, 1963; Vol. 2, p. 226–228.
24. Holemann, A.; Stocker, B. L.; Seeberger, P. H. *J. Org. Chem.* **2006**, *71*, 8071–8088.

27 2-Acetamido-3,4,6-tri-*O*-acetyl-2-deoxy-1-*O*-*p*-nitrophenoxycarbonyl-α-D-glucopyranose

Torben Seitz, Caroline Maierhofer,
*Daniel Matzner,[†] and Valentin Wittmann**

CONTENTS

Preparation of glycoconjugates is of current interest for the investigation of multivalent carbohydrate–protein interactions, and methods for the stereospecific attachment of sugars to scaffold molecules are highly demanded.[1] Glycosyl *p*-nitrophenyl carbonates are suitable precursors for bioconjugation of carbohydrates because they undergo fast reaction with amines producing glycosyl carbamates with retention of configuration at the anomeric center in virtually theoretical yields.[2–6] Glycosyl carbamates also find application as prodrugs.[7]

    The title compound **2** was prepared[5] from 2-acetamido-3,4,6-tri-*O*-acetyl-2-deoxy-D-glucopyranose (**1**) that is obtained through selective anomeric deprotection of 2-acetamido-1,3,4,6-tetra-*O*-acetyl-2-deoxy-D-glucopyranose according to

---

* Corresponding author; e-mail: valentin.wittmann@uni-konstanz.de.
† Checker, under the supervision of Prof. Dr. G. Mayer; e-mail: gmayer@uni-bonn.de.

Zhang and Kováč.[8] Treatment of **1** with p-nitrophenyl chloroformate and triethylamine in dichloromethane at 0°C exclusively led to formation of the α-glycosyl carbonate. Since carbonate **2** slowly decomposes on silica gel, it was crucial to keep the
silica gel contact time short during column chromatography by using a short column
and fast elution. Compound **2** was employed for the synthesis of various multivalent
wheat germ agglutinin ligands.[5–6]

EXPERIMENTAL METHODS

GENERAL METHODS

Solvents were purified by distillation prior to use. CH_2Cl_2 was dried over CaH_2.
Thin-layer chromatography (TLC) was performed on Merck silica gel 60 F_{254} aluminum sheets. Compound spots were visualized by UV light ($\lambda = 254$ nm) and immersion in a solution of p-anisaldehyde (3.7 mL) in a mixture of EtOH (135 mL), H_2SO_4
(5 mL), and glacial acetic acid (15 mL) followed by heating. Preparative flash column
chromatography (FC) was carried out on Merck silica gel 60 (40–63 μm). NMR
spectra were recorded at 300 K on a Bruker Avance 400 instrument. Chemical shifts
are reported in ppm relative to the $CDCl_3$ signal ($\delta_H = 7.26$, $\delta_C = 77.16$). Assignments
of proton and carbon signals were carried out with the aid of DQF-COSY and
HMQC experiments. Optical rotation was measured with a Jasco P-2000 polarimeter. Elemental analyses were performed by the microanalytical facility at the
Universität Konstanz with a vario EL instrument from Elementar Analysensysteme
(Hanau, Germany).

2-ACETAMIDO-3,4,6-TRI-*O*-ACETYL-2-DEOXY-1-*O*-*P*-
NITROPHENOXYCARBONYL-α-D-GLUCOPYRANOSE (2)

To a solution of **1**[8,9] (1.04 g, 3 mmol) and dry Et_3N (832 μL, 6 mmol) in dry
CH_2Cl_2 (10 mL), a solution of p-nitrophenyl chloroformate (1.33 g, 6.6 mmol) in dry
CH_2Cl_2 (1 mL) was added dropwise at 0°C. A precipitate formed, and after 3 h at
0°C, when TLC (EtOAc) showed that the reaction was complete, the solvent was
evaporated, and the residue was dissolved in EtOAc and washed sequentially with
10% aq. citric acid and water. The combined aqueous layers were reextracted with
EtOAc, and the combined EtOAc solutions were dried ($MgSO_4$) and concentrated.
Short column FC (gradient from petroleum 2:1 ether–EtOAc→EtOAc) gave **2** as
white solid. Yield 1.41 g (92%); m.p. 70°C–71°C ($CHCl_3$); $[\alpha]_D^{24}$ +108 (c 1, $CHCl_3$);
R_f 0.51 (EtOAc). ^1H NMR ($CDCl_3$) δ: 8.31–8.27 (m, 2 H, Ar-H), 7.44–4.41 (m, 2 H,
Ar-H), 6.16 (d, 1 H, $J_{1,2}$ 3.6 Hz, H-1), 5.86 (d, 1 H, $J_{NH,2}$ 8.6 Hz, NH), 5.31 (dd, 1 H, $J_{3,2}$
10.5 Hz, $J_{3,4}$ 9.6 Hz, H-3), 5.25 (t, 1 H, $J_{4,3} \approx J_{4,5} \approx 9.7$ Hz, H-4), 4.53 (ddd, 1 H, $J_{2,1}$ 3.6
Hz, $J_{2,NH}$ 8.6 Hz, $J_{2,3}$ 10.5 Hz, H-2), 4.28 (dd, 1 H, $J_{6a,6b}$ 12.5 Hz, $J_{6a,5}$ 4.0 Hz, H-6a),
4.17–4.10 (m, 2 H, H-6b, H-5), 2.08 (s, 3 H, CH_3), 2.06 (s, 3 H, CH_3), 2.05 (s, 3 H,
CH_3), 1.97 (s, 3 H, CH_3). ^{13}C NMR ($CDCl_3$) δ: 171.8, 170.7, 170.3, 169.2 (4 C=O),
155.0 (O-C(O)-O), 150.7, 145.8, 125.6, 121.6 (CAr), 95.9 (C-1), 70.4 (C-5), 70.3 (C-3),

67.3 (C-4), 61.5 (C-6), 51.5 (C-2), 23.1, 20.8, 20.6, (4 CH$_3$); MS (ESI): calcd for C$_{21}$H$_{24}$N$_2$NaO$_{13}$ [M+Na]$^+$ m/z: 535.1, found: 535.0. Anal. calcd for C$_{21}$H$_{24}$N$_2$O$_{13}$: 49.22; H, 4.72, N, 5.47. Found: C, 49.49, H, 4.91, N, 5.22.

ACKNOWLEDGMENTS

This work was funded by the University of Konstanz and the Konstanz Research School Chemical Biology (KoRS-CB). T.S. acknowledges a fellowship from KoRS-CB.

2-Acetamido-3,4,6-tri-*O*-acetyl-2-deoxy-1-*O*-*p*-nitrophenoxycarbonyl-α-ᴅ-glucopyranose (**2**)
^{1}H NMR (400 MHz) in CDCl$_3$

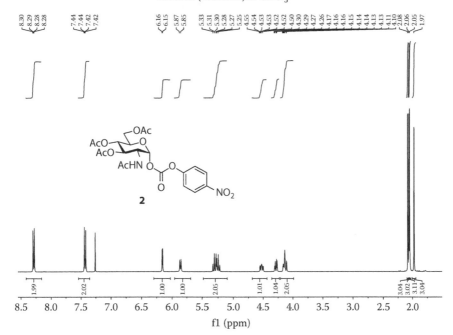

f1 (ppm)

2-Acetamido-3,4,6-tri-*O*-acetyl-2-deoxy-1-*O*-*p*-nitrophenoxycarbonyl-α-D-glucopyranose (**2**)
^{13}C NMR (100 MHz) in CDCl$_3$

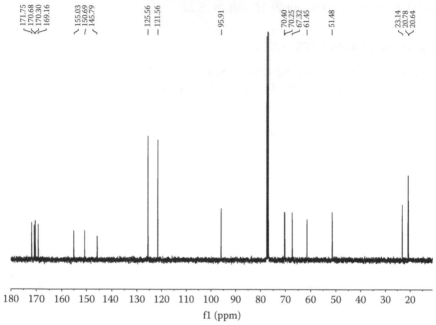

REFERENCES

1. Renaudet, O.; Roy, R. *Chem. Soc. Rev.* **2013**, *42*, 4515–4517 and cited references.
2. Azoulay, M.; Escriou, V.; Florent, J.-C.; Monneret, C. *J. Carbohydr. Chem.* **2001**, *20*, 841–853.
3. Cherif, S.; Leach, M. R.; Williams, D. B.; Monneret, C. *Bioorg. Med. Chem. Lett.* **2002**, *12*, 1237–1240.
4. André, S.; Specker, D.; Bovin, N. V.; Lensch, M.; Kaltner, H.; Gabius, H.-J.; Wittmann, V. *Bioconjugate Chem.* **2009**, *20*, 1716–1728.
5. Schwefel, D.; Maierhofer, C.; Beck, J. G.; Seeberger, S.; Diederichs, K.; Möller, H. M.; Welte, W.; Wittmann, V. *J. Am. Chem. Soc.* **2010**, *132*, 8704–8719.
6. Braun, P.; Nägele, B.; Wittmann, V.; Drescher, M. *Angew. Chem. Int. Ed.* **2011**, *50*, 8428–8431.
7. Madec-Lougerstay, R.; Florent, J.-C.; Monneret, C. *J. Chem. Soc. Perkin Trans. 1* **1999**, 1369–1376.
8. Zhang, J.; Kováč, P. *J. Carbohydr. Chem.* **1999**, *18*, 461–469.
9. Jha, R.; Davis, J. T. *Carbohydr. Res.* **1995**, *277*, 125–134.

28 Short Synthesis of (2-Acetamido-3,4,6-tri-*O*-acetyl-2-deoxy-β-D-glucopyranosyl) benzene and 4-(2-Acetamido-3,4,6-tri-*O*-acetyl-2-deoxy-β-D-glucopyranosyl) bromobenzene

Torben Seitz, Carsten Fleck,[†]
and Valentin Wittmann*

CONTENTS

* Corresponding author; e-mail: valentin.wittmann@uni-konstanz.de.
† Checker, under supervision of Wolfgang Maison; e-mail: maison@chemie.uni-hamburg.de.

C-Glycosyl arenes are widespread in nature.[1,2] In carbohydrate chemistry, they have found interest because of their stability against chemical and enzymatic hydrolysis.[3–5] However, up to date, there are only a few reports on the synthesis of *C*-glycosyl arenes derived from *N*-acetylglucosamine (GlcNAc).[6–8] Compounds containing electron-rich arenes can be obtained by Friedel–Crafts-like reactions of glycosyl donors, such as trichloroacetimidates.[7] This approach, however, is not suitable for electron-poor arenes. A frequently used method for the preparation of *C*-glycosyl compounds is the addition of organometallic compounds to sugar lactones followed by reductive dehydroxylation.[9–11] The approach has also been applied to the synthesis of *C*-glycosyl compounds derived from GlcNAc via 2-azido-lactone **1**. However, preparation of **2** (R = H) was reported not to be possible.[8] Although we were able to manage the conversion of **1** into **2** (R = Br) in a yield of 55%, the complete synthesis is very time consuming.

Here, we report a short access to β-*C*-glucosyl benzenes **5** and **6** starting from hydrobromide **4**, which is available in one step from D-glucosamine hydrochloride.[12] The syntheses are slight modifications of a procedure for the preparation of **5** published by Yoshimura et al.[13] Hydrobromide **4** was treated with a large excess of Grignard reagent to generate the unprotected *C*-glucosyl compound, which was subsequently reacetylated to give **5**. The product was purified by crystallization without chromatography and obtained in a yield of 33% comparable to the published one. From the mother liquor, a small amount (7%) of the pure α-isomer and an anomeric mixture (α/β 1:1.25) (10%) could be obtained by column chromatography with EtOAc as eluent. Similarly, treatment of **4** with monolithiated 1,4-dibromobenzene and acetylation gave *C*-glucosyl bromobenzene **6** in a yield of 12% after crystallization. Although the yields of **5** and **6** are rather moderate, they are quite attractive when compared to the even lower estimated overall yield of the 11-step synthesis[8,14,15] via lactone **1**. In addition, the short access via hydrobromide **4** saves a number of synthetic steps. Compound **6** is an important starting material for further modification of the bromophenyl moiety, for example, by palladium-catalyzed cross-coupling reactions.[16]

EXPERIMENTAL METHODS

GENERAL METHODS

Reagents (reagent grade) were purchased from Aldrich and used without further purification. THF was distilled over KNa. Diethyl ether (p.a. grade) was dried over molecular sieves 4 Å. 3,4,6-Tri-O-acetyl-2-amino-2-deoxy-α-D-glucopyranosyl bromide hydrobromide **4** was synthesized as described.[12] Thin-layer chromatography was performed on Merck silica gel 60 F_{254}–coated aluminum sheets with detection by UV light ($\lambda = 254$ nm). NMR spectra were recorded on a Bruker Avance III 400 instrument. Chemical shifts are reported in ppm relative to the DMSO signal ($\delta_H = 2.5$, $\delta_C = 39.5$). Assignments of the signals in the NMR spectra were performed using 2D spectroscopy (DQF-COSY and HSQC). Optical rotation was measured with a Jasco P-2000 polarimeter. Elemental analyses were performed by the micro-analytical facility at the Universität Konstanz with a vario EL instrument from Elementar Analysensysteme (Hanau, Germany). High-resolution ESI mass spectra were recorded on a Bruker MicrOTOF-Q instrument operating in positive mode. Samples were dissolved in MeCN/H_2O mixtures or pure MeOH and were injected directly via syringe.

(2-ACETAMIDO-3,4,6-TRI-O-ACETYL-2-DEOXY-β-D-GLUCOPYRANOSYL)BENZENE (5)

Under N_2, 3,4,6-tri-O-acetyl-2-amino-2-deoxy-α-D-glucopyranosyl bromide hydro-bromide (**4**) (2.0 g, 4.44 mmol) was suspended in dry Et_2O (20 mL). After cooling to 0°C, a 3 M solution of PhMgBr in Et_2O (16 mL, 49 mmol) was slowly added, and the mixture was stirred at 0°C for 2 h. After addition of PhMgBr, a brown solid was formed. The reaction was allowed to warm up to r.t. and stirred over-night. The suspension was again cooled to 0°C, the reaction was quenched with 1 M HCl (10 mL), and the product was extracted with water (3 × 100 mL). The combined aqueous phases were concentrated under reduced pressure, upon which the color turned from slightly yellow to dark brown, and pyridine (30 mL) and Ac_2O (15 mL) were added. The mixture was stirred overnight, and the solvents were removed under reduced pressure. The residue was dissolved in EtOAc, washed with 1 M HCl (3 × 50 mL) and sat. aq. $NaHCO_3$ (50 mL), and dried ($MgSO_4$). EtOAc was removed under reduced pressure, and the residue was crystallized from a mini-mum of i-PrOH. Yield 590 mg (33%); m.p. 257°C; $[\alpha]_D^{24}$ −35.3 (c 1, $CHCl_3$); R_f 0.38 (EtOAc). ^{1}H NMR (DMSO-d_6) δ: 7.90 (d, 1 H, $J_{NH,2}$ 9.3 Hz, NH), 7.32-7.26 (m, 5 H, Ar-H), 5.18 (t, 1 H, $J_{2,3} \approx J_{3,4} \approx 9.3$ Hz, H-3), 5.01 (t, 1 H, $J_{3,4} \approx J_{4,5} \approx 9.3$ Hz, H-4), 4.51 (d, 1 H, $J_{1,2}$ 10.3 Hz, H-1), 4.15 (dd, 1 H, $J_{6a,6b}$ 12.5 Hz, $J_{6a,5}$, 5.4 Hz, H-6a), 4.07 (dd, 1 H, $J_{6b,6a}$ 12.5 Hz, $J_{6b,5}$ 5.4 Hz, H-6b), 4.03-3.92 (m, 2 H, H-2, H-5), 2.01 (s, 6 H, 2 CH_3), 1.91 (s, 3 H, CH_3), 1.54 (s, 3 H, CH_3). ^{13}C NMR (DMSO-d_6) δ: 170.1, 169.8,

169.4, 168.6 (4 C=O), 137.9, 128.1, 127.9, 127.3 (C^{Ar}), 79.4 (C-1), 74.9 (C-5), 74.1 (C-3), 68.8 (C-4), 62.4 (C-6), 53.9 (C-2), 22.3, 20.6, 20.5, 20.4 (4 CH_3); MS (ESI) calcd for $C_{20}H_{26}NO_8$ [M+H]$^+$ m/z: 408.2, found: 408.4; HRMS (ESI): m/z calcd for $C_{20}H_{25}NO_8$ [M+Na]$^+$ 430.1472, found 430.1471; Anal. calcd for $C_{20}H_{25}NO_8$: C, 58.96; H, 6.18, N, 3.44. Found: C, 58.69; H, 6.38, N, 3.52.

4-(2-ACETAMIDO-3,4,6-TRI-O-ACETYL-2-DEOXY-β-D-GLUCOPYRANOSYL)BROMOBENZENE (6)

1,4-Dibromobenzene (11.54 g, 49 mmol) was dissolved in dry THF (200 mL) and cooled to −78°C. n-BuLi (2.5 M in hexane, 19.6 mL, 49 mmol) was added dropwise, and the reaction was stirred for 20 min. Glucosyl bromide 4 (2 g, 4.45 mmol) dissolved in dry THF (100 mL) was added slowly. The reaction mixture has a high viscosity at −78°C and is a suspension. The mixture was allowed to warm up to r.t. and stirred overnight. The suspension was cooled to 0°C, and the reaction was quenched with 1 M HCl (10 mL). Et$_2$O (300 mL) was added, and the product was extracted with water (3 × 100 mL). The aqueous solution was concentrated under reduced pressure, upon which the color turned from slightly yellow to dark brown, and pyridine (120 mL) and Ac$_2$O (60 mL) were added to the oily residue. The mixture was stirred overnight at r.t. and concentrated. The residue was dissolved in EtOAc and washed with 1 M HCl (3 × 50 mL) and sat. aq. NaHCO$_3$ (50 mL), the organic phase was dried (MgSO$_4$) and concentrated, and the product 6 was crystallized from a minimum of i-PrOH. Yield 249 mg, (12%); m.p. 262°C; $[\alpha]_D^{25}$ −47.8 (c 1, CHCl$_3$); R_f 0.42 (petroleum ether/EtOAc 1:3). ^{1}H NMR (DMSO-d_6) δ: 7.95 (d, 1 H, $J_{NH,2}$ 9.3 Hz, N-H), 7.53-7.51 (m, 2 H, Ar-H), 7.28-7.26 (m, 2 H, Ar-H), 5.18 (t, 1 H, $J_{2,3} \approx J_{3,4} \approx 9.8$ Hz, H-3), 5.01 (t, 1 H, $J_{3,4} \approx J_{4,5} \approx 9.8$ Hz, H-4), 4.50 (d, 1 H, $J_{1,2}$ 10.3 Hz, H-1), 4.15 (dd, 1 H, $J_{6a,6b}$ 12.7 Hz, $J_{6a,5}$ 5.3 Hz, H-6a), 4.06 (dd, 1 H, $J_{6b,6a}$ 12.7 Hz, $J_{6b,5}$ 2.4 Hz, H-6b), 3.99-3.91 (m, 2 H, H-2, H-5), 2.01 (s, 6 H, 2 CH$_3$), 1.91 (s, 3 H, CH$_3$), 1.56 (s, 3 H, CH$_3$). ^{13}C NMR (DMSO-d_6) δ: 170.2, 169.8, 169.4, 168.7 (4 C=O), 137.4, 130.9, 129.4, 121.3 (C^{Ar}), 78.6 (C-1), 74.9 (C-5), 73.9 (C-3), 68.7 (C-4), 62.4 (C-6), 53.9 (C-2), 22.4, 20.6, 20.5, 20.4 (4 CH$_3$); MS (ESI): calcd for $C_{20}H_{25}BrNO_8$ [M+H]$^+$ m/z: 488.1, found: 488.0; HRMS (ESI): m/z calcd for $C_{20}H_{24}BrNO_8$ [M+Na]$^+$ 508.0578, found 508.0568; Anal. calcd for $C_{20}H_{24}BrNO_8$: C, 49.40; H, 4.97, N, 2.88. Found: C, 49.64, H, 4.95, N, 3.15.

ACKNOWLEDGMENTS

This work was funded by the University of Konstanz and the Konstanz Research School Chemical Biology (KoRS-CB). T.S. acknowledges a fellowship from KoRS-CB.

(2-Acetamido-3,4,6-tri-O-acetyl-2-deoxy-β-D-glucopyranosyl)benzene (5)
^{1}H NMR (400 MHz) in DMSO-d_6

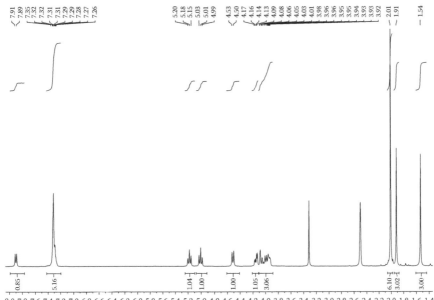

(2-Acetamido-3,4,6-tri-O-acetyl-2-deoxy-β-D-glucopyranosyl)benzene (5)
^{13}C NMR (100 MHz) in DMSO-d_6

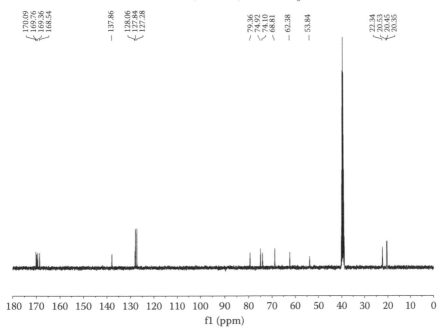

4-(2-Acetamido-3,4,6-tri-O-acetyl-2-deoxy-β-D-glucopyranosyl)bromobenzene (**6**)
^{1}H NMR (400 MHz) in DMSO-d_6

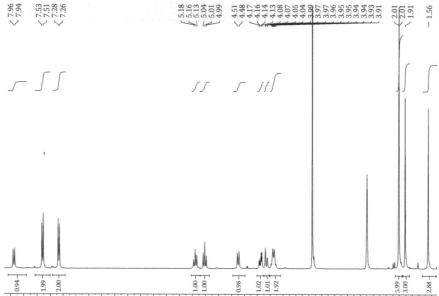

4-(2-Acetamido-3,4,6-tri-O-acetyl-2-deoxy-β-D-glucopyranosyl)bromobenzene (**6**)
^{1}H NMR (400 MHz) in DMSO-d_6

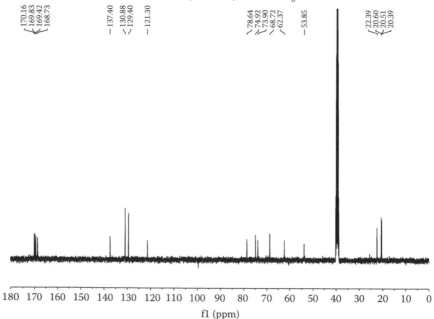

REFERENCES

1. Haynes, L. J. *Adv. Carbohydr. Chem.* **1965**, *20*, 357–369.
2. Hacksell, U.; Daves Jr, G. D. *Progress Med. Chem.* **1985**, *22*, 1–65.
3. Postema, M. H. D. *C-Glycoside Synthesis*; CRC Press: Boca Raton, FL, **1995**.
4. Levy, D. E.; Tang, C. *The Chemistry of C-Glycosides*; Elsevier: New York, **1995**.
5. Jaramillo, C.; Knapp, S. *Synthesis* **1994**, 1–20.
6. Yoshimura, J. *Bull. Chem. Soc. Jpn.* **1961**, *34*, 8–12.
7. Castro-Palomino, J. C.; Schmidt, R. R. *Liebigs Ann.* **1996**, 1623–1626.
8. Ayadi, E.; Czernecki, S.; Xie, J. *Chem. Commun.* **1996**, 347–348.
9. Lewis, M. D.; Cha, K.; Kishi, Y. J. *J. Am. Chem. Soc.* **1982**, *104*, 4976–4978.
10. Kraus, G. A.; Molina, M. T. *J. Org. Chem.* **1988**, *53*, 752–753.
11. Czernecki, S.; Ville, G. *J. Org. Chem.* **1989**, *54*, 610–612.
12. Fodor, G.; Ötvös, L. *Chem. Ber.* **1956**, *89*, 701–708.
13. Yoshimura, J.; Muramatsu, N.; Sato, T. *Nippon Kagaku Zasshi* **1958**, *79*, 1503–1506.
14. Czernecki, S.; Ayadi, E. *Can. J. Chem.* **1995**, *73*, 343–350.
15. Ayadi, E.; Czernecki, S.; Xie, J. *J. Carbohydr. Chem.* **1996**, *15*, 191–199.
16. de Meijere, A.; Diederich, F. *Metal-Catalyzed Cross-coupling Reactions*; 2nd edn.; Wiley-VCH: Weinheim, Germany, **2004**.

29 Isopropyl 2,3,4,6-Tetra-O-benzoyl-1-thio-β-D-galactopyranoside, a Convenient Galactosyl Donor

Peng Xu, Sayantan Bhaduri,† and Pavol Kováč*

CONTENTS

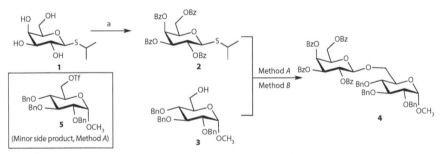

Reagents and conditions: (a) BzCl, pyridine, CH₂Cl₂, 0°C. Method A: Tf₂O–Me₂S₂, 4 Å MS, CH₂Cl₂, 0°C, 89%. Method B: NIS, AgOTf, 4 Å MS, CH₂Cl₂, −10°C, 87%.

* Corresponding author; e-mail: kpn@helix.nih.gov.
† Checker, under supervision of Nicola Pohl; e-mail: npohl@indiana.edu.

Isopropyl 1-thio-β-D-galactopyranoside[1] (**1**), or isopropyl thiogalactopyrano-side (IPTG), as the compound is often referred to by commercial suppliers of carbohydrate derivatives, is known more to molecular biologists than to carbohydrate chemists. IPTG is a molecular biology reagent, largely because of its ability to induce expression of genes under the control of the lac operon.[2] Carbohydrate chemists are more familiar with other alkyl and aryl 1-thioglycosides and use them, in their fully protected form, as glycosyl donors in carbohydrate synthesis. Their popularity lies in their versatility in glycosylation reactions combined with long shelf life. Unfortunately, due to the odorous nature of most thiols, preparation of many useful 1-thioglycosides is neither a pleasant nor a desirable task in an academic laboratory, the latter due to health and environmental concerns. IPTG has been available commercially for many years, but at pricing that made the routine use of the compound as a synthetic intermediate cost prohibitive. This may soon change as IPTG has recently become available commercially as a rather inexpensive commodity.* Consequently, while to date IPTG and its variously protected derivatives have been only seldom used in carbohydrate synthesis (e.g., References [3–5]), the substance may soon become one of the standard tools in synthetic carbohydrate chemistry, opening avenues toward an array of complex galactosyl donors and acceptors.

The advantage of benzoylated glycosyl donors over their acetylated counterparts has been duly recognized many years ago,[6,7] and here we report the preparation of the crystalline isopropyl 2,3,4,6-tetra-*O*-benzoyl-1-thio-β-D-galactopyranoside (**2**) from the commercially available **1**. The preparation of **2** and its activation with N-iodosuccinimide/trimethylsilyl trifluoromethanesulfonate (NIS/TMSOTf) reagent was previously reported in connection with the synthesis of L-methyl β-D-galactopyranoside but **2** was not fully characterized, and the original working protocol[8] is not readily available worldwide. Its use as a convenient glycosyl donor was tested here through activation with the NIS/AgOTf[9] and Me_2S_2/Tf_2O.[10] The use of the latter, a very useful promoter, deserves the following comment. According to the original protocol,[10] the promoter is prepared from Me_2S_2 and Tf_2O with a slight excess of the former. Although free Tf_2O is not present in the reagent, this promoter is a rather potent triflating reagent. When the reagent was used to promote glycosylation of methyl 2,3,4-tri-*O*-benzyl-α-D-glucopyranoside (**3**), the corresponding 6-triflyl derivative (**5**) was formed as a very minor by-product (see section "Experimental Methods"). The same 6-*O*-triflate was formed in substantial amount (>50%, thin-layer chromatography [TLC]) within the same short reaction time (10 min, 0°C) when the promoter was added to a solution of **3** in the absence of a glycosyl donor. While the extent of triflation of glycosyl acceptor during Me_2S_2/Tf_2O-promoted glycosylations is normally preparatively unimportant, this side reaction might decrease the yield of glycosylation with less reactive glycosyl acceptors and/or donors.

Reactions of 1-thioglycoside **2** with **3** gave the expected coupling product in over 80% yields via both Methods A and B.

---

* Crystalline IPTG of excellent quality, which we verified, can be obtained from Carbosynth Ltd. at <$2.00/g in 100 g batches, and the price goes further down when larger quantity is ordered.

EXPERIMENTAL METHODS

GENERAL METHODS

Optical rotations were measured at ambient temperature in $CHCl_3$ with a Jasco automatic polarimeter, Model P-2000. All reactions were monitored by TLC on silica gel 60–coated glass slides (Analtech, Inc.). Column chromatography was performed by elution from columns of silica gel with Biotage Isolera chromatograph. Solvent mixtures less polar than those used for TLC were used at the onset of separations. Nuclear magnetic resonance (NMR) spectra were measured at 600 MHz (^{1}H) and 150 MHz (^{13}C) with a Bruker Avance 600 spectrometer in $CDCl_3$ as solvent. ^{1}H and ^{13}C chemical shifts are referenced to signals of TMS (0 ppm) and $CDCl_3$ (77.0 ppm). Assignments of NMR signals were made by homonuclear and heteronuclear 2D correlation spectroscopy using software supplied with the spectrometer. When reporting assignments of NMR signals of the disaccharide, sugar residues are serially numbered, beginning with the one bearing the glucose residue, and are identified by a Roman numeral superscript in listings of signal assignments. Isopropyl 1-thio-β-D-galactopyranoside was purchased from Carbosynth, Ltd. Unless otherwise stated, all other chemicals (best available quality) were purchased from Sigma Chemical Company and used as supplied. Me_2S_2/Tf_2O promoter was prepared as described[10] and used within 24 h. Dichloromethane used in glycosylation reactions had been stirred with indicator-free Drierite (10–20 mesh) for several hours and kept at room temperature over dying reagent. Solutions in organic solvents were dried with Na_2SO_4 and concentrated at 2 kPa/<40°C.

Isopropyl 2,3,4,6-Tetra-*O*-benzoyl-1-thio-β-D-galactopyranoside (2)

In a 100 mL flask, isopropyl 1-thio-β-D-galactopyranoside (**1**) (3.0 g, 12.6 mmol) was dissolved in a mixture of CH_2Cl_2 (45 mL) and pyridine (22 mL). The clear solution formed was cooled to 0°C (ice-water bath) and, with stirring, benzoyl chloride (7.3 mL, 62.9 mmol) was added slowly over 5 min. The cloudy mixture formed was stirred overnight, during which time the temperature gradually increased to room temperature. TLC (hexane/EtOAc 3:1) showed that all starting material was converted into one product. MeOH (15 mL) was added, to destroy excess of BzCl and, after stirring for 1 h, the solution was concentrated to dryness (40°C/5 mmHg) to remove most of pyridine. The residue was dissolved in CH_2Cl_2 (200 mL) and washed successively with 1N HCl aq. (75 mL), sat. $NaHCO_3$ (75 mL), and sat. NaCl solution (75mL). The organic phase was dried, concentrated to ~15 mL, and added dropwise into hexane (300 mL) under stirring. The white precipitate formed was filtered, washed with hexane (10 mL×3), and sucked dry. After further drying (60°C/13 Pa) overnight, a pure (TLC, NMR) compound **2** (7.7 g, 93%) was obtained as a white powder. Analytically, the pure material was prepared by crystallization from EtOAc–MeOH,* m.p. 175°C–175.5°C; $[\alpha]_D$ +113 (*c* 1.2, $CHCl_3$).

---

* Compound **2** (7.0 g) was dissolved in EtOAc (18 mL) with heating. After cooling to room temperature, MeOH (36 mL) was added slowly with manual stirring. The white crystals formed were filtered, washed with EtOAc/MeOH 1:2, and sucked dry. Further drying at 60°C/13 Pa overnight gave the analytically pure sample.

¹H NMR (CDCl₃): δ (ppm) 8.09 (d, 2H, Ar-H), 8.02 (d, 2H, Ar-H), 7.95 (d, 2H, Ar-H), 7.78 (d, 2H, Ar-H), 7.62 (t, 1H, Ar-H), 7.56 (t, 1H, Ar-H), 7.53–7.35 (m, 8H, Ar-H), 7.23 (m, 2H, Ar-H), 6.04 (d, 1H, $J_{3,4}$ = 3.4 Hz, H-4), 5.82 (t, 1H, $J_{2,3}$ = 10.0 Hz, H-2), 5.66 (dd, 1H, $J_{3,4}$ = 3.4 Hz, H-3), 4.96 (d, 1H, $J_{1,2}$ = 10.0 Hz, H-1), 4.65 (dd, 1H, $J_{6a,6b}$ = 11.3 Hz, H-6a), 4.43 (dd, 1H, H-6b), 4.36 (dd, 1H, $J_{5,6a}$ = 6.8 Hz, $J_{5,6b}$ = 6.0 Hz, H-5), 3.28 (heptet, 1H, J = 6.8 Hz, CH-iPr), 1.32 (d, 6H, J = 6.8 Hz, CH_3-iPr×2). ¹³C NMR (CDCl₃): δ (ppm) 166.0, 165.6, 165.5, 165.3, 133.6, 133.3, 130.0, 129.8, 129.7, 129.4, 129.3, 129.0, 128.8, 128.6, 128.4, 128.3, 128.3, 84.1 (C-1), 75.0 (C-5), 72.8 (C-3), 68.4 ×2 (C-2, C-4), 62.4 (C-6), 35.8 (CH-iPr), 23.9 (CH_3-iPr), 23.8 (CH_3-iPr). ESI-MS (m/z): [M+Na]⁺ calcd for $C_{37}H_{34}NaO_9S$, 677.1816; found, 677.1816. Anal. calcd for $C_{37}H_{34}O_9S$: C 67.88, H 5.23; found, C 68.11, H 5.38.

Methyl 2,3,4,6-Tetra-*O*-benzoyl-β-ᴅ-galactopyranosyl-(1→6)-2,3,4-tri-*O*-benzyl-α-ᴅ-glucopyranoside (4)

Method A

Promoter preparation:[10] Trifluoromethanesulfonic anhydride (0.168 mL, 1 mmol) was added to a solution of dimethyl disulfide (0.1 mL, 1.1 mmol) in dry CH_2Cl_2 (0.75 mL) at 0°C and the mixture was stirred at 0°C for 30 min. A ~1 M solution of the promoter was obtained. Gradually, a yellow to brown color developed over a period of several hours.

A solution of the freshly prepared promoter (0.9 mL, ~0.9 mmol) was added at 0°C dropwise with stirring to a mixture of compound **2** (392 mg, 0.6 mmol), **3** (232 mg, 0.5 mmol), and 4 Å MS (700 mg) in anhydrous CH_2Cl_2 (5 mL). After 10 min at 0°C, TLC (hexane/EtOAc 3:1) showed disappearance of the acceptor **3** and formation of one major (R_f ~ 0.3) faster moving product. The reaction was quenched with Et₃N (0.3 mL) and CH_2Cl_2 (20 mL) was added. After filtration through a Celite pad, the filtrate was washed successively with 1N HCl aq. (10 mL), sat. NaHCO₃ (10 mL), and sat. aq. NaCl solution (10 mL). The organic phase was dried, concentrated, and the residue was chromatographed (40 g silica gel, 20:1 → 15:1 toluene/EtOAc).

A small amount of material (a few mg) eluted first was methyl 2,3,4-tri-*O*-benzyl-6-*O*-trifluoromethanesulfonyl-α-ᴅ-glucopyranoside (**5**).* ¹H NMR (CDCl₃) δ (ppm) 7.24–7.38 (m, 15H, Ar-H), 5.02 (d, 1H, J = 10.9 Hz, 3-OBn), 4.93 (d, 1H, J = 11.0 Hz, 4-OBn), 4.82 (d, 1H, J = 10.9 Hz, 3-OBn), 4.79 (d, 1H, J = 12.2 Hz, 2-OBn), 4.66 (d, 1H, J = 12.2 Hz, 2-OBn), 4.60 (d, 1H, $J_{1,2}$ = 3.5 Hz, H-1), 4.56 (d, 1H, J = 11.0 Hz, 4-OBn), 4.55 (dd, 1H, H-6a), 4.46 (dd, 1H, $J_{6a,6b}$ = 10.7 Hz, H-6b), 4.02 (t, 1H, $J_{3,4}$ = 9.3 Hz, H-3), 3.86 (ddd, 1H, $J_{5,6a}$ = 1.7 Hz, $J_{5,6b}$ = 5.4 Hz, H-5), 3.53 (dd, 1H, $J_{2,3}$ = 9.6 Hz, H-2), 3.42 (dd, 1H, $J_{4,5}$ = 9.9 Hz, H-4), 3.38 (s, 3H, 1-OCH₃). ¹³C NMR (CDCl₃): δ (ppm) 138.3, 137.8, 137.4, 128.6, 128.5, 128.5, 128.2, 128.1, 127.9, 127.8, 118.5 (q, $^1J_{CF}$ = 318.9 Hz, –CF₃), 98.0 (C-1), 81.8 (C-3), 79.7 (C-2), 76.4 (C-4), 75.8 (3-OBn), 75.1 (4-OBn), 74.9 (C-6), 73.5 (2-OBn), 68.2

---

* The same material (TLC, MS, NMR) was formed (>50% according to TLC) when a solution of glycoside **3** (~25 mg) in CH_2Cl_2 (0.3 mL) was treated with the promoter (~0.1 mL) for 10 min at 0°C.

(C-5), 55.5 (1-OCH$_3$). ESI–MS (*m/z*): [M+NH$_4$]$^+$ calcd for C$_{29}$H$_{35}$F$_3$NO$_8$S, 614.2030; found, 614.2031.

The title compound **4** (463 mg, 89%) was obtained as syrup, which became a white fluffy, amorphous solid after lyophilization of its solution in benzene.

[α]$_D$ +68 (c 1.3, CHCl$_3$); ^{1}H NMR (CDCl$_3$) δ (ppm) 8.08 (d, 2H, Ar-H), 8.01 (d, 2H, Ar-H), 7.88 (d, 2H, Ar-H), 7.76 (d, 2H, Ar-H), 7.61 (t, 1H, Ar-H), 7.55 (t, 1H, Ar-H), 7.48 (t, 2H, Ar-H), 7.44–7.18 (m, 21H, Ar-H), 7.12 (d, 2H, Ar-H), 5.96 (d, 1H, H-4^{II}), 5.84 (dd, 1H, $J_{2,3}$= 10.4 Hz, H-2^{II}), 5.59 (dd, 1H, $J_{3,4}$= 3.5 Hz, H-3^{II}), 4.89 (d, 1H, J= 11.0 Hz, 3^I-OBn), 4.75 (d, 1H, $J_{1,2}$= 8.0 Hz, H-1^{II}), 4.72 (d, 1H, J= 12.0 Hz, 5^I-OBn), 4.69 (d, 1H, J= 11.0 Hz, 3^I-OBn), 4.66 (dd, 1H, $J_{6a,6b}$= 11.2 Hz, H-6a^{II}), 4.58 (d, 1H, J= 12.0 Hz, 5^I-OBn), 4.55 (d, 1H, J= 11.2 Hz, 4^I-OBn), 4.49 (d, 1H, $J_{1,2}$= 3.5 Hz, H-1^I), 4.39 (dd, 1H, H-6b^{II}), 4.37 (d, 1H, J= 11.2 Hz, 4^I-OBn), 4.24 (dd, 1H, J $J_{5,6a}$= 6.5, $J_{5,6b}$= 6.8 Hz, H-5^{II}), 4.20 (dd, 1H, J= 4.2 Hz, 12.8 Hz, H-6a^I), 3.89 (t, 1H, $J_{3,4}$= 9.3 Hz, H-3^I), 3.75 (m, 2H, H-5^I, H-6b^I), 3.39 (dd, 1H, $J_{2,3}$= 9.6 Hz, H-2^I), 3.36 (t, 1H, $J_{4,5}$= 9.0 Hz, H-4^I), 3.20 (s, 3H, 1^I-OCH$_3$). ^{13}C NMR (CDCl$_3$): δ (ppm) 166.03, 165.6, 165.5, 165.1, 138.7, 138.2, 138.1, 133.6, 133.3, 133.1, 129.7, 129.7, 129.4, 129.2, 129.0, 128.7, 128.6, 128.5, 128.4, 128.4, 128.3, 128.3, 128.1, 127.9, 127.8, 127.7, 127.5, 127.5, 102.0 (C-1^{II}), 97.9 (C-1^I), 81.9 (C-3^I), 79.8 (C-2^I), 77.4 (C-4^I), 75.5 (3^I-OBn), 74.7 (4^I-OBn), 73.3 (2^I-OBn), 71.6 (C-3^{II}), 71.3 (C-5^{II}), 69.7 (C-2^{II}), 69.5 (C-5^I), 68.6 (C-6^I), 68.0 (C-4^{II}), 61.8 (C-6^{II}), 55.0 (1^I-OCH$_3$). ESI-MS (*m/z*): [M+NH^4]$^+$ calcd for C$_{62}$H$_{62}$NO$_{15}$, 1060.4114; found, 1060.4117. Anal. calcd for C$_{62}$H$_{58}$O$_{15}$: C 71.39, H 5.60; found, C 71.60, H 5.51.

Method B

Compound **2** (360 mg, 0.55 mmol) and **3** (232 mg, 0.5 mmol) were placed in a 25 mL pear-shaped flask and dissolved in anhydrous CH$_2$Cl$_2$ (5 mL). 4 Å MS (400 mg) was added and the mixture was cooled to −10°C under Ar. After stirring for 20 min at −10°C, NIS (135 mg, 0.6 mmol) was added followed by AgOTf (38.5 mg, 0.15 mmol). The color of the reaction mixture turned into dark red in 5 min. TLC (hexane/EtOAc 3:1) showed absence of the acceptor (**3**) and formation of a major faster moving product (*R$_f$* ~ 0.3). The reaction was quenched with Et$_3$N (0.1 mL) and CH$_2$Cl$_2$ (20 mL) was added. After filtration through a Celite pad, the filtrate was washed successively with 1N HCl aq. (10 mL), 10% Na$_2$S$_2$O$_3$ aq. (10 mL), sat. NaHCO$_3$ (10 mL), and sat. NaCl aq. solution (10 mL). The organic phase was dried and concentrated, and the residue was chromatographed (40 g silica gel, 4:1 → 3:1 hexane–acetone), to give the title compound **4** (455 mg, 87%), which was identical in all aspects with the compound described earlier.

ACKNOWLEDGMENT

This research was supported by the Intramural Research Program of the NIH and NIDDK.

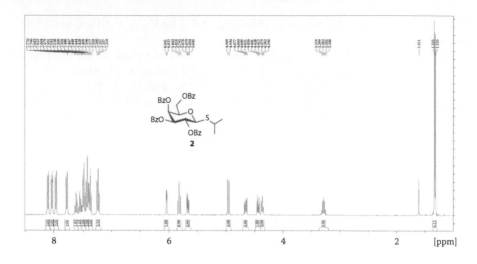

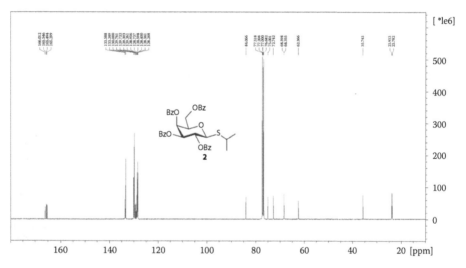

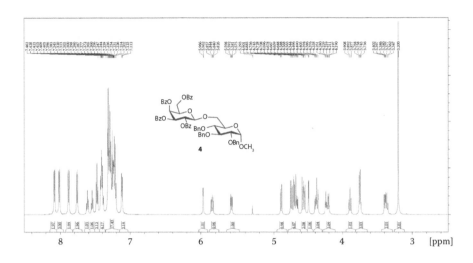

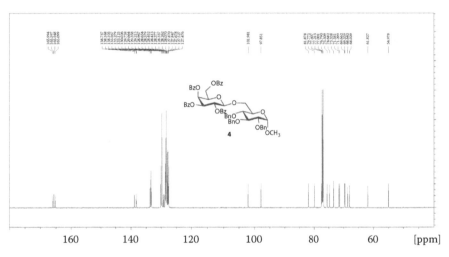

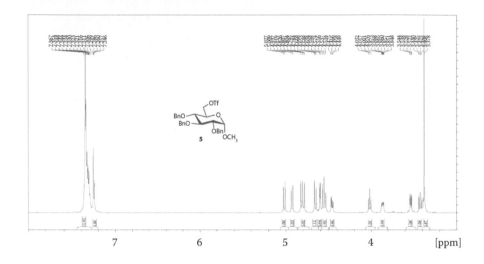

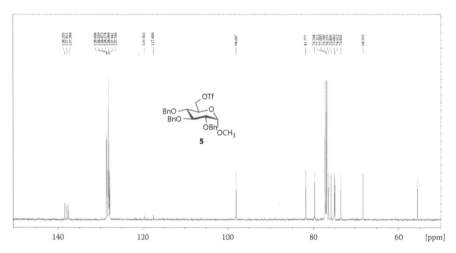

REFERENCES

1. Schmidt, R. R.; Stumpp, M. *Justus Liebigs Ann. Chem.* **1983**, 1249–1256.
2. Hansen, L. H.; Knudsen, S.; Sorensen, S. J. *Curr. Microbiol.* **1998**, *36*, 341–347.
3. Gu, G.; Yang, F.; Du, Y.; Kong, F. *Carbohydr. Res.* **2001**, *336*, 99–106.
4. Li, A.; Kong, F. *Bioorg. Med. Chem.* **2005**, *13*, 839–853.
5. Chen, L.; Shi, S.-D.; Liu, Y.-Q.; Gao, Q.-J.; Yi, X.; Liu, K.-K.; Liu, H. *Carbohydr. Res.* **2011**, *346*, 1250–1256.
6. Garegg, P. J.; Norberg, T. *Acta Chem. Scand.* **1979**, *B33*, 116–118.
7. Fugedi, P.; Garegg, P. J. *Carbohydr. Res.* **1986**, *149*, C9–C12.
8. Shi, S.; Chen, L.-q.; Liu, Y.-Q.; Gao, Q.-J.; Wu, K.-Q. *Guangzhou Huaxue (in Chinese)* **2010**, *35*, 1–6.
9. Konradson, P.; Udodong, U. E.; Fraser-Reid, B. *Tetrahedron Lett.* **1990**, *31*, 4313–4316.
10. Tatai, J.; Fugedi, P. *Org. Lett.* **2007**, *9*, 4647–4650.

30 Facile Access to (1→3)-Glucosamine Linkages

Synthesis of Propyl 4,6-Di-O-benzyl-2-deoxy-2-trichloroacetamido-1-thio-β-D-glucopyranoside

*Chinmoy Mukherjee, Tze Chieh Shiao,[†]
and Nicola L.B. Pohl*[*]*

CONTENTS

(1→3)-Glucosamine linkages are common in a variety of natural products.[1,2] However, the synthesis of the corresponding glucosamine building blocks often requires multiple steps to carry out various mandatory protecting group manipulations. To reduce the number of these steps, researchers often use either

---

[*] Corresponding author; e-mail: npohl@indiana.edu.
[†] Checker, under supervision of R. Roy; e-mail: roy.rene@uqam.ca.

4,6-O-benzylidene or 4,6-O-di-*tert*-butyl silyl groups to leave the 3-OH free for gly-cosylation.[3-5] However, these protecting groups conformationally lock the resulting pyranoses[3] and also render them largely incompatible with acidic and/or basic reaction conditions. Benzyl ethers are regarded as one of the most stable and compatible protecting groups in many reaction conditions.

Selective incorporation of benzyl ethers in amino sugars often requires a number of synthetic steps.[6-8] A new regioselective benzylation method using a 2-NHTCA-protected glucosamine substrate significantly reduces the number of steps to achieve the target building block in high yield.[9] In the presence of sodium hydroxide and 15-crown-5, compound **2** can easily be benzylated at the 4- and 6-position hydroxyls directly from the acetylated precursor to afford compound **3** in 76% yield. Compound **2** was prepared from its acetylated precursor **1**[10] through a simple thioglycosidation reaction with 1-propanethiol[11] in the presence of boron trifluoride diethyl etherate as a Lewis acid catalyst.

EXPERIMENTAL METHODS

GENERAL METHODS

Reactions were performed using oven-dried glassware under argon using anhydrous solvents. Sodium hydroxide (Fisher Scientific) pellets were ground to fine powder prior to use. 1-Propanethiol, boron trifluoride diethyl etherate, benzyl bromide, and 15-crown-5 were purchased from Sigma-Aldrich, United States. Ambient temperature in the laboratory was usually 20°C. All reactions were performed under argon. Thin-layer chromatography (TLC) was performed using glass-backed Silica Gel HL TLC plates w/UV254 from Sorbent Technologies. Visualization of TLC plates was performed by UV, 5% sulfuric acid/ethanol, or *p*-anisaldehyde/ethanol. NMR spectra were recorded on a Varian I400 (400 MHz for ^{1}H, 100 MHz for ^{13}C) and a Varian I500 (500 MHZ for ^{1}H, 125 MHz for ^{13}C). Chemical shifts are reported in parts per million (ppm) on the δ scale. ^{1}H NMR and ^{13}C NMR spectra taken for solution in $CDCl_3$ were referenced to the solvent peak at 7.260 ppm (1H) and 77.16 ppm (^{13}C). The assignments of ^{1}H NMR peaks were made primarily from 2D ^{1}H-^{1}H COSY and ^{1}H-^{13}C HSQC spectra. High-resolution mass spectra (HRMS, ESI mode) were obtained in an LCT KC366 instrument. Melting points were recorded with an electrothermal instrument, and specific optical rotations were recorded with a Perkin Elmer (Model 343) spectrometer. Combustion analyses were performed in Atlantic Microlab, Inc, Georgia, United States.

Propyl 3,4,6-Tri-O-acetyl-2-deoxy-2-trichloroacetamido-1-thio-β-D-glucopyranoside (2)

To a solution of **1** (2.30 g, 4.67 mmol) in dry CH_2Cl_2 (25 mL) was added 1-propa-nethiol (0.63 mL, 0.53 g, 7.00 mmol); the mixture was cooled to 0°C and $BF_3 \cdot OEt_2$ (1.44 mL, 1.66 g, 11.7 mmol) was added. The mixture was stirred at ambient tem-perature for 2 h, diluted with CH_2Cl_2 (60 mL), and washed successively with ice-cold water (40 mL), satd. aq. $NaHCO_3$ (2 × 40 mL), and brine (40 mL). The organic layer was dried (Na_2SO_4), filtered, and concentrated under reduced pressure; the resulting residue was crystallized from hot ethanol. The solids were filtered and dried under reduced pressure to afford the title product **2** (1.85 g, 3.64 mmol, 78%) as colorless crystals. R_f=0.42 (hexane/EtOAc, 3/2, v/v); M.p. 151°C–153°C; $[\alpha]_D^{20}$ −39.3 (c 1.0, $CHCl_3$). ¹H NMR (400 MHz, $CDCl_3$): δ=7.05 (d, J=9.3 Hz, 1 H, N*H*CO), 5.35 (dd, J=10.1, 9.7 Hz, 1 H, H-3), 5.10 (dd, J=9.8, 9.7 Hz, 1 H, H-4), 4.66 (d, J=10.3 Hz, 1 H, H-1), 4.23 (dd, J=12.4, 5.2 Hz, 1 H, H-6$_a$), 4.15–4.06 (m, 2 H, H-2, H-6$_b$), 3.76 (ddd, J=10.0, 5.2, 2.3 Hz, 1 H, H-5), 2.73–2.58 (m, 2 H, S*CH*$_{2ab}$), 2.06, 2.02, 2.01 (3 s, 9 H, 3 COC*H*$_3$), 1.64–1.57 (m, 2 H, *CH*$_{2ab}$CH$_3$), 0.95 (t, J=7.3 Hz, 1 H, CH$_2$C*H*$_3$). ¹³C NMR (100 MHz, $CDCl_3$): δ = 171.2 (*C*O), 170.8 (*C*O), 169.3 (*C*O), 162.0 (NH*C*O), 92.2 (*C*Cl$_3$), 84.0 (C-1), 76.1 (C-5), 73.3 (C-3), 68.6 (C-4), 62.5 (C-6), 54.8 (C-2), 32.3 (S*C*H$_2$), 23.1 (*C*H$_2$CH$_3$), 20.8 (*C*OCH$_3$), 20.7 (*C*OCH$_3$), 20.6 (*C*OCH$_3$), 13.5 (CH$_2$*C*H$_3$). ESIHRMS Calcd for $C_{17}H_{24}Cl_3NNaO_8S$ (M+Na)⁺: 530.0186; found, 530.0192. Anal. Calcd. for $C_{17}H_{24}Cl_3NO_8S$ (507.03): C, 40.13; H, 4.75; N, 2.75; found, C, 40.36; H, 4.68; N, 2.81.

Propyl 4,6-Di-O-benzyl-2-deoxy-2-trichloroacetamido-1-thio-β-D-glucopyranoside (3)

To a solution of the acetylated glucosamine derivative **2** (100 mg, 0.20 mmol) in anhydrous Tetrahydrofuran (THF) (2 mL) was added powdered NaOH (94 mg, 2.358 mmol) and 15-crown-5 (23 μL, 26 mg, 0.118 mmol), and the mixture was stirred at ambient temperature under argon for 1 h. BnBr (70.0 μL, 101 mg, 0.589 mmol) was added to the slurry dropwise and the reaction mixture was stirred for 6 h. The reaction mixture was then neutralized with acetic acid and concentrated under reduced pressure, water (15 mL) was added, and the mixture was extracted with ethyl acetate (3 × 30 mL). The combined organic layer was washed with brine (30 mL), dried (Na_2SO_4), filtered, and concentrated. The crude mixture was chromatographed (4:1 hexane–EtOAc) to afford the title compound **3** as a white solid (84 mg, 0.149 mmol, 76%). R_f=0.5 (2:1 hexane–EtOAc); M. p. 125°C–126.5°C (EtOH); $[\alpha]_D^{20}$ −25.5 (c 1.0, $CHCl_3$). ¹H NMR (500 MHz, $CDCl_3$): δ = 7.20–7.06 (m, 10 H, Ar-*H*), 6.69 (d, J=7.8 Hz, 1 H, N*H*CO), 4.58 (d, J=11.8 Hz, 1 H, PhC*H*$_2$), 4.51–4.47 (m, 2 H, H-1, PhC*H*$_2$), 4.44 (d, J=12.1 Hz, 1 H, PhC*H*$_2$), 4.39 (d, J=12.1 Hz, 1 H, PhC*H*$_2$), 3.78 (ddd, J=9.6, 8.2, 4.1 Hz, 1 H, H-3), 3.60 (dd, J=10.6, 2.1 Hz, 1 H, H-6$_a$), 3.58–3.50 (m, 2 H, H-2, H-6$_b$), 3.41 (dd, J=9.5, 8.1 Hz, 1 H, H-4), 3.35 (ddd, J=9.6, 4.2, 2.0 Hz, 1 H, H-5), 2.70 (d, J=4.2 Hz, 1 H, O*H*), 2.55 (ddd, J=14.4, 7.9, 6.5 Hz, 1 H, S*CH*$_{2a}$), 2.50–2.43 (m, 1 H, S*CH*$_{2b}$), 1.53–1.41 (m, 2 H, *CH*$_{2ab}$CH$_3$), 0.80 (t, J=7.3 Hz, 3 H, CH$_2$C*H*$_3$). ¹³C NMR (125 MHz, $CDCl_3$): δ = 162.4 (NH*C*O), 138.1 (2 C, Ar-*C*), 128.7–127.8 (10 C, Ar-*C*), 92.5 (*C*Cl$_3$), 83.2 (C-1), 79.4 (C-5), 78.4 (C-4), 75.6 (C-3), 74.8 (Ph*C*H$_2$), 73.6 (Ph*C*H$_2$), 69.1 (C-6), 57.7 (C-2), 32.3 (S*C*H$_2$), 23.3 (*C*H$_2$CH$_3$), 13.6 (CH$_2$*C*H$_3$). ESIHRMS Calcd for $C_{25}H_{30}Cl_3NNaO_5S$ (M+Na)⁺: 584.0808; found, 584.0795. Anal. Calcd. for $C_{25}H_{30}Cl_3NO_5S$ (561.09): C, 53.34; H, 5.37; N, 2.49; found, C, 53.47; H, 5.30; N, 2.47.

ACKNOWLEDGMENTS

This work was supported in part by the National Institutes of Health (1R01GM090280). Daniel E. Kabotso is gratefully acknowledged for technical assistance.

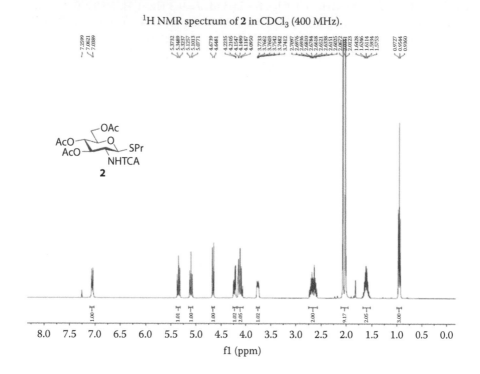

^{1}H NMR spectrum of **2** in CDCl$_3$ (400 MHz).

^{13}C NMR spectrum of **2** in CDCl$_3$ (100 MHz).

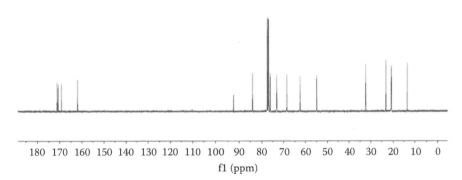

^{1}H NMR spectrum of **3** in CDCl$_3$ (500 MHz).

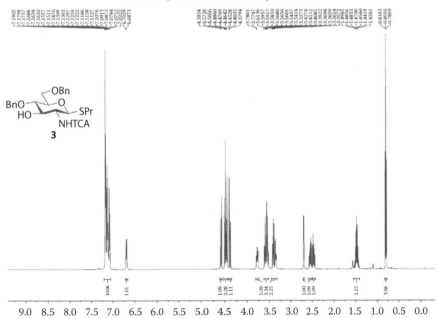

^{13}C NMR spectrum of **3** in CDCl$_3$ (125 MHz).

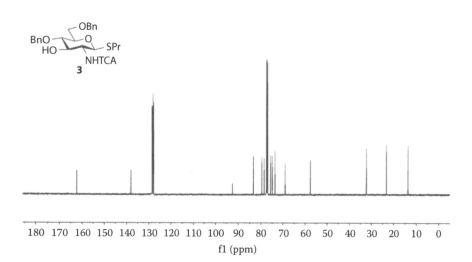

180 170 160 150 140 130 120 110 100 90 80 70 60 50 40 30 20 10 0

f1 (ppm)

REFERENCES

1. Shida, K.; Misonou, Y.; Korekane, H.; Seki, Y.; Noura, S.; Ohue, M.; Honke, K.; Miyamoto, Y. *Glycobiology* **2009**, *19*, 1018–1033.
2. Meyer, K.; Palmer, J. W. *J. Biol. Chem.* **1934**, *107*, 629–634.
3. Crich, D. *J. Org. Chem.* **2011**, *76*, 9193–9209.
4. Dinkelaar, J.; Gold, H.; Overkleeft, H. S.; Codée, J. D. C.; van der Marel, G. A. *J. Org. Chem.* **2009**, *74*, 4208–4216.
5. Lu, X.; Kamat, M. N.; Huang, L.; Huang, X. *J. Org. Chem.* **2009**, *74*, 7608–7617.
6. Nashed, M. A.; Slife, C. W.; Kiso, M.; Anderson, L. *Carbohydr. Res.* **1980**, *82*, 237–252.
7. Ogawa, S.; Matsunaga, N.; Li, H.; Palcic, M. M. *Eur. J. Org. Chem.* **1999**, 631–642.
8. Horlacher, T.; Oberli, M. A.; Werz, D. B.; Kröck, L.; Bufali, S.; Mishra, R.; Sobek, J. et al. *ChemBioChem* **2010**, *11*, 1563–1573.
9. Mukherjee, C.; Lin, L.; Pohl, N. L. B. *Adv. Synth. Catal.* **2014**, *356*, 2247–2256.
10. Palmacci, E. R.; Plante, O. J.; Hewitt, M. C.; Seeberger, P. H. *Helv. Chim. Acta* **2003**, *86*, 3975–3990.
11. Goswami, M.; Ellern, A.; Pohl, N. L. B. *Angew. Chem. Int. Ed.* **2013**, *52*, 8441–8445.

31 Stereoselective Synthesis of 7-Deoxy-1,2;3,4-di-*O*-isopropylidene-D-*glycero*-α-D-*galacto*-heptopyranose

Purav P. Vagadia, Stephen P. Brown,[†]
Deanna L. Zubris, Nicholas A. Piro,
Walter J. Boyko, W. Scott Kassel,
and Robert M. Giuliano*

CONTENTS

* Corresponding author; e-mail: robert.giuliano@villanova.edu.
† Checker, under the supervision of Amos B. Smith III; e-mail: smithab@sas.upenn.edu.

The addition of organometallic reagents to 1,2;3,4-di-O-isopropylidene-α-D-*galacto*-1,6-dialdo-hexopyranose **1** provides higher-carbon heptopyranoses that have been used in the synthesis of sialoside probes for neuraminidase inhibition,[1,2] immunologically relevant disaccharides,[3] substrates for glycosidases used in kinetic resolution,[4] trisaccharides used to assess monoclonal antibody binding,[5] and other targets containing chain-extended sugars.[6] The addition of methylmagnesium halides to **1**, as well as the addition of methyllithium and methylmanganese halides,[7] gives mixtures of the α-D-*glycero* **2** and β-L-*glycero* **3** diastereomers. An extensive study by Tolsitkov and coworkers revealed very high diastereoselectivity for the addition of methylmanganese iodide to **1** to give a 97:3 ratio in favor of the β-L-*glycero* isomer **3**.[7] In spite of the variety of conditions that have been examined, a highly stereoselective route to **2** based on the addition of an organometallic reagent to **1** has not been achieved. Methylation conditions typically favor the formation of **3**, which is predicted on the basis of chelation control involving metal binding with the aldehyde and ring oxygens and alkylation from the less hindered side of the carbonyl group.[7] The highest ratios of **2:3** are only on the order of 2.5:1 to 3:1 and have been achieved using methylmagnesium chloride or methyllithium in diethyl ether. Herein, we wish to report a highly stereoselective synthesis of **2** using methyl(triisopropoxy)titanium[8] $CH_3Ti(Oi\text{-}Pr)_3$ as the methylating agent.

Our choice of methyl(triisopropoxy)titanium[9] was based on our previous work[10] on the addition of organometallic reagents to pentodialdo-1,4-furanoses, in which we observed high Felkin–Anh stereoselectivity when methyl(triisopropoxy)titanium[9]

TABLE 31.1

Comparison of ¹H NMR Data of 2 with Data Reported for 2 and 3

Compound 3[5]	Compound 2[5]	Compound 2 (This Work, Method b)
δ 5.58 (d, 1H, J=4.6 Hz, H-1)	δ 5.55 (d, 1H, J=5 Hz, H-1)	5.56 (d, 1H, J=5.1 Hz, H-1)
δ 4.59 (dd, 1H, J=8.0, 2.1 Hz, H-3)	δ 4.62 (dd, 1H, J=8.0, 2.2 Hz, H-3)	4.62 (dd, 1H, J=7.8, 2.4 Hz, H-3)
δ 4.32 (dd, 1H, H-2)	δ 4.30 (dd, 1H, H-2)	4.32 (dd, 1H, J=4.8, 2.4 Hz, H-2)
δ 4.26 (dd, 1H, J=2.0 Hz, H-4)	δ 4.46 (dd, 1H, J=2.0 Hz, H-4)	4.47 (dd, 1H, J=7.5, 2.4 Hz, H-4)
δ 3.99 (m, 1H, H-6)	δ 3.96 (m, 1H, H-6)	3.97 (m, 1H, H-6)
δ 3.48 (dd, 1H, J=6.8 Hz, H-5)	δ 3.50 (dd, 1H, J=7.2 Hz, H-5)	3.51 (dd, 1H, J=7.5, 1.8 Hz, H-5)
δ 1.50 (CH₃ singlet)	δ 1.51 (CH₃ singlet)	δ 1.52 (CH₃ singlet)
δ 1.44 (CH₃ singlet)	δ 1.45 (CH₃ singlet)	δ 1.47 (CH₃ singlet)
δ 1.32 (CH₃ singlet)	δ 1.36 (CH₃ singlet)	δ 1.36 (CH₃ singlet)
	δ 1.32 (overlap with δ 1.29)	δ 1.33 (CH₃ singlet)
δ 1.26 (d, 3H, J=6 Hz, C-7)	δ 1.29 (overlap with δ 1.32)	1.31 (d, 3H, J=6.3 Hz, C-7)
		2.29 (d, 1H, J=6.0 Hz, OH)

Source: Lemieux, R.U. et al., *Can. J. Chem.*, *60*, 81–86, 1982.

was used. The conversion, as described in the section "Experimental Methods," gave **2** in a ratio of approximately 19:1, as determined by integration of the C-6 hydroxyl group proton at 2.34 and 2.73 ppm, respectively, in the ^{1}H Nuclear Magnetic Resonance (NMR) spectrum of the crude product. Unchanged aldehyde **1** (~5%) was also present.

Pure, title compound **2** was isolated by flash chromatography and its isomeric identity and purity were proven by x-ray analysis of the crystalline *p*-bromobenzoate **4**. As noted by Lemieux,[5] many of the proton chemical shifts for **2** and **3** are nearly identical; however, the H-4 resonances are sufficiently resolved in spectra of the two diastereomers[5] (δ 4.47 in **2** and δ 4.26 in **3**, Table 31.1). A comparison of ^{1}H NMR data of **2** with those reported previously[5] for **2** and **3** is given in Table 31.1, which proves that the identity of the material described here is consistent with that reported by Lemieux and coworkers.[5] Nevertheless, we cannot explain the difference in the $[\alpha]_D$ values found for the two materials synthesized independently. Herein, we describe a gram-scale, stereoselective synthesis of the title compound **2** and further illustrate the Felkin–Anh (nonchelation) selectivity in reactions of dialdopyranose sugars with methyl(triisopropoxy)titanium.

EXPERIMENTAL METHODS

GENERAL METHODS

Chlorotitanium triisopropoxide, 97% (CAS number 20717-86-6), was purchased from Strem Chemicals and stored in a nitrogen glove box at room temperature. All other reagents were purchased from Sigma-Aldrich, including anhydrous diethyl ether and anhydrous dichloromethane (Sure/Seal™). Thin-layer chromatography (TLC) was performed on aluminum-backed silica gel 60 F_{254} plates (Merck KGaA 64271 Darmstadt). Visualization of spots was facilitated by dipping the TLC plates in Hanessian's stain [3 g $Ce(SO_4)_2$, 10 g $(NH_4)Mo_7O_{24}\cdot4H_2O$, H_2O, 450 mL, conc. H_2SO_4, 50 mL] followed by heating with a heat gun. Column chromatography was performed on silica gel 60 Å (Fluka, 40–63 μm particle size). NMR spectra were recorded on a Varian Mercury 300 MHz spectrometer in $CDCl_3$. ^{1}H and ^{13}C spectra (300 and 75 MHz, respectively) were referenced to Tetra Methyl Silane (TMS) (δ=0.0 ppm) and $CDCl_3$ (δ=76.9 ppm). Signal-nuclei assignments were confirmed using Distortion less Enhancement by polarization Transfer (DEPT) and 2D experiments. Optical rotations were measured at 23°C using Perkin–Elmer 241 and Jasco p-2000 polarimeters.

7-Deoxy-1,2;3,4-di-*O*-isopropylidene-D-*glycero*-α-D-*galacto*-heptopyranose (2)

Method A

In a nitrogen glove box, chlorotitanium triisopropoxide (23.3 g, 89.4 mmol, ~6 equiv.)* was added to a Schlenk flask equipped with a Teflon plug and a magnetic stir bar. The Schlenk flask was fitted with a rubber septum and removed from the

---

* An excess of chlorotitanium triisopropoxide and methyllithium over the 6 equivalents of each may be needed for the methylation of **1** to go to completion.

glove box. The Schlenk flask was then connected to a Schlenk line, and argon was introduced into the flask. Anhydrous diethyl ether (55 mL) was dispensed into the flask using a syringe. The mixture was stirred at room temperature until the solids dissolved, then chilled to −50°C (dry ice/acetone bath). Under an argon counter-flow, the septum was replaced with a dropping funnel (itself capped with a septum). Methyllithium (57 mL of 1.6 M methyllithium in diethyl ether, 91 mmol, ~6 equiv.) was added to the dropping funnel via cannula transfer. Methyllithium was added dropwise at a rate that the temperature of the cooling bath did not rise above −50°C. When the addition was complete, the light yellow reaction mixture was stirred for 1 h (the reaction warmed to −25°C during this time) and a white precipitate formed. Under an argon counterflow, the dropping funnel was removed and replaced with a stopper. While maintaining the cooling bath at −25°C, diethyl ether was removed *in vacuo* (using the Schlenk line, with a liquid nitrogen–cooled trap to collect the solvent). When the solvent was removed, methyl(triisopropoxy)titanium thus formed was kept under argon, and the flask's stopper was replaced with a septum.

Methyl(triisopropoxy)titanium was suspended in anhydrous dichloromethane (35 mL), and the Schlenk flask was placed in a dry ice/acetone bath (−78°C). A solution of 1,2;3,4-di-*O*-isopropylidene-α-D-galacto-1,6-dialdo-hexopyranose **1** (3.8 g, 14.7 mmol),* prepared by TEMPO/BAIB oxidation,[11] in anhydrous dichloromethane (15 mL) was added dropwise via a syringe while maintaining the cooling bath at −78°C. The cooling bath was left in place to allow the temperature to rise slowly over a period of 22 h under argon (a needle was affixed to an argon balloon and was inserted through the septum of the Schlenk flask in order to avoid pressure buildup). The temperature of the bath at this point had reached −5°C. In order to quench the reaction, a few drops of water were added slowly with stirring to the reaction mixture. Once vigorous effervescence subsided, additional water (95 mL) was added, followed by diethyl ether (200 mL). The entire mixture (including the solids formed) was transferred into an Erlenmeyer flask (500 mL) equipped with a large stirring bar and the mixture was stirred vigorously for 10 min. After filtration, the solids were washed with diethyl ether (125 mL), the filtrate was transferred into a separatory funnel, the layers were separated, and the aqueous phase was washed with diethyl ether (2×25 mL). Combined organic extracts were washed successively with H_2O (50 mL) and saturated NaCl solution (20 mL), dried ($MgSO_4$), and filtered through a pad of silica gel. The silica gel was rinsed with additional diethyl ether (100 mL) and the filtrate was concentrated to yield crude **2** as a light yellow oil. Chromatography (1:1 ethyl acetate/hexanes) yielded 2.63 g of **2** as a colorless oil (65.1%). Small amounts of impurities were evident in the NMR spectrum, but the material produced correct analytical figures.

$[\alpha]_D$ = −37.8 (*c* 0.4, $CHCl_3$), lit. [5][†] $[\alpha]_D$ = −34.8 (*c* 0.3, $CHCl_3$), for nearly pure substance.

---

* Aldehyde **1** is unstable and it should be prepared freshly before use. See also reference 12.
† The optical rotation for **2** reported by Lemieux and coworkers[5] is for a white solid, mp 57–58°C, that was obtained by a sequence consisting of benzoylation of **2** and debenzoylation of the benzoate **4**. Gateau-Olesker and coworkers[13] independently synthesized **2** and obtained it in a non crystalline form.

^{1}H NMR (CDCl$_3$): δ 1.31 (d, 3H, $J_{7,6}$=6.3 Hz, (C-7) H$_3$, 1.33, 1.36, 1.47, 1.52 (4s, 12H, 4×isopropylidene CH$_3$), 2.29 (d, 1H, $J_{OH,6}$=6.0 Hz, OH), 3.51 (dd, 1H, $J_{5,6}$=7.5 Hz, $J_{5,4}$=1.8 Hz, H-5), 3.96 (m, 1H, H-6), 4.32 (dd, 1H, $J_{2,1}$=5.4 Hz, $J_{2,3}$=2.4 Hz, H-2), 4.47 (dd, 1H, $J_{4,3}$=8.1 Hz, $J_{4,5}$=2.4 Hz, H-4), 4.62 (dd, 1H, $J_{3,4}$=7.8 Hz, $J_{3,2}$=2.4 Hz, H-3), 5.56 (d, 1H, $J_{1,2}$=5.1 Hz, H-1). ^{13}C NMR (CDCl$_3$): δ 109.2, 108.5 (quaternary C of isopropylidene), 96.4 (C-1), 71.2 (C-5), 70.7 (C-4), 70.6 (C-3), 70.4 (C-2), 66.4 (C-6), 25.9, 25.9, 24.9, 24.4 (4×isopropylidene CH$_3$), 20.3 (C-7)

Anal. Calcd for C$_{13}$H$_{22}$O$_6$: C, 56.92; H, 8.08. Found: C, 57.19; H, 8.29.

Method B

The 4-bromobenzoate **4** (0.98 g, 2.1 mmol) was dissolved in anhydrous methanol (10 mL). Sodium methoxide solution (0.26 g, 25 wt% in methanol) was added, and the solution was stirred at room temperature. TLC analysis (1:1 ethyl acetate: hexanes) indicated the presence of some starting material; therefore, additional 0.23 g of sodium methoxide solution (overall 0.49 g, 25 wt% in methanol, 2.3 mmol, 1.1 equiv.) was added and the reaction was stirred overnight. After 12 h, all starting material was consumed and the reaction mixture had turned yellow. The mixture was neutralized by adding, in small portions, Dowex 50WX8-200 (H$^+$), until neutral pH was achieved. The mixture was filtered and the filtrate was concentrated to yield a solid residue containing 4-bromomethyl benzoate as the major impurity. Chromatography (2:3 ethyl acetate/hexanes) yielded 0.54 g (91.5%) of **2** as colorless oil, which was dried overnight in a vacuum oven at 50°C.

$[\alpha]_D$ = −66.9 (*c*, 0.3, CHCl$_3$), lit. [5]* $[\alpha]_D$ = −34.8 (*c* 0.3, CHCl$_3$), for nearly pure substance.

^{1}H NMR (CDCl$_3$): δ 1.31 (d, 3H, $J_{7,6}$=6.3 Hz, (C-7) H$_3$, 1.33, 1.36, 1.47, 1.52 (4s, 12H, 4×isopropylidene CH$_3$), 2.29 (d, 1H, $J_{OH,6}$=6.0 Hz, OH), 3.51 (dd, 1H, $J_{5,6}$=7.5 Hz, $J_{5,4}$=1.8 Hz, H-5), 3.97 (m, 1H, H-6), 4.32 (dd, 1H, $J_{2,1}$=4.8 Hz, $J_{2,3}$=2.4 Hz, H-2), 4.47 (dd, 1H, $J_{4,3}$=7.5 Hz, $J_{4,5}$=2.4 Hz, H-4), 4.62 (dd, 1H, $J_{3,4}$=7.8 Hz, $J_{3,2}$=2.4 Hz, H-3), 5.56 (d, 1H, $J_{1,2}$=5.1 Hz, H-1).

^{13}C NMR (CDCl$_3$): δ 109.3, 108.5 (quaternary C of isopropylidene), 96.4 (C-1), 71.0 (C-5), 70.8 (C-4), 70.6 (C-3), 70.4 (C-2), 66.9 (C-6), 25.9, 25.9, 24.9, 24.4 (4×isopropylidene CH$_3$), 20.3 (C-7).

^{1}H NMR spectra of compound **2** prepared by method A (top spectrum) and method B (bottom spectrum) as well as the ^{13}C NMR spectrum for **2** prepared by method B are provided.

6-*O*-(4-Bromobenzoyl)-7-deoxy-1,2;3,4-di-*O*-isopropylidene-D-*glycero*-α-D-*galacto*-heptopyranose (4)

To a solution of **2** (2.63 g, 9.58 mmol) in anhydrous pyridine (80 mL) was added 4-bromobenzoyl chloride (3.16 g, 14.4 mmol, 1.5 equiv.). The reaction was stirred at room temperature overnight. TLC analysis indicated majority of the starting

---

* The optical rotation for **2** reported by Lemieux and coworkers[5] is for a white solid, mp 57–58°C, that was obtained by a sequence consisting of benzoylation of **2** and debenzoylation of the benzoate **4**. Gateau-Olesker and coworkers[12] independently synthesized **2** and obtained it in a noncrystalline form.

material was consumed (1:1 Ethyl acetate/hexanes). Excess pyridine was removed under reduced pressure to yield white solids. Hexanes (10 mL) were added to the solids and, after brief manual swirling, the mixture was filtered through a medium-porosity sintered glass funnel. Solids remaining in the flask were combined with the bulk in the funnel using nine additional portions of hexanes (10 mL, 100 mL of hexane overall), rubbing the combined solids with hexanes after addition of each portion, and the combined filtrates were concentrated. If more solids appeared in the filtrate, the filtration was repeated (to remove any solid). Concentration of the clear solution yielded 2.6 g of a yellow oil. Crystallization from methanol (15 mL) gave 1.27 g of a white solid, $[\alpha]_D = -68.7$ (c 0.3, $CHCl_3$); mp 93.5°C–95.5°C.

^{1}H NMR ($CDCl_3$): δ 1.26 (s, 3H, isopropylidene CH_3), 1.34 (s, 3H, isopropylidene CH_3), 1.43 (s, 3H, isopropylidene CH_3), 1.44 (d, 3H, $J=6.3$ Hz, (C-7)H$_3$), 1.53 (s, 3H, isopropylidene CH_3), 3.86 (dd, 1H, $J=7.5$ Hz, 1.8 Hz, H-5), 4.33 (m, 2H, H-2, H-4), 4.61 (dd, 1H, $J=8.1$ Hz, 2.4 Hz, H-3), 5.24 (dq, 1H, $J=7.5$ Hz, 6.3 Hz, H-6), 5.56 (d, 1H, $J=5.1$ Hz, H-1), 7.57 (d, 1H, $J=8.1$ Hz, aromatic), 7.89 (d, 1H, $J=8.1$ Hz, aromatic).

^{13}C NMR ($CDCl_3$): δ 164.7 (ester carbonyl), 131.7, 131.1, 129.4, 128.0 (aromatic), 109.5, 108.7 (quaternary C of isopropylidene), 96.4 (C-1), 70.7 (C-2), 70.7 (C-3), 70.6 (C-4), 70.1 (C-6), 69.8 (C-5), 26.1, 26.0, 24.9, 24.4 (4×isopropylidene CH_3), 17.0 (C-7).

Anal. Calcd for $C_{20}H_{25}BrO_7$: C, 52.53; H, 5.51; Br, 17.47. Found: C, 52.43; H, 5.36; Br, 17.39.

X-Ray Structural Analysis

A single colorless shard ($0.15 \times 0.17 \times 0.24$ mm^3) was mounted using NVH immersion oil onto a nylon fiber and cooled to the data collection temperature of 100(2) K. Data were collected on a Brüker-AXS Kappa APEX II CCD diffractometer with 0.71073 Å Mo-Kα radiation. Unit cell parameters were obtained from 60 data frames, 0.5° ϕ, from three different sections of the Ewald sphere yielding $a = 13.3815(10)$, $b = 19.0003(14)$, $c = 25.1987(19)$ Å, $V = 6406.8(8)$ Å^3. 141,249 reflections ($R_{int} = 0.0838$) were collected (14,146 unique) over $\theta = 1.342°$–$27.122°$. The systematic absences in the data were consistent with the noncentrosymmetric, orthorhombic space group $P2_12_12_1$. The data set was treated with SADABS absorption corrections based on redundant multiscan data (Sheldrick, G. Bruker-AXS, 2012); $T_{max}/T_{min} = 1.12$. The asymmetric unit contains three independent molecules yielding $Z' = 3$ and $Z = 12$. All nonhydrogen atoms were refined with anisotropic displacement parameters. All hydrogen atoms were treated as idealized contributions with methyl groups allowed to rotate freely. To confirm the absolute structure, the data were passed through one round of refinement as a racemic twin with the twin ratio parameter allowed to refine freely; this parameter corresponds to the Flack x parameter and refined to 0.004(5), which confirms that the correct absolute configuration was identified and that the crystal is enantiomerically pure. The goodness of fit on F^2 was 1.021 with $R_1 = 0.0326$ (for $I > 2\sigma (I)$) and $wR_2 = 0.0618$ (all data). The largest difference peak and hole are 0.0287 and -0.372 e/Å^3.

ACKNOWLEDGMENTS

We thank Villanova University for supporting this project.

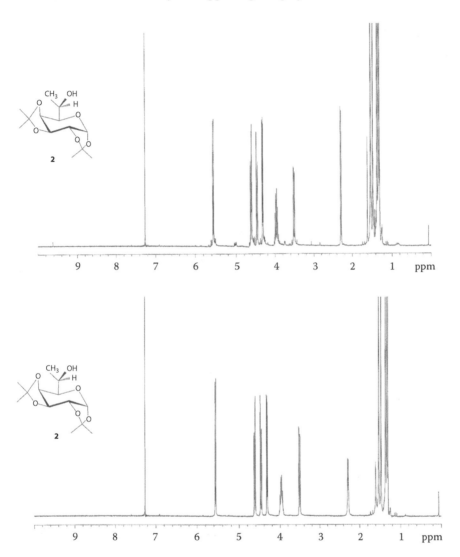

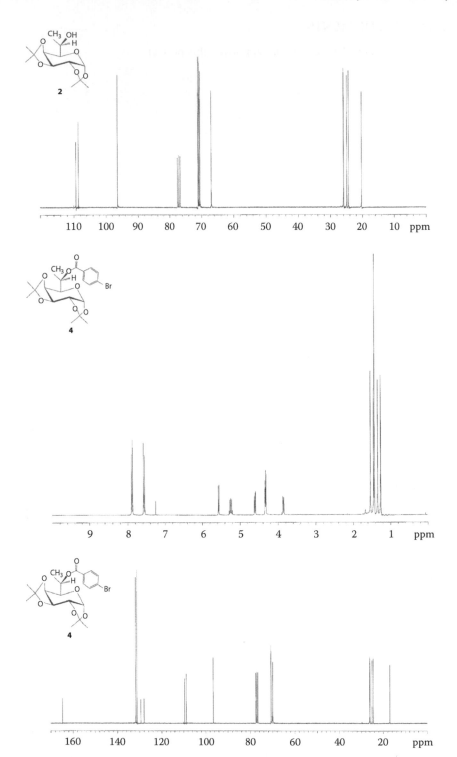

REFERENCES

1. Sabesan, S.; Neira, S.; Wasserman, Z. *Carbohydr. Res.*, **1995**, *267*, 239–261.
2. Sabesan, S.; Neira, S.; Davidson, F.; Duus, J. O.; Bock, K. *J. Am. Chem. Soc.*, **1994**, *116*, 1616–1634.
3. Martin, P.; Lequart, V.; Cecchelli, R.; Boullanger, P.; Lafont, D.; Banoub, J. *Chem. Lett.*, **2004**, *33*, 696–697.
4. Grabowska, U.; MacManus, D. A.; Biggadike, K.; Bird, M. I.; Davies, S.; Gallagher, T.; Hall, L. D.; Vulfson, E. N. *Carbohydr. Res.*, **1998**, *305*, 351–361.
5. Lemieux, R. U.; Wong, T. C.; Thogersen, H. *Can. J. Chem.*, **1982**, *60*, 81–86.
6. Gyorgydeak, Z.; Pelyvas, I. F. *Monosaccharide Sugars: Chemical Synthesis, Chain Elongation, Degradation, and Epimerization*, Academic Press, San Diego, CA, **1998**, pp. 193–197.
7. Kasatkin, A. N.; Podlipchuk, R. K.; Biktimirov, R. K.; Tolsitkov, G. A. *Russ. Chem. Bull.*, **1993**, *42*, 1078–1082.
8. Rausch, M. D.; Gordon, H. B., U. *J. Organomet. Chem.*, **1974**, *74*, 85–90.
9. Weidmann, B.; Seebach, D. *Helv. Chim. Acta.*, **1980**, *63*, 2451–2454.
10. Giuliano, R. M.; Villani, Jr., F. J. *J. Org. Chem.* **1995**, *60*, 202–211.
11. De Mico, A.; Margarita, R.; Parlanti, L.; Vescovi, A.; Piancatelli, G. *J. Org. Chem.*, **1997**, *62*, 6974–6977.
12. Brunjes, M.; Sourkouni-Argirusi, G.; Krishning, A. *Adv. Synth. Catal.*, **2003**, *345*, 635–642.
13. Gateau-Olesker, A.; Sepulchre, A. M.; Vass, G.; Gero, S. D. *Tetrahedron*, **1977**, *33*, 393–397.

32 Synthesis of 4-Methoxyphenyl α-D-Rhamnopyranoside

*Tze Chieh Shiao, Sylvain Rocheleau,[†]
and René Roy**

CONTENTS

D-Rhamnose is a key component of several bacterial polysaccharides and lipopolysaccharides.[1] Commercial D-rhamnose's cost is prohibitive for large-scale synthesis. Although several chemical syntheses of D-rhamnosides have been reported,[2–5] the methods are generally not readily applicable for accessing large quantities of building blocks toward complex oligosaccharide synthesis. We previously reported a convenient synthesis of D-rhamnoside **3** in five steps from D-mannose, albeit with no synthetic details.[6,7] We wish to present herein the high-yielding detailed synthetic procedures for its preparation on a multigram scale from 4-methoxyphenyl α-D-mannopyranoside **1**[8,9] in 87% overall yield, together with specifically functionalized D-rhamnoside **2**, which are useful building blocks for oligosaccharide synthesis.

Thus, starting from unprotected D-mannopyranoside **1**,[8,9] its direct conversion to the 6-deoxy-6-iodo-derivative **2** using the Garegg–Samuelsson's procedure (I₂, Ph₃P, imidazole, THF)[10] occurred uneventfully in 93% yield according to the general Scheme. Radical dehalogenation using tri-*n*-butyltin hydride in the presence

---

* Corresponding author; e-mail: roy.rene@uqam.ca.
† Checker, under supervision of Nicolas Moitessier; e-mail: Nicolas.moitessier@mcgill.ca.

of 2,2'-azobisisobutyronitrile (AIBN) in refluxing toluene provided the desired D-rhamnopyranoside **3** in 94% yield after silica gel column chromatography. Notably, we report herein its Nuclear Magnetic Resonance (NMR) spectra both in CD_3OD and in $CDCl_3$, the latter showing a first-order spectrum.

EXPERIMENTAL METHODS

GENERAL METHODS

The reaction was carried out under argon using freshly distilled solvent. After work-up, the organic layer was dried over anhydrous $MgSO_4$ and concentrated at reduced pressure. The progress of reactions was monitored by thin-layer chromatography using silica gel 60 F_{254} coated plates (E. Merck). NMR spectra were recorded on Varian Inova AS600 and Bruker Avance III HD 600 MHz spectrometers. Proton and carbon chemical shifts (δ) are reported in ppm relative to the chemical shift of residual $CHCl_3$, which was set at 7.26 ppm (^1H) and 77.16 ppm (^{13}C). Coupling constants (J) are reported in Hertz (Hz), and the following abbreviations are used for peak multiplicities: singlet (s), doublet (d), doublet of doublets (dd), doublet of doublet with equal coupling constants (t_{ap}), triplet (t), and multiplet (m). Analysis and assignments were made using COSY and HSQC experiments. High-resolution mass spectra (HRMS) were measured with liquid chromatography–mass spectrometry–time of flight (Agilent Technologies) in positive and/or negative electrospray mode by the analytical platform of UQAM.

4-Methoxyphenyl 6-Deoxy-6-iodo-α-D-mannopyranoside (2)[6]

A suspension of mannopyranoside **1**[8,9] (3.00 g, 10.48 mol, 1.00 equiv.) in dry THF (80 mL) was refluxed under argon atmosphere (the solution becomes homogeneous). Triphenylphosphine (PPh_3, 5.50 g, 20.96 mmol, 2.00 equiv.) and imidazole (1.78 g, 26.20 mmol, 2.50 equiv.) were rapidly added into the solution. Stirring at reflux was continued until a clear solution was obtained, and a solution of iodine (5.32 g, 20.96 mmol, 2.00 equiv.) in dry THF (40 mL) was added dropwise via a syringe. The mixture was stirred at reflux for 2 h and, after cooling to room temperature, the white precipitate of triphenylphosphine oxide was removed by filtration and washed with THF. The filtrate was concentrated under reduced pressure keeping a minimum amount of THF. The concentrated THF solution was next directly applied on top of a dry silica gel column, and chromatography ($CHCl_3 \rightarrow$ 95:5 $CHCl_3$–MeOH), to afford the 6-iodo derivative **2** as a yellow solid (3.85 g, 93%); mp, 133°C–135°C ($CHCl_3$); $R_f = 0.36$, 9:1 $CHCl_3$–MeOH; $[\alpha]_D^{24}$ +14 ($c = 1.0$, MeOH); ^1H NMR (CD_3OD) δ 7.07 (d, 2H, $J_{H,H} = 9.1$ Hz, H-arom), 6.83 (d, 2H, $J_{H,H} = 9.1$ Hz, H-arom), 5.32 (d, 1H, $J_{1,2} = 1.8$ Hz, H-1), 3.99 (dd, 1H, $J_{2,3} = 3.4$ Hz, $J_{1,2} = 1.8$ Hz, H-2), 3.86 (dd, 1H, $J_{3,4} = 8.7$ Hz, $J_{2,3} = 3.4$ Hz, H-3), 3.73 (s, 3H, OCH_3), 3.58–3.52 and 3.28–3.25 ppm (m, 4H, H-4, H-5, H-6a et H-6b); ^{13}C NMR (CD_3OD) δ 156.6, 151.9 (C_q-arom), 119.3, 115.5 (CH-arom), 101.1 (C-1), 74.5 (C-5), 72.5 (C-3), 72.1 (C-2), 72.0 (C-4), 56.0 (OCH_3), 6.3 ppm (C-6).

ESI$^+$-HRMS: [M+Na]$^+$ calcd for $C_{13}H_{18}IO_6Na$, 418.9962; found, 418.9955. Anal. Calcd. for $C_{13}H_{17}IO_6$: C, 39.41; H, 4.33. Found: C, 39.37; H, 4.39.

4-Methoxyphenyl α-D-Rhamnopyranoside (4-Methoxyphenyl 6-Deoxy-α-D-Mannopyranoside) (3)[7]

To a solution of iodide **2** (4.85 g, 12.24 mmol, 1.00 equiv.) in dry toluene were added tributyltin hydride (Bu$_3$SnH, 5.02 mL, 18.40 mmol, 1.50 equiv.) and AIBN (0.20 g, 1.22 mmol, 0.10 equiv.). The reaction mixture was stirred under reflux for 1 h under argon atmosphere and, after cooling to room temperature, concentrated. The residue was chromatographed (CHCl$_3$ → 95:5 CHCl$_3$–MeOH), to give the desired deoxy compound **3** as a white solid (3.24 g, 94%). mp: 101°C–103°C (AcOEt-PE),[6] (Lit. 11 mp = 101°C–102°C); R_f = 0.67, 9:1 CH$_3$CN–H$_2$O; R_f = 0.29, 9:1 CHCl$_3$–MeOH; $\left[\alpha\right]_D^{24}$ +96 (c = 1.0, CHCl$_3$), (Lit. 11 $\left[\alpha\right]_D^{24}$ +117 (c = 0.37, MeOH); ^{1}H NMR (CD$_3$OD) δ 6.97 (d, 2H, $J_{H,H}$ = 9.1 Hz, H-arom.), 6.83 (d, 2H, $J_{H,H}$ = 9.1 Hz, H-arom.), 5.28 (d, 1H, $J_{1,2}$ = 1.8 Hz, H-1), 3.98 (dd, 1H, $^3J_{2,3}$ = 3.5 Hz, $J_{1,2}$ = 1.8 Hz, H-2), 3.82 (dd, 1H, $J_{3,4}$ = 9.5 Hz, $J_{2,3}$ = 3.4 Hz, H-3), 3.73 (s, 3H, OCH$_3$), 3.68 (dq, 1H, $J_{4,5}$ = 9.5 Hz, $J_{5,6}$ = 6.2 Hz, H-5), 3.44 (dd, 1H, $J_{3,4}$ = $J_{4,5}$ = 9.5 Hz, H-4) and 1.22 (d, 3H, $^3J_{5,6}$ = 6.2 Hz, H-6); ^{13}C NMR (CD$_3$OD) δ 156.4, 151.0 (C$_q$-arom.), 118.8, 115.6 (CH-arom.), 100.7 (C-1), 73.9 (C-3), 72.2 (C-4), 72.2 (C-2), 70.4 (C-5), 56.0 (OCH$_3$) and 18.0 ppm (C-6). ^{1}H NMR (CDCl$_3$) δ 6.91 (d, 2H, $J_{H,H}$ = 9.1 Hz, H-arom.), 6.73 (d, 2H, $J_{H,H}$ = 9.1 Hz, H-arom.), 5.38 (d, 1H, $J_{1,2}$ = 1.7 Hz, H-1), 4.84 (d, 1H, $J_{3,OH}$ = 6.1 Hz, OH-3), 4.49 (d, 1H, $J_{2,OH}$ = 4.9 Hz, OH-2), 4.42 (d, 1H, $J_{4,OH}$ = 5.0 Hz, OH-4), 4.17 (dd, 1H, $^3J_{2,3}$ = 3.2 Hz, $J_{1,2}$ = 1.7 Hz, H-2), 4.02 (dd, 1H, $J_{3,4}$ = 9.1 Hz, $J_{2,3}$ = 3.2 Hz, H-3), 3.72 (s, 3H, OCH$_3$), 3.79 (dq, 1H, $J_{4,5}$ = 9.4 Hz, $J_{5,6}$ = 6.2 Hz, H-5), 3.44 (ddd, 1H, H-4) and 1.27 (d, 3H, $^3J_{5,6}$ = 6.2 Hz, H-6); ^{13}C NMR (CDCl$_3$) δ 154.9, 150.0 (C$_q$-arom.), 117.6, 114.6 (CH-arom.), 98.6 (C-1), 72.9 (C-3), 71.6 (C-4), 71.0 (C-2), 68.0 (C-5), 55.5 (OCH$_3$) and 17.5 ppm (C-6). ESI$^+$-HRMS: [M+Na]$^+$ calcd for $C_{13}H_{18}O_6Na$, 293.0996; found, 293.1004. Anal. Calcd. for $C_{13}H_{18}O_6$: C, 57.77; H, 6.71. Found: C, 56.25; H, 6.68.

ACKNOWLEDGMENTS

This work was supported by the Natural Sciences and Engineering Research Council of Canada and a Canadian Research Chair in Therapeutic Chemistry to R.R.

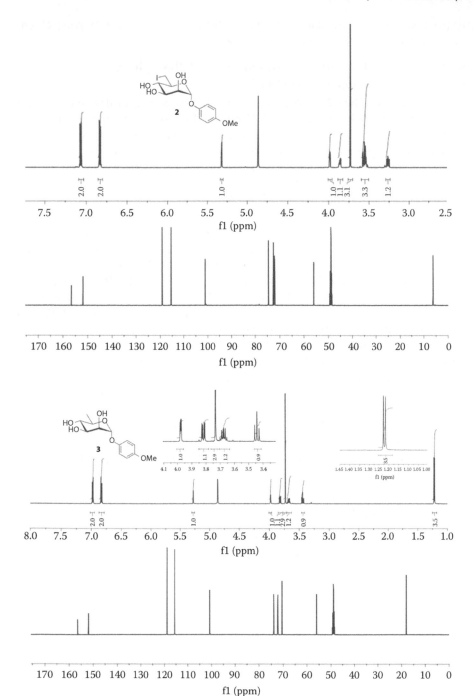

REFERENCES

1. (a) Comegna, D.; Bedini, E.; Parrilli, M. *Tetrahedron*, **2008**, *64*, 3381–3391; (b) Zunk, M.; Kiefel, M. J. *Tetrahedron Lett.*, **2011**, *52*, 1296–1299.
2. (a) Tsvetkov, Y. E.; Backinowsky, L. V.; Kochetkov, N. K. *Carbohydr. Res.*, **1989**, *193*, 75–90; (b) Ramm, M.; Lobe, M.; Hamburger, M. *Carbohydr. Res.*, **2003**, *338*, 109–112; (c) Bedini, E.; Carabellese, A.; Corsaro, M. M.; De Castro, C.; Parrilli, M. *Carbohydr. Res.*, **2004**, *339*, 1907–1915; (d) Bedini, E.; Carabellese, A.; Barone, G.; Parrilli, M. *J. Org. Chem.*, **2005**, *70*, 8064–8070; (e) Ley, S. V.; Owen, D. R.; Wesson, K. E. *J. Chem. Soc. Perkin Trans. 1*, **1997**, 2805–2806; (f) Zou, W.; Sen, A. K.; Szarek, W. A.; MacLean, D. B. *Can. J. Chem.*, **1993**, *71*, 2194–2200; (g) Crich, D.; Bowers, A. A. *Org. Lett.*, **2006**, *8*, 4327–4330; (h) Crich, D.; Li, L. *J. Org. Chem.*, **2009**, *74*, 773–781.
3. Dhénin, S. G. Y.; Moreau, V.; Morel, N.; Nevers, M.-C.; Volland, H.; Créminon, C.; Djedaïni-Pilard, F. *Carbohydr. Res.*, **2008**, *343*, 2101–2110.
4. Bundle, D. R.; Gerken, M.; Peters, T. *Carbohydr. Res.*, **1988**, *174*, 239–251.
5. Dhénin, S. G. Y.; Moreau, V.; Nevers, M.-C.; Créminon, C.; Djedaïni-Pilard, F. *Org. Biomol. Chem.*, **2009**, *7*, 5184–5199.
6. Fauré, R.; Shiao, T. C.; Damerval, S.; Roy, R. *Tetrahedron Lett.*, **2007**, *48*, 2385–2388.
7. Fauré, R.; Shiao, T. C.; Lagnoux, D.; Giguère, D.; Roy, R. *Org. Biomol. Chem.*, **2007**, *5*, 2704–2708.
8. Jaworek, C. H.; Iacobucci, S.; Calias, P.; d'Alarcao, M. *Carbohydr. Res.*, **2001**, *331*, 375–391.
9. Weingart, R.; Schmidt, R. R. *Tetrahedron Lett.*, **2000**, *41*, 8753–8758.
10. (a) Garegg, P. J.; Samuelsson, B. *J. Chem. Soc., Chem. Commun.*, **1979**, 978–980: (b) Garegg, P. J.; Johansson, R.; Ortega, C.; Samuelsson, B. *J. Chem. Soc. Perkin Trans. 1*, **1982**, 681–683.
11. Balthaser, B. R.; McDonald, F. *Org. Lett.*, **2009**, *11*, 4850–4853.

33 Synthesis of *N*-(4-Methoxybenzyl)-2-(α-D-glucopyranosyl) acetamide

Jennie G. Briard, Tze Chieh Shiao,†
and Robert N. Ben*

CONTENTS

* Corresponding author; e-mail: rben@uottawa.ca.
† Checker under supervision of René Roy: e-mail: roy.rene@uqam.ca.

The phenomenon of ice recrystallization has been recognized for the past 40 years. It is a thermodynamically driven process whereby large ice crystals grow larger at the expense of smaller ice crystals through the process of Ostwald ripening.[1-4] During the cryopreservation of biological materials, ice recrystallization can cause extensive cell damage resulting in reduced post-thaw cell viability.[5-9] Conventional cryoprotectants promote dehydration of a cell during freezing but fail to protect against cellular damage caused by ice recrystallization, which would be a beneficial property.[5,10-12] C-linked glycoproteins, synthetic analogues of antifreeze glycoproteins, have been shown to possess potent ice recrystallization inhibition (IRI) activity and also function as cryoprotectants.[13-15] However, these compounds are large-molecular-weight materials. Recently, it was discovered that structurally different classes of carbohydrate-based small molecules possess IRI activity[16-20] and that a crucial balance between hydrophobic and hydrophilic interactions is essential for IRI activity.[19,21] Therefore, the general synthesis of small molecule IRIs like carbon-linked N-(4-methoxybenzyl)-2-(α-D-glucopyranosyl)acetamide (4) is an efficient method to couple hydrophobic and hydrophilic functionalities when designing IRI active small molecules.

EXPERIMENTAL METHODS

GENERAL METHODS

All anhydrous reactions were performed in flame-dried or oven-dried glassware under a positive pressure of dry argon or nitrogen. All solvents used for anhydrous reactions were distilled. All reactions were monitored using analytical thin-layer chromatography (TLC) with 0.2 mm pre-coated silica gel aluminum plates 60 F254 (E. Merck). Components were visualized by illumination with a short-wavelength (254 nm) ultraviolet light and/or staining (orcinol stain solution, 0.1% in 5% H_2SO_4). All flash chromatography was performed with E. Merck silica gel 60 (230–400 mesh). Optical rotations were measured with an Anton Paar MCP500 polarimeter. ^{1}H (400 or 500 MHz) and ^{13}C NMR (100 or 125 MHz) spectra were recorded at ambient temperature with a Bruker Avance 400 or Bruker Avance 500 spectrometer using $CDCl_3$ or D_2O as solvent. Chemical shifts are reported in ppm and referenced to residual solvent peak. Splitting patterns are designated as follows: s, singlet; d, doublet; dd, doublet of doublets; t, triplet; q, quartet; m, multiplet; and br, broad. Assignments were made by COSY experiments. Low-resolution mass spectrometry (LRMS) was performed on a Micromass Quattro-LC electrospray spectrometer with a pump rate of 20 µL/min using electrospray ionization (ESI). 2-(6-Chloro-1H-benzotriazol-1-yl)-1,1,3,3-tetramethylammonium hexafluorophosphate (HCTU) was purchased from Chem-Impex International, Inc.

N-(4-Methoxybenzyl)-2-(2,3,4,6-tetra-O-acetyl-α-D-glucopyranosyl)acetamide (3)

Sodium periodate (2.0 g, 9.36 mmol) and ruthenium trichloride (10 mg, 0.05 mmol, 3 mol% of 1) were added to a mixture of allyl glycoside 1[14,15] (0.58 g, 1.56 mmol), dicholoromethane (DCM) (4 mL), acetonitrile (4 mL), and distilled water (6 mL), and the mixture was stirred for 16 h when TLC (8:2 DCM–EtOAc) showed that the reaction was complete (**1** R_f=0.7, **2** R_f=0.2). The product was extracted into DCM (40 mL) and the extract was washed with 1% HCl (40 mL). The aqueous phase was re-extracted with DCM (20 mL); the organic layers were combined and washed with 1% HCl (2×40 mL). Saturated sodium chloride was added as needed to separate layers. The organic phase was dried over MgSO$_4$, concentrated under reduced pressure, and purified by chromatography (8:2 → 6:4 DCM–EtOAc), to give (2,3,4,6-tetra-*O*-acetyl-α-D-glucopyranosyl)acetic acid[14] (**2**, 0.43 g, 71%). ^{1}H NMR (400 MHz, CDCl$_3$) δ 5.24 (t, 1H, $J_{2,3}$=$J_{3,4}$=8.72 Hz, H-3), 5.14 (dd, 1H, $J_{1,2}$=5.57, $J_{2,3}$=9.0 Hz, H-2), 4.98 (t, 1H, $J_{3,4}$=$J_{4,5}$=8.71 Hz, H-4), 4.67 (m, 1H, H-1), 4.22 (dd, 1H, $J_{5,6}$=5.27, $J_{6,6'}$=12.2 Hz, H-6), 4.09 (dd, 1H, $J_{5,6'}$=2.76, $J_{6,6'}$=12.25 Hz, H-6'), 4.92 (m, 1H, H-5), 2.78 (dd, 1H, $J_{7,1}$=9.26, $J_{7,7'}$=15.5 Hz, H-7), 2.67 (dd, 1H, $J_{7',1}$=5.41, $J_{7,7'}$=5.56 Hz, H-7'), 2.04 (m, 12H, 4×CH$_3$). ^{13}C NMR (100 MHz, CDCl$_3$) δ 175.32 (COOH), 170.98 (CO), 170.26 (CO), 169.73 (CO), 169.61 (CO), 70.28 (C-1), 70.18 (C-5), 69.66 (C-3), 69.39 (C-4), 68.41 (C-2), 62.11 (C-6), 33.03 (C-7), 20.86 (CH$_3$), 20.83 (CH$_3$), 20.78 (CH$_3$). LRMS (ESI): *m/z* calcd. for C$_{16}$H$_{22}$NaO$_{11}$ [M+Na]$^+$ 413.33, found 412.93.

A mixture of the foregoing compound **2** (0.43 g, 1.11 mmol), HCTU (0.46 g, 1.11 mmol), and dry *N,N*-diisopropylethylamine (DIPEA) (0.19 mL, 1.11 mmol) in DCM (10 mL) was stirred for 20 min. 4-Methoxybenzylamine (0.36 mL, 2.78 mmol) and dry DIPEA (0.19 mL, 1.11 mmol) were added causing formation of an off-white precipitate. The mixture was stirred for 16 h, when TLC (2% MeOH in DCM) showed that the reaction was complete (**3**, R_f=0.26). DCM (~30 mL) was added, and the mixture was washed successively with 10% HCl (2×40 mL), H$_2$O (2×40 mL), sodium bicarbonate (2×40 mL), and brine (2×40 mL). After drying (MgSO$_4$) and concentration, chromatography (2% MeOH in DCM) gave **3** (0.29 g, 51%), [α]$_D$ +89 (*c* 0.4, CH$_2$Cl$_2$); ^{1}H NMR (400 MHz, CDCl$_3$) δ 7.18 (d, 2H, $J_{9,10}$=8.65 Hz, H-10), 6.84 (d, 2H, $J_{9,10}$=8.68 Hz, H-9), 6.14 (t, 1H, $J_{NH,8}$=5.71 Hz, NH), 5.21 (t, 1H, $J_{2,3}$=$J_{3,4}$=8.22 Hz, H-3), 5.08 (dd, 1H, $J_{1,2}$=5.21, $J_{2,3}$=8.44 Hz, H-2), 4.95 (t, 1H, $J_{3,4}$=$J_{4,5}$=8.2 Hz, H-4), 4.66 (m, 1H, H-1), 4.37 (m, 2H, H-8), 4.17 (dd, 1H, $J_{5,6}$=5.57, $J_{6,6'}$=12.13 Hz, H-6), 4.09 (dd, 1H, $J_{5,6'}$=3.24, $J_{6,6'}$=12.08 Hz, H-6'), 3.94 (m, 1H, H-5), 3.78 (s, 3H, OCH$_3$), 2.65 (dd, 1H, $J_{1,7}$=9.86, $J_{7,7'}$=15.56 Hz, H-7), 2.5 (dd, 1H, $J_{1,7'}$=4.24, $J_{7,7'}$=15.49 Hz, H-7'), 2.01 (m, 12H, 3×CH$_3$). ^{13}C NMR (100 MHz, CDCl$_3$) δ 170.75 (CO-amide), 170.05 (CO), 169.61 (CO), 169.59 (CO), 168.90 (CO), 159.26 (C-13), 130.24 (C-11), 129.14 (C-9), 114.31 (C-10), 70.34 (C-1), 69.89 (C-5), 69.60 (C-3), 69.49 (C-4), 68.39 (C-2), 62.20 (C-6), 55.48 (OCH$_3$), 43.28 (C-8), 34.54 (C-7), 20.85 (CH$_3$), 20.82 (CH$_3$), 20.81 (CH$_3$), 20.79 (CH$_3$). LRMS (ESI): *m/z* calcd. for C$_{24}$H$_{31}$NNaO$_{11}$ [M+Na]$^+$ 532.49, found 532.03. Anal. Calcd for C$_{24}$H$_{31}$NO$_{11}$: C, 56.58; H, 6.13; N, 2.75. Found: C, 56.28; H, 6.10; N, 2.81.

N-(4-Methoxybenzyl)-2-(α-D-glucopyranosyl)acetamide (4)

A solution of **3** (0.29 g, 0.56 mmol) in 16 mL sodium methoxide (1 M in methanol) and 4 mL H$_2$O was stirred for 1 h when TLC (1:10 MeOH–DCM) showed that the reaction was complete (**4** appears at the baseline, **3** R_f=0.5). The mixture was neutralized with Amberlite (Fluka Amberlite IR120 H$^+$) cation-exchange resin (H$^+$-form), filtered, and lyophilized to yield product as a white powder (0.11 g, 60%), [α]$_D$ +70 (*c* 0.5, MeOH); ^{1}H NMR (400 MHz, D$_2$O) δ 7.28 (d, 2H, $J_{9,10}$=8.53 Hz, H-10),

6.98 (d, 2H, $J_{9,10} = 8.59$ Hz, H-9), 4.48 (m, 1H, H-1), 4.4 (d, 1H, $J_{8,8'} = 15.13$ Hz, H-8), 4.27 (d, 1H, $J_{8,8'} = 14.86$ Hz, H-8′), 3.82 (s, 3H, OCH$_3$), 3.76 (dd, 1H, $J_{3,4} = 6.29$, $J_{2,3} = 9.81$ Hz, H-3), 3.66 (dd, 1H, $J_{1,2} = 4.38$, $J_{2,3} = 12.26$ Hz, H-2), 3.55 (m, 3H, H-5, H-6), 3.39 (t, 1H, $J_{3,4} = J_{4,5} = 9.33$ Hz, H-4), 2.68 (m, 2H, H-7). ^{13}C NMR (400 MHz, D$_2$O) δ 173.15 (CO-amide), 158.12 (C-13), 130.58 (C-11), 128.93 (C-9), 114.17 (C-10), 73.45 (C-1), 73.16 (C-5), 73.10 (C-3), 70.43 (C-4), 69.73 (C-2), 60.50 (C-6), 55.40 (OCH$_3$), 42.52 (C-8), 32.43 (C-7). LRMS (ESI): m/z calcd. for C$_{16}$H$_{23}$NNaO$_7$ [M+Na]$^+$ 364.35, found 364.00.

ACKNOWLEDGMENTS

This work was supported by the Natural Sciences and Engineering Research Council of Canada, Canadian Blood Services (CBS), Canadian Institutes of Health Research, Canadian Foundation of Innovation, and GreenCentre Canada. J.G.B. acknowledges CBS for a graduate fellowship program award.

^{1}H NMR of compound **2** (0–10 ppm).

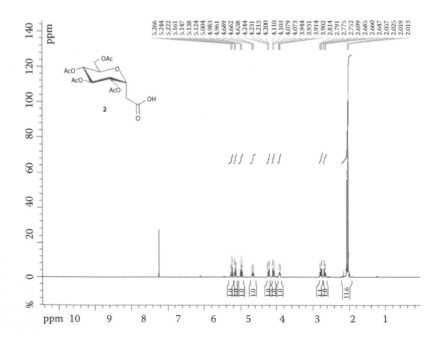

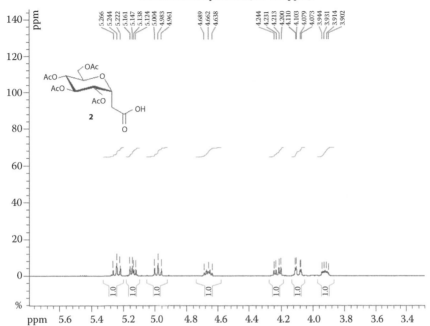

^{1}H NMR of compound **2** (3.4–5.6 ppm).

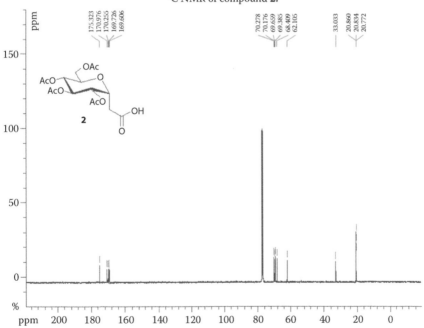

^{13}C NMR of compound **2**.

^{1}H NMR of compound **3** (0–10 ppm).

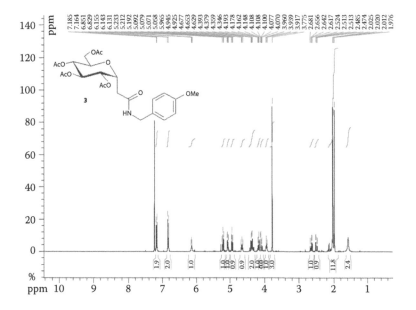

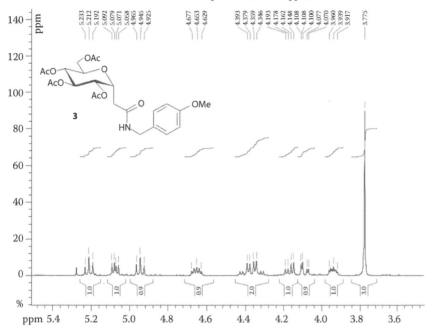

^{1}H NMR of compound **3** (3.6–5.4 ppm).

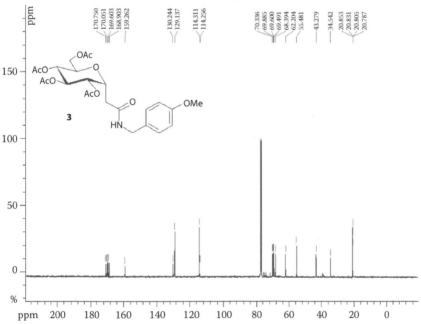

^{13}C NMR of compound **3**.

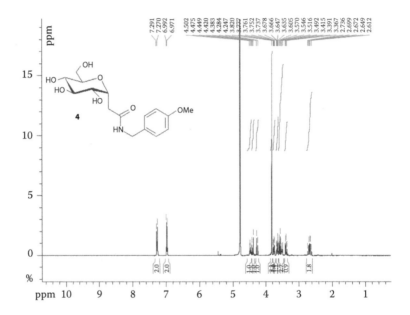

^{1}H NMR of compound 4 (0–10 ppm).

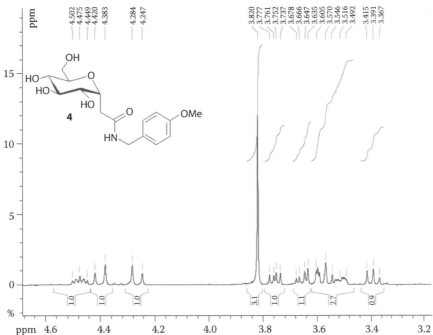

¹H NMR of compound **4** (3.2–4.6 ppm).

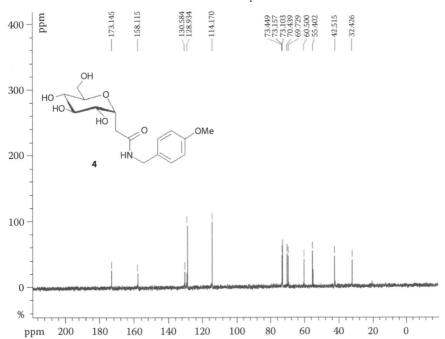

¹³C NMR of compound **4**.

REFERENCES

1. Sutton, R.L.; Lips, A.; Piccirillo, G. and Sztehlo, A. *J. Food Sci.*, **1996**, *61*, 741–745.
2. Pronk, P.; Infante, F.C.A. and Witkamp, G.J. *J. Cryst. Growth*, **2005**, *275*, e1355–e1361.
3. Hagiwara, T.; Hartel, R. and Matsukawa, S. *Food Biophys.*, **2006**, *1*, 74–82.
4. Fletcher, N.H. *The Chemical Physics of Ice*, 1st edn., Cambridge University Press, London, U.K., **1970**, 104–129.
5. Mazur, P. *Am. J. Physiol. Cell Phys.*, **1984**, *247*, C125–C142.
6. Sakai, A. and Otsuka, K. *Plant Physiol.*, **1967**, *42*, 1680–1694.
7. Shimada, K. and Asahina, E. *Cryobiology*, **1975**, *12*, 209–218.
8. Farrant, J.; Walter, C.; Lee, H. and McGann, L. *Cryobiology*, **1977**, *14*, 273–286.
9. Fowler, A. and Toner, M. *Ann. NY. Acad. Sci.*, **2005**, *1066*, 119–135.
10. Chao, H.; Davies, P.L. and Carpenter, J.F. *J. Exp. Biol.*, **1996**, *199*, 2071–2076.
11. Mazur, P. *Science*, **1970**, *168*, 939–949.
12. Meryman, H.T. *Cryobiology*, **1971**, *8*, 173–183.
13. Eniade, A.; Purushotham, M.; Ben, R.N.; Wang, J.B. and Horwath, K. *Cell Biochem. Biophys.*, **2003**, *38*, 115–124.
14. Czechura, P.; Tam, R.Y.; Dimitrijevic, E.; Murphy, A.V. and Ben, R.N. *J. Am. Chem. Soc.*, **2008**, *130*, 2928–2929.
15. Leclère, M.; Kwok, B.K.; Wu, L.K.; Allan D.S. and Ben, R.N. *Bioconjug. Chem.*, **2011**, *22*, 1804–1810.
16. Tam, R.Y.; Ferreira, S.S.; Czechura, P.; Chaytor, J.L. and Ben, R.N. *J. Am. Chem. Soc.*, **2008**, *130*, 17494–17501.
17. Chaytor J.L. and Ben, R.N. *Bioorg. Med. Chem. Lett.*, **2010**, *20*, 5251–5254.
18. Balcerzak, A.K.; Ferreira, S.S.; Trant, J.F. and Ben, R.N. *Bioorg. Med. Chem. Lett.*, **2012**, *22*, 1719–1721.
19. Capicciotti, C.J.; Leclère, M.; Perras, F.A.; Bryce, D.L.; Paulin, H.; Harden, J.; Liu, Y. and Ben, R.N. *Chem. Sci.*, **2012**, *3*, 1408–1416.
20. Balcerzak, A.K.; Febbraro, M. and Ben, R.N. *RSC Adv.*, **2013**, *3*, 3232–3236.
21. Trant, J.F.; Biggs, R.A.; Capicciotti, C.J. and Ben, R.N. *RSC Adv.*, **2013**, *3*, 26005–26009.

Index

A

Acetals
benzaldehyde dimethyl, II, 191
1,3-benzylidene, II, 232
4,6-*O*-benzylidene, I, 199, 205; II, 9, 10, 162, 175, 189, 191
chlorobenzylidene, II, 72
diphenylmethylene, II, 10
dithio-, II, 270
hemi-, 24, 184
kinetic formation, II, 231
mono-, II, 176
p-methoxybenzylidene, II, 11
pseudo-C2-symmetric *bis*-, II, 231
stannylene, II, 59, 61
2-Acetamido-4,6-*O*-benzylidene-2-deoxy-D-glucopyranose
preparation, I, 200
spot visualization, I, 200
4-*O*-(2-Acetamido-6-*O*-benzyl-2-deoxy-3,4-*O*-isopropylidene-β-D-talopyranosyl)-2,3:5,6-di-*O*-isopropylidene-*aldehydo*-D-glucose dimethyl acetal, III, 74
4-*O*-(2-Acetamido-3,6-di-*O*-benzyl-2,4-dideoxy-α-L-*erythro*-hex-4-enopyranosyl)-2,3:5,6-di-*O*-isopropylidene-*aldehydo*-D-glucose dimethyl acetal, III, 76
(2-Acetamido-3,4,6-tri-*O*-acetyl-2-deoxy-β-D-glucopyranosyl)benzene, III, 221
4-(2-Acetamido-3,4,6-tri-*O*-acetyl-2-deoxy-β-D-glucopyranosyl)bromobenzene, III, 222
2-Acetamido-3,4,6-tri-*O*-acetyl-2-deoxy-1-*O*-*p*-nitrophenoxycarbonyl-α-D-glucopyranose, III, 216
Acetobromoglucose, II, 184
Acetolysis
deoxysugars, of, I, 3
Acetone elimination, III, 73
N-Acetylneuraminic acid
methyl α-glycoside, II, 197
N-Acetyl-D-galactopyranose, III, 173
N-Acetyl-D-galactosamine, III, 174
N-Acetylglucosamine
benzylidenation of, I, 200

N-Acetylneuraminic acid
acetylated glycal, preparation, I, 245
benzyl glycosides NMR data, I, 251, 255
chloride, NMR data, I, 246, 249
Acylation of carbohydrate derivatives
selective, II, 61
Alcohols
oxidation of, II, 21
Aldonolactones
synthesis from aldose hemiacetals, III, 33
Alkenylamines, II, 67
precursors, II, 68
Vasella-reductive amination procedure, II, 48
Alkylation of carbohydrate derivatives
selective, II, 61
Allyl 2-acetamido-2-deoxy-4,6-di-*O*-pivaloyl-β-D-galactopyranoside, III, 174
Allyl 2-acetamido-2-deoxy-3,6-di-*O*-pivaloyl-β-D-glucopyranoside, III, 175
Allyl glycoside
from anomeric orthoester, II, 150
4-Amino-4-deoxy-L-arabinose (Ara4N), III, 193
2-Aminoethyl diphenylborinate, II, 62
***p*-{N-[4-(4-Aminophenyl α-D-glucopyranosyl)-2,3-dioxocyclobut-1-enyl]amino} phenyl α-D-glucopyranoside,** III, 108
***p*-{N-[4-(4-Aminophenyl β-D-glucopyranosyl)-2,3-dioxocyclobut-1-enyl]amino} phenyl β-D-glucopyranoside,** III, 109
N-(3-Aminopropyl)-2,3,4,6-tetra-*O*-benzyl-D-gluconamide, III, 29
N-(3-Aminopropyl)-2,3,4-tri-*O*-benzyl-D-xylonamide, III, 28
1,6-Anhydrosugars, II, 98
Anomeric
azide, II, 257
S-deacetylation, III, 89
dihalides, II, 245
Anthracenylmethylene, II, 10
Arabinofuranose
1,2,3,5-tetra-*O*-benzoyl-α-D-, I, 234
1,2,3,5-tetra-*O*-benzoyl-α,β-D-, I, 234
Arabinofuranose 1,2,5-orthobenzoate, III, 147
α-D-Arabinofuranoside
methyl 2,3,5-tri-*O*-benzoyl-, III, 147, 149
p-tolyl 2,3,5-tri-*O*-benzoyl-1-thio-
preparation, I, 343
α-D-Arabinofuranosyl bromide
2,3,5-tri-*O*-benzoyl-, III, 149